AF559847

System Based Integrated Nutrient Management

System Based Integrated Nutrient Management

B. GANGWAR
V.K. SINGH

2012

New India Publishing Agency
Pitam Pura, New Delhi-110 088

Published by
Sumit Pal Jain *for*
New India Publishing Agency
101, Vikas Surya Plaza, CU Block, L.S.C. Mkt.,
Pitam Pura, New Delhi- 110 088, (India)
Phone: 011-27341717, Fax: 011-27341616
E-mail: info@nipabooks.com
Web: www.nipabooks.com

ISBN : 978-93-81450-05-5

Typeset at: Harminder Singh Kharb *for* Typographiya
Printed at: Jai Bharat Printing Press, Delhi

भारतीय कृषि अनुसंधान परिषद
कृषि अनुसंधान भवन - II, पूसा, नई दिल्ली 110 012
INDIAN COUNCIL OF AGRICULTURAL RESEARCH
KRISHI ANUSANDHAN BHAVAN-II, PUSA, NEW DELHI 110 012

डॉ. अरविन्द कुमार
उप महानिदेशक (शिक्षा)
Dr. Arvind Kumar
Deputy Director General (Education)

Phone : 011-25841760 (O)
011-25843644 (R)
Fax : 011-25843932
E-mail : ddgedn@icar.org.in
Website : www.icar.org.in

12th March, 2012

Foreword

In order to meet the needs of the ever growing population with constraint of low per capita land availability India, enchancing productivity of arable lands is the only viable alternative. Efforts to raise the crop yields through the chemical fertilizer use were successful to some extent but yield stagnation of major food crops led to deciline in growth rate of productivity in recent years. Decline in factor productivity, owing to deterioration of soil health and emergence of wide spread deficiencies of various soil nutrients, is considered to be the major reason for such a situation. Results from long-term experiments (LETs) have revealed a net negative balance of N, P and K even with recommended fertilization due to greater drain of the native soil fertility, thus, rendering them unable to support high productivity levels in succeeding years.

The consistent increase in fertilizer dose shall neither be economically feasible nor sustainable on ecological grounds and the only option left is to up-scale the efficienty of applied nutrients. Rational nutrients management strategies, particularly Integrated Nutrient Management (INM), assumes greater significance in sustaining the productivity of key production systems.

The INM approach aims at efficient and judicious use of all the major plant nutrient's sources like mineral, organic and biological in an integrated manner, so as to attain maximum economic returns without any deleterious effect on soil properties and also maintaining ecological balance, thus, off setting yield plateauing trends in different crops and cropping systems. INM is a judicious way to mobilize all available, accessible and affordable plant nutrient sources in the working capital asset and within the investment capacity of farmer, having paramount significance for sustained higher crop productivity without deteriorating soil health.

The information based on the cropping system based INM in soil-plant system is lacking and the compilation of relevant literature in the form of a publication, "System based integrated nutrient management" by Dr. B. Gangwar and Dr. V.K. Singh highlights the options and optimization needs of available nutrient resources, including management strategies to improve nutrient use efficiency and its impact on soil physical, chemical and biological environment. The authors have also documented the methodology for nutrient budgetting, economic evaluation, crop simulation modelling and its application for integrated nutrient management which shall be of immense practical value to all the stake holders. I trust that the book would be useful to all concerned and serve as a knowledge resource with respect to nutrient management and its related issues for sustainability.

(Arvind Kumar)

Preface

The nutrient management is considered to be one of the major contributors (about 40%) in achieving national food targets. The demand of increased use of fertilizer will remain linearly related with the food goal in future. To achieve the projected targets of about 350 million tonnes of food grains by 2030 AD country will be need fertilizer to the tune of 100 million tonnes. In fact, our experiences during the green revoluation era have clearly shown that the imbalance of nutrient use especially under cereal based systems has resulted in many fold soil and environment related problems. With the increase in fertilizer use coupled with increasing cropping intensity under the changing scenario of climatic concern naturally require the special attention and call for efficient management of nutrients in integrated manner. Our estimate clearly shows that the chemical fertilizer alone cannot suffice the nutrient requirement of different crops and cropping system. In this context, the integrated nutrient management involving different organic sources like FYM, vermi-compost, crop residue, green manure, Bio-fertilizer and *in-situ* role of legumes along with balanced nutrient use (major and micro-nutrients) deserves due attention. In fact, based on series of experiments and studies conducted all over the country with reference to nutrient management for various crop as applicable to varying soil environment representing different agro-climatic zones have been generated at national level but the system based information's on integrated nutrient management is still lacking and yet to be documented. Therefore, the present attempt was made to compile the system based information for major cropping systems in the form of a book.

In this publication chapter 1, 3 and 4 deals with general issues and management options for integrated nutrient management with special reference to irrigated eco-system, while chapter 2 focused on crop residue management. The chapter 5 and 10 are enlightens the soil-test based nutrient management for sustainable soil health, while chapters 6 and 7 are related to nutrient economy through integrated farming system and inclusion of legumes under cereal based cropping systems. The chapter 8 is focused on integrated nutrient management in rice-wheat cropping system, while chapter 9 on oilseed based, 11 on soybean based, 13 on vegetable and chapter 23 on seed spices based cropping systems. The issues related to SSNM, protected agriculture, soil chemical, biological and microbial diversity are discussed in chapter 12, 14 and 18, respectively. The aspects related to system based nutrient budgeting, soil carbon management and sequestration, balanced crop nutrition in relation to crop diseases, economics and nutrient modeling have been duly discussed in chapters from 19 to 25. It is hoped that all related issues of system based integrated nutrient management have been duly discussed and presented in this publication.

During the process of compilation of this information, we received the kind encouragement from Dr. A.K. Singh, Deputy Director General (Natural Resource Management) and Dr. Arvind Kumar, Deputy Director General (Education), Indian Council of Agricultural Research, New Delhi and therefore, we extend our sincere gratitude to them. In fact, this publication is the improved version of lectures delivered during the winter school "System based integrated nutrient management for sustained productivity and soil health" by selected resource persons/ subject matter specialists. We place our sincere thanks to all the contributors for their timely action for improving their write ups as per requirement. The help extended by Mr. Rajesh Kumar, Senior stenographer (PA) in setting the text matter and designing the cover page is thankfully acknowledged. We assume that our efforts in the form of this publication will be use full to all the stake holders involved in nutrient management and others.

Editors

Contributors

A.K. Pandey, Head, principal Scientist, Indian Institute of Vegetable Reserch, Varanasi-221305

A.K. Vyas, Senior. Scientist, Division of Agronomy, Indian Agricultural Research, Modipuram, Meerut-250110, U.P.

Akath Singh, Senior. Scientist, Project Directorate for farming System Research, Modipuram, Merrut-250110, U.P.

Awani Kumar Singh, Senior. Scientist, Center for Protected Cultivation Technology (CPCT), I.A.R.I. (Pusa), New Delhi-110012

B. Gangawar, Director, Project Director for Farming Systems Research, Modipuram Meerut-250110, U.P.

B.G. Shivakumar, Scientist, Division of Agronomy, Indian Agricultural Research Institute, New Delhi-110012

B.M. Sharma, Principal Scientist, Division of Soil Science and Agricultural Chemistry, I.A.R.I., New Delhi-110012

B.S. Dwivedi, Head, Division of Soil Science and Agricultural Chemistry, I.A.R.I., New Delhi-110012

Balraj Singh, Principal Scientist, Center for protected Cultivation Technology (CPCT), I.A.R.I. (Pusa), New Delhi-110012

Chandra Bhanu, Scientist, Project Directorate for farming Systems Research, Modipuram, Merrut.

D. Singh, Project Directorate for Farming Systems Research, Modipuram, Meerut

D.M. Hegde, Ex-Director, Directorate of Oilseeds Research, Rajendranagar, Hyderabad-500030

J.P. Singh, Principal Scientist, Project Directorate for Farming Systems Research, Modipuram Meerut-250110, U.P.

K.K. Singh, Principal Scientist, Project Directorate for Farming Systems Research, Modipuram, Modipuram (Merrut), India

K.P Tripathi, Senior. Scientist, Project Directorate of Cropping Systems Research, Merrut

M. Shamim, Scientist, Project Directorate for Farming Systems Reserch, Modipuram

M.A. Khan, Senior Scientist, Centeral Potato Research Institute Campus, Modipuram, Meerut

N.C. Upadhayay, Principal Scientist, Central Potato Research Institute Campus, Modipuram, Meerut

N.K. Jat, Scientist, Project Directorate for Farming Systems Research, Modipuram, Merrut, U.P.

O.P. Aishwath, Senior Scientist, National Research Centre on Seed Spices, Tabiji-305206, Ajmer, Rajasthan

Rakesh Kumar, Research Associate, Center for Protected Cultivation Technology (CPCT), I.A.R.I. (Pusa), New Delhi-110012

S.P. Mazumdar, Scientist, Project Directorate for Farming System Research Modipuram, Meerut

S.P. Singh, Senior Scientist, Project Directorate for Farming Systems Research, Modipuram Merrut

S.S. Pal, Principal Scientist, Project Directorate for Farming System Research, Modipuram, Meerut

V.K. Singh, National Fellow (ICAR), Project Directorate for Farming Systems Research, Modipuram, Meerut-250110, U.P.

Yadvinder Singh, INSA Seniour Scientist, Department of Soils, Punjab Agricultural University, Ludhiana

Contents

System Based Integrated Nutrient Management, 2012
© B. Gangwar & V.K. Singh (eds.), pp. 1-28
New India Publishing Agency, New Delhi (India)
e-mail : info@nipabooks.com; website : www.nipabooks.com

CHAPTER 1

System Based Integrated Nutrient Management

B. GANGWAR AND V.K. SINGH

With the advent of modern crop varieties, better irrigation facilities and greater use of fertilizers and other inputs country has changed within the last 50 years from a region of food scarcity to a region of food sufficiency. The green revolution technologies involving greater use of synthetic agro-chemicals such as fertilizers and pesticides with adoption of nutrient- responsive, high yielding varieties of crops have boosted the production per unit area under different crops and cropping systems. However, this increase in production has slowed down and in some cases there are indications of decline in growth of productivity and production (Table 1). These production scenarios may be envisaged as corresponding decrease in soil organic matter and continuous increase in multi nutrient deficiency (Dwivedi *et al.*, 2006). If such situation is allowed to continue for another few decades, there are chances that todays' productive land may become unproductive. The declines in yield and production fatigue have been noticed in various cropping systems of different agro-eco region (Gill and Singh, 2009). Farmers' participatory surveys conducted in Indo-Gangetic Plain (IGP) region indicates that farmers' have resorted higher doses of nutrients each year to obtain the same yield as obtained in previous year (Dwivedi *et al.*, 2001). The probable reasons behind these are continuous mining of nutrients, inadequate use of organic sources, imbalance fertilization and decline of soil organic matter.

Table 1 : Compound growth rate (% per year) of production and productivity of principal crops (1980-81 to 2006-07)

Crop	Production			Productivity		
	1980-81 to 1989-90	1990-91 to 1999-2000	2001-2007	1980-81 to 1989-90	1990-91 to 1999-2000	2001-2007
Rice	3.62	1.74	1.35	3.19	1.34	1.53
Wheat	3.57	3.27	0.58	3.10	1.83	-0.53
Total cereals	3.03	1.86	1.21	2.90	1.59	2.18
Total pulses	1.52	0.59	3.23	1.61	0.93	1.41
Total food grains	2.85	1.66	1.40	2.74	1.52	2.00

Source: Agricultural Statistics at a Glance, 2007

Presently, soil degradation is a major concern in agriculture because of non-judicious use of agricultural inputs and over exploitation of natural resources. Imbalance between nutrient addition v/s removals, the later being on higher side, is a matter of concern. Thus, extra mining of nutrients will have to be checked in order to maintain the soil health. On the other hand, subsidized fertilizer availability has enable the farmers to apply fertilizer at maximum level which is especially the N driven fertilizer management. But, during the same period, use of organic sources including green manures under different cropping systems has declined substantially (Kannaiyan, 1998). Diagnostic surveys on nutrient management practices prevailing in RWCS dominated areas of Indo-Gangetic Plain (IGP) revealed that nearly one-third of the farmers practicing RWCS apply as high as 180 kg fertilizer N ha^{-1} to each rice and wheat crop as against local recommendation of 120 kg N ha^{-1}. The use of P is sub-optimal and other secondary and micro- nutrient including organic manures are meagre (Dwivedi *et al.*, 2001). In these areas, indiscriminate use of fertilizer N needs to be curbed because a further increase in fertilizer application rate may not be only uneconomic but also environmentally unsafe (Yadav *et al.*, 2000). Unbalanced use of fertilizer N may enhance nitrate leaching beyond root zone, posing a potential threat of the pollution of ground water used for drinking purposes in rural areas (Singh *et al.*, 1995). All these factors led to imbalance nutrient use ratio between N: P_2O_5: K_2O. For example, at presently this ratio in state like Punjab is 18.4: 5.9: 1 as against the recommended balanced fertilization ratio of 4:2:1 (FAI, 2010). Under these circumstances, it is very pertinent to narrow down the crop nutrition ratio in order to sustain the productivity.

Integrated nutrient management (INM) in an approach which takes cares of efficient and beneficial nutrition. The basic objective of INM is to restore organic matter in soil to enhance nutrient use efficiency and to maintain soil quality in terms of physical, biological and chemical properties. Lack of awareness among farmers about negative effects of high intensive farming without concern to conserve the natural resources or soil fertility has induced the production fatigue over the years. In view of this, there is a need for alternative practices of managing the nutrients more judiciously, efficiently, and in balanced proportions. Since, crop rotation is the key component of integrated sustainable farming system, the system based integrated nutrient management is of essential desire for sustained productivity and soil health.

In this chapter on attempt has been made to compile the available information on the different aspects of INM under various cropping systems for enhancing crop productivity and maintaining soil health.

Productivity of predominant cropping systems and yield gaps

Not withstanding the projections highlighting the need for increasing agricultural production, a vast untapped potential in respect of the yield for all the crops prevails in different agro-climatic zones of the country. It is worth noting here that the gaps between potential and realisable and between realisable and averaged realised yield in the country generally ranged around 50 to 100%, respectively. The agro-climatic zone wise yield potential computed in terms of REY at Project Directorate for Farming Systems Research, Modipuram indicates that highest productivity ranging between 8.46-12.04 t ha^{-1} for rice-wheat system, 8.61-11.35 t ha^{-1} for maize-wheat system and 5.84-7.46 t ha^{-1} for pearl millet-wheat system in Trans Gangetic Plains. While lowest system productivity of rice-wheat system (4.72-8.73 t ha^{-1}) and maize-wheat system (4.35-9.34 t ha^{-1}) was in Western Himalayan region. Assessing yield gap in terms of REY for various cropping systems varies from 0.88 to 7.54 t ha^{-1} in different agro-climatic zone (Fig. 1). Further, yield gap between state average and on-station research average varied from cropping system to cropping system and from region to region within a system (Gill *et al.*, 2008). The important viable options for vertical productivity improvement are efficient input management such as balanced nutrient application, use of quality seed, and efficient crop protection measures in conjunction with increasing the irrigation facilities/infrastructure. Among the different factors fertilizer ofcourse plays a key note to bridge such gap as it alone can contribute 50% for achieving the production target. The apparent example like: average wheat productivity in Bihar and Uttar Pradesh are 43% and 61% to that of Punjab. On the other hand, the average rate of fertilizer use in Bihar and U.P. as percentage of the Punjab is 46% and 66%, respectively (Tiwari, 2002). These facts call for scientific use of fertilizers in a integrated and balance manner for enhanced productivity.

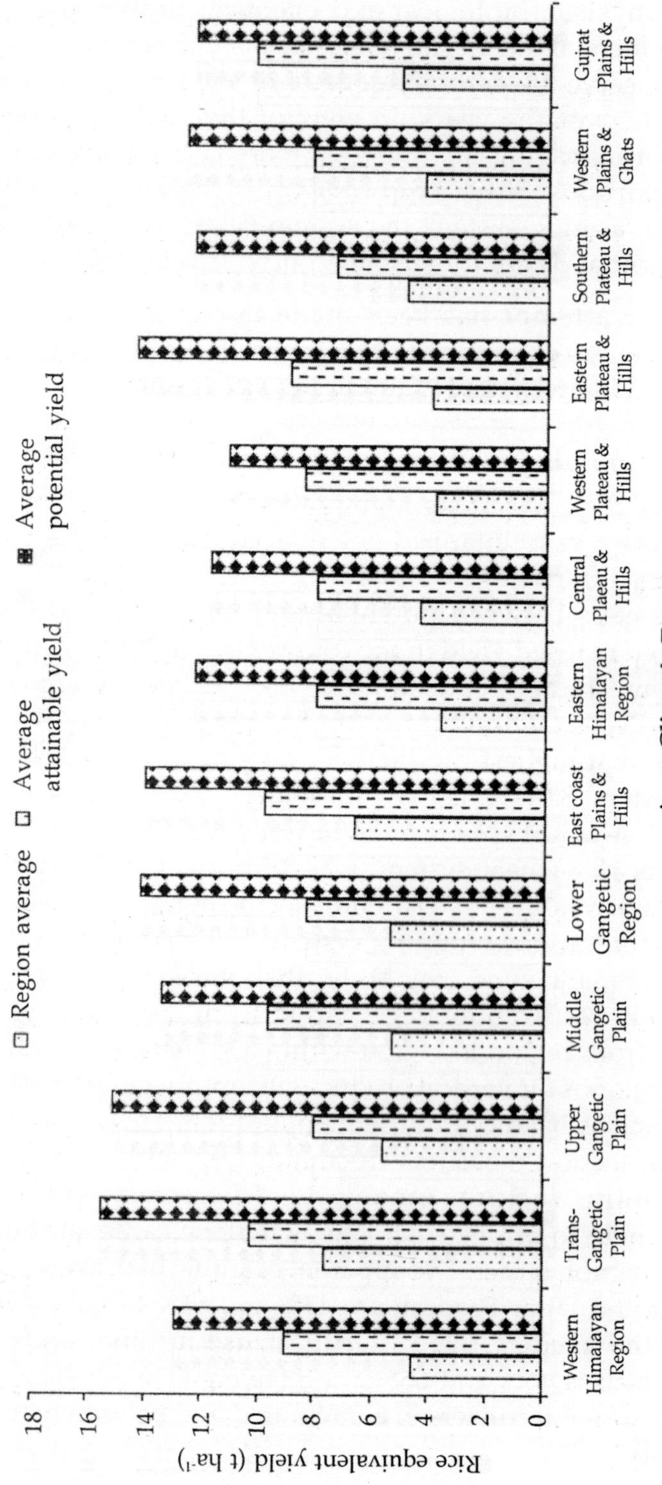

Fig. 1 : Yield gap between agro-climatic zone and attainable yield, and between attainable and potential yield.

Critical role of balance nutrient application

Long-term studies conducted under AICRP-CS revealed that the indiscriminate use of straight fertilizer; say N only had deletions effects on crop productivity and soil fertility (Yadav *et al.*, 2000), leads removal of other nutrients apart from N and as long as the soil has an adequate reserve of other nutrients (Ladha *et al.*, 2003). The further addition of N will fail to produce any increase in crop yield. To site a specific example, on an average, a crop of wheat removes from the soil 25, 9, 33, 5.3, 4.7 and 3.7 kg of N, P, K, Ca, Mg and S, respectively and 624, 70, 56, 24, 48 and 2 g of Fe, Mn, Zn, Cu, B and Mo, respectively tonne^{-1} of harvested grains. It is, thus, evident that balanced application of major and micronutrients is essential for getting more yields per unit area of land. During the initial years of introduction of the HYVs, only macronutrient deficiencies were discovered as an obstacle to their high yields. With passes of time the situation has worsened with increasing use of high analysis fertilizers free from secondary and micronutrients, decreasing use of organic manures and neglected recycling of crop residues. As a result, multi-nutrient deficiencies (macro+micro) are being observed in recent years (Dwivedi *et al.*, 2006; Sanyal and Chatterjee, 2007). Soil exhaustion has spread to such an extent that N deficiency is now universal across the Indian soils, phosphorus deficiency affects is 65-70% of these soils, while potassium fertility is either low or medium in more than 50% of soils (Table 2).

Table 2 : Extent of macro-nutrient deficiency in India

Nutrient	Number of sample analysed	% sample		
		Low	Medium	High
N	3650004	63	26	11
P	3650004	42	38	20
K	3650004	13	37	50
S	27000	40	35	25

Source: Motsara, 2002

The deficiency of sulphur is also alarming (Singh and Gangwar, 2011). Further, Table 3 indicates that deficiency of zinc, iron, manganese, copper and boron is also coming up to a great extent (Table 3, Fig. 2). In fact, inadequate uses of organic amendments have caused gradual depletion of soil organic carbon pool, thereby adversely affecting the physical, chemical and biological properties of the soils. Furthermore, this has led to a reducing overall soil indigenous nutrient supplying capacity. Therefore, it is of urgent need of optimum fertilizer prescription, along with integrated organic incorporation, to ensure the correction of all existing nutrient deficiencies as well as to fulfil the crop demand. Here it is pertinent to mention that full advantages of organic along with inorganic only be harness when all deficient nutrients are fulfilled in right amount and right time.

Table 3 : Extent of micro- nutrient deficiency in India

Nutrient	Number of sample analysed	% sample deficient
Zn	251660	49
Fe	251660	12
Mn	251660	5
Cu	251660	3
B	36825	33
Mo	36825	13

Source: Singh, 2001

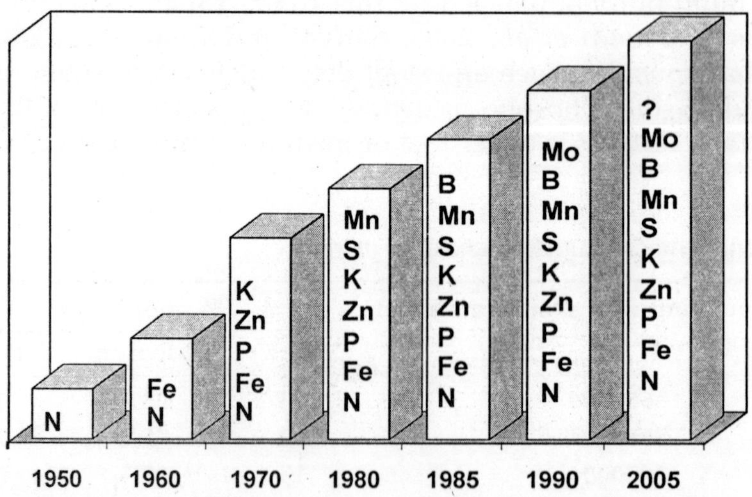

Fig. 2 : Progressive expansion in the occurance of nutreint deficiencies in India
Source: (Tiwari, 2006)

Potential of organic sources in India

The system based integrated nutrient management is solely depends on type of organic and inorganic sources used, their combination and method of application. In fact, options of organic sources are very limited and pre-dominantly farm yard manure, composts, green manure, press mud, industrial or city waste, ash fly, crop residue and bio fertilizers are being used in different forms. Recently vermin-composting has emerged as a new viable option for nutrient enriched organic source. Although the exact assessment of these sources are quite difficult but the estimates made by Tandon (1997) reveals that the total trappable N + P_2O_5 + K_2O from different sources in the country are 6.05 mt at present and may be increased up to 7.25 mt by 2025 AD (Table 4). Further, statistics on projections on population growth, food demands and fertilizer needs reveal that by 2021 AD to feed 1420 million mouths in India, food demand and fertilizer consumption would be 280 and 24 million tonnes (N+P_2O_5+K_2O), respectively. Similar figures for 2051 AD are projected at 1780 million people, 350 million tonnes of food and 31 million tonnes of fertilizers nutrients (N+P_2O_5+K_2O) against the current consumption of about 15-16 million tonnes. If it is assumed that 50% of the waste material finds its way back to the farm, potential contribution of agricultural crop residue, rural and urban wastes altogether could be 5.4 and 7.2 million tonnes of N+P_2O_5+K_2O in the years of 2021 and 2051, respectively. If these projections are exactly met, then by 2021 AD the shortage of 18.6 million tonnes has to be primarily met through fertilizers and other sources such as green manure, organic manures and bio-fertilizers. According to Tandon (1997), animal dung and crop residues together have a potential supply of 9.0-14.4 million tonne of N + P_2O_5 + K_2O of which hardly 2.7-4.3 million tonne are available for agricultural use.

Table 4 : Projections of trappable nutrients from different organic sources for agriculture in India

Resources	Year		
	2000	2010	2025
Generators			
Human population (Million)	1000	1120	1300
Livestock population (Million)	498	537	596
Tappable resources (mt)			
Human Excreta (Dry)	13	15	17
Livestock dung (Dry)	113	119	128
Crop residue	99	112	162
Nutrient potential, mt (N+P_2O_5+ K_2O)			
Human Excreta (Dry)	2	2.24	2.6
Livestock dung (Dry)	6.64	7.00	7.54
Crop residues	6.21	7.1	20.27
Tappable nutrients, mt (N+P_2O_5+ K_2O)			
Human Excreta	1.6	1.8	2.1
Livestock dung	2	2.1	2.26
Crop residues	2.05	2.34	3.39
Total	5.05	6.24	7.25

Tappable = 30% of dung, 80% of excreta, 33% of crop residues
Source: Tandon, 1997

In view of above projections, it looks imperative that bulk of the nutrient supply has to be met through fertilizers. Here lies the need of an objective analysis, ingenuity and discretion as to how this precious organic sources of nutrients which would be best constitute 15-20% of the total NPK requirement be used with advantage to protect the soil from degradation and to maintain its productivity under intensive agriculture. In this context, judicious use of resources (Organic and inorganic) in system mode may answer, how best integrated use of organic sources can be made for harnessing the maximum production benefits. This approach should be a priori and major consideration for achieving sustainable system productivity.

Integrated nutrient management strategies

Although integrated plant nutrient supply system is an age-old concept but its importance was not realised earlier due to low nutrient turn over in soil-plant system or almost all the nutrient needs of the then sustenance agriculture were met through organic sources which also supplied secondary and micro nutrients besides major nutrients. The IPNS has now assume great importance firstly, because of the present negative nutrient balance and secondary, neither the chemical fertilizers alone nor can the organic resources exclusively achieve the production sustainability of the soil as well as crops under highly intensive cropping systems. In present era of intensive agriculture, importance of system based integrated nutrient management is being further magnified because of inadequate and imbalanced use of fertilizers causing negative nutrient balance, depletion of soil fertility and decline in fertilizer use efficiency. It is important to mention here that the approach for integrated nutrient management in system perspective varies as per different crops grown under different cropping system, their nutrient need and ability to harness the residual advantages of applied. Some of the viable options of INM under different cropping systems are being discussed here as under:

a) INM under cereal based systems

Crop responses to organic and biological nutrient carriers are not as spectacular as to fertilisers, but the supplementary and complementary use of these sources is known to enhance the use efficiency of applied fertilisers, besides improving soil physico-chemical properties and preventing emergence of micronutrient deficiencies. Moreever, the system based nutrient management take care of residual use of available nutrients to succeeding crops. Studies carried out with cereal based cropping systems under AICRP-

IFS has established that 25-50% fertiliser NPK dose of *kharif* crops can be substituted with the use of FYM under different situations (Table 5).

Table 5: Productivity of cereal-cereal cropping systems as influenced by integrated use of FYM and inorganic fertilizers under AICRP-IFS (mean over 2005-06 to 2009-10)

Treatments		Grain yield (kg ha^{-1})		
Kharif	**Rabi**	**Kharif**	**Rabi**	**Rice equivalent yield**
Rice-wheat				
Ludhiana				
100% NPK	100% NPK	5361	4639	10315
50% NPK+ 50% N (FYM)	100% NPK	5689	4961	10986
75% NPK+ 25% N (FYM)	75% NPK	5853	4798	10977
Kanpur				
100% NPK	100% NPK	4496	3360	8084
50% NPK+ 50% N (FYM)	100% NPK	5054	3694	8999
75% NPK+ 25% N (FYM)	75% NPK	4770	3376	8375
Kalyani				
100% NPK	100% NPK	3904	3021	7131
50% NPK+ 50% N (FYM)	100% NPK	4070	3342	7639
75% NPK+ 25% N (FYM)	75% NPK	4285	3095	7590
Rice-rice				
Chiplima				
100% NPK	100% NPK	5335	5517	10852
50% NPK+ 50% N (FYM)	100% NPK	5848	5657	11505
75% NPK+ 25% N (FYM)	75% NPK	5786	5632	11418
Bhubaneswar				
100% NPK	100% NPK	5358	5919	11277
50% NPK+ 50% N (FYM)	100% NPK	5558	6329	11887
75% NPK+ 25% N (FYM)	75% NPK	5502	6385	11887
Karmana				
100% NPK	100% NPK	4601	4432	9033
50% NPK+ 50% N (FYM)	100% NPK	5228	4536	9764
75% NPK+ 25% N (FYM)	75% NPK	5042	4594	9636

Contd...

Rice-mustard				
Rudrur				
100% NPK	100% NPK	4869	810	5679
50% NPK+ 50% N (FYM)	100% NPK	4996	846	5842
75% NPK+ 25% N (FYM)	75% NPK	5208	795	6003
Maize-wheat				
100% NPK	100% NPK	4849	5157	9462
50% NPK+ 50% N (FYM)	100% NPK	5517	5806	10700
75% NPK+ 25% N (FYM)	75% NPK	5095	5257	9769
Rice-maize				
Kathalgare				
100% NPK	100% NPK	6590	4372	10156
50% NPK+ 50% N (FYM)	100% NPK	6459	4685	10280
Pearl millet-wheat				
S. K. Nager				
100% NPK	100% NPK	1071	2305	2753
50% NPK+ 50% N (FYM)	100% NPK	1277	2938	3437
75% NPK+ 25% N (FYM)	75% NPK	1430	2872	3508
Hisar				
100% NPK	100% NPK	3557	5852	7674
50% NPK+ 50% N (FYM)	100% NPK	3715	5999	7922
Junagadh				
100% NPK	100% NPK	1450	3325	3895
50% NPK+ 50% N (FYM)	100% NPK	1899	4061	4860
75% NPK+ 25% N (FYM)	75% NPK	1659	3500	4208
Sorghum-wheat				
Parbhani				
100% NPK	100% NPK	2679	2748	4841
50% NPK+ 50% N (FYM)	100% NPK	2712	3052	5122
Ranchi				
100% NPK	100% NPK	2850	3437	5570
50% NPK+ 50% N (FYM)	100% NPK	2957	3845	6006

Source : AICRP-IFS Reports

The long-term sustainability of intensively cultivated rice-wheat system with integrated use of FYM was also reported by earlier workers (Yadav *et al.*, 2000; Ladha *et al.*, 2003; Katyal *et al.*, 1999). Studies conducted at Modipuram revearls that use of industrial waste like decomposed sulphitation press mud under rice-wheat system also had similar advantage in terms of yield stabilization of rice and wheat crop (Fig. 3).

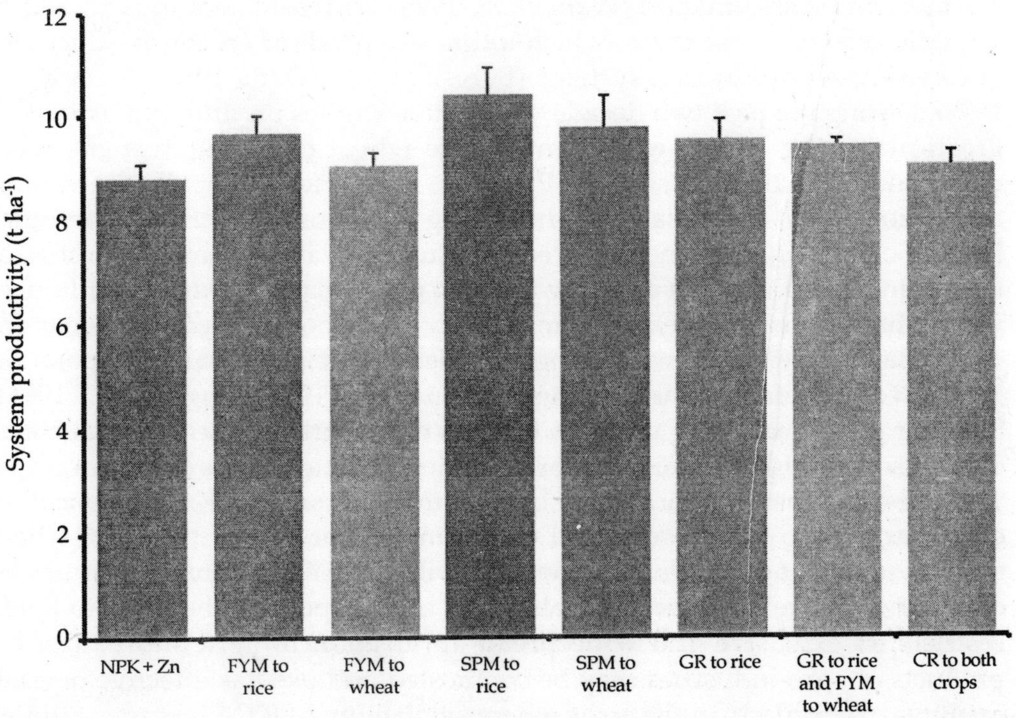

Fig. 3 : Systems productivity as infleunced by different organic sources. Bar indicates standard error of mean, n=4.
Source: Singh and Mishra, 2010

Since FYM has limitations as its competitive use for domestic fuel, legume green manuring especially during fallow period like summer under rice- wheat system is another viable option for sustained productivity. Results from AICRP-IFS indicated that a saving of 25 to 50% N can be made under rice-wheat, rice-rice, rice-maize, maize-wheat, pearl millet-wheat and sorghum-wheat system (Table 6). Growing legumes as green manure can provide the equivalent of 40-60 kg of urea N ha^{-1} to the following rice crops, with added improvement in rice equivalent yield (REY) of rice- wheat system. The increase in REY over rice–wheat system with inclusion of a green manure crop in post wheat summer season was ascribed to incorporation of additional amounts of plant nutrients through green biomass amounting to 23.8 to 27.4 t ha^{-1} in case of *Sesbania* green manure and 25.4 to 28.5 t ha^{-1} in case of mungbean green manure (Singh *et al.*, 2001). The benefits of green manuring in terms of yield increase and soil fertility regeneration in rice-wheat system have already been well documented (Singh *et al.*, 1991). In fact, a green manure hastens the nutrient transformation (Wani *et al.*, 1995), increases root growth and N use efficiency of cereal crops, which ultimately result in greater productivity of cereal-based production systems (Buresh and De-Datta 1991; Yadav *et al.*, 1998). During the past two decade, exhaustive studies on different aspects of green-manuring in rice-wheat and maize-wheat cropping systems were undertaken on alluvial soils of Punjab with some practically important conclusions as (i) Sesbania green-manuring was more promising than other legume crops, (ii) efficiency of green-manure N was comparable to that of urea N in rice, (iii) benefit accruing through green manure was higher in rice-wheat than in maize-wheat system, and (iv) incorporation of green-manure crop a day before rice transplanting gave better efficiency than incorporation 1 to 2 weeks before transplanting (Meelu *et al.*, 1992; Singh *et al.*, 1994). Summer green-manuring involves the loss of a short crop season that could otherwise be utilised for raising short-duration summer pulses or fodder crops. These results seem to indicate that organic manure (such as FYM) application can contribute to maintaining and enhancing the productivity of rice-wheat systems in the long run. But the non availability of FYM because of burning of cow dung for fuel and the high labor cost of transporting the FYM to fields restricts its extensive and widespread application in agriculture. The bi-products of agro-industries may be composted and used as effective organic manures, particularly in the areas where availability of FYM is scarce. Studies at Modipuram indicated that a decomposed sulphitation pressmud (SPM) gave even better performance over FYM in increasing the yields of rice and wheat (Singh and Mishra, 2010). The FYM and composts applied to rice produce a better yield advantage than those applied to wheat.

Disposal of rice and wheat straw in TGP and UGP has emerged as a great problem. In these combine-harvested areas farmers opt to burn the residues *in situ*, losing precious nutrients on one hand and polluting environment on the other. Recycling of these residues back to fields helps to build stable organic matter in the soil, as also to sustain crop yield levels. Studies in AICARP suggested that in some areas, incorporation of crop residues made it possible to curtail 25% of fertilizer NPK requirement of rice (Table 7).

Table 6 : Effect of integrated use of green manuring and inorganic fertilizers on productivity of cereal based cropping system under AICRP-IFS (mean over 2005-06 to 2009-10)

Treatments		Grain yield (kg ha^{-1})		
Kharif	Rabi	Kharif	Rabi	Rice equivalent yield
Rice-wheat				
Ludhiana				
100% NPK	100% NPK	5361	4639	10315
50% NPK+ 50% N (GM)	100% NPK	6027	4780	11132
75% NPK+ 25%N (GM)	75% NPK	6096	4401	10796
Varanasi				
100% NPK	100% NPK	4496	3360	8084
50% NPK+ 50% N (GM)	100% NPK	4759	3512	8510
75% NPK+ 25%N (GM)	75% NPK	4625.2	3275	8123
Kalyani				
100% NPK	100% NPK	3904	3021	7131
50% NPK+ 50% N (GM)	100% NPK	4065	3527	7831
75% NPK+ 25%N (GM)	75% NPK	4230	3257	7708
Rice-rice				
Rajendra Nagar				
100% NPK	100% NPK	4744	5166	9909
50% NPK+ 50% N (GM)	100% NPK	4865	5433	10299
75% NPK+ 25%N (GM)	75% NPK	4949	5292	10241
Chiplima				
100% NPK	100% NPK	5335	5517	10852
50% NPK+ 50% N (GM)	100% NPK	6046	6061	12108
Bhubaneswar				
100% NPK	100% NPK	5358	5919	11277
50% NPK+ 50% N (GM)	100% NPK	6081	6737	12818
75% NPK+ 25%N (GM)	75% NPK	5871	6129	12000
Karmana				
100% NPK	100% NPK	4601	4432	9431
50% NPK+ 50% N (GM)	100% NPK	4993	4494	9488
75% NPK+ 25%N (GM)	75% NPK	4970	4441	9411
Rice-mustard				
Rudrur				
100% NPK	100% NPK	4869	810	6308
50% NPK+ 50% N (GM)	100% NPK	5371	815	6818

Contd...

Maize -wheat				
Ranchi				
100% NPK	100% NPK	4849	5157	9462
50% NPK+ 50% N (GM)	100% NPK	5517	5806	10700
75% NPK+ 25%N (GM)	75% NPK	5095	5257	9769
Rice-maize				
Kathalagere				
100% NPK	100% NPK	6590	4372	10156
50% NPK+ 50% N (GM)	100% NPK	6459	4685	10280
Pearl millet-wheat				
S. K. Nager				
100% NPK	100% NPK	1071	2305	2753
50% NPK+ 50% N (GM)	100% NPK	1453	2940	3583
75% NPK+ 25%N (GM)	75% NPK	1530	2994	3690
Hisar				
100% NPK	100% NPK	3557	5852	7674
50% NPK+ 50% N (GM)	100% NPK	3715	5999	7922
Junagadh				
100% NPK	100% NPK	1450	3325	3895
50% NPK+ 50% N (GM)	100% NPK	1585	3643	4264
75% NPK+ 25%N (GM)	75% NPK	1507	3351	3962
Sorghum-wheat				
Ranchi				
100% NPK	100% NPK	2850	3437	5570
50% NPK+ 50% N (GM)	100% NPK	2854	3458	5591

GM= Green manure
Source : AICRP-IFS Reports

The advantages of cereal crop residues, however, appear on their continuous incorporation over years, as these are relatively resistant to decomposition. Application of 10-20 kg N ha^{-1} at the time of incorporation of residues hastened the rate of decomposition, and consequently increased the beneficial effect in terms of grain yield and soil fertility build-up (Table 8). Since 70-80% of K taken up by these crops is retained in straw component, residue recycling may be the best option to replenish K to the soil and avoid the mining of soil K reserves. The effective ways of residue recycling, so as to harness this potential source is still a important researchable issue for sustainability of rice-wheat system.

Table 7 : Productivity of cereal-cereal cropping systems as influenced by integrated use of crop residue and inorganic fertilizers under AICRP-IFS (mean over 2005-06 to 2009-10)

Treatments		Grain yield (kg ha^{-1})		
Kharif	**Rabi**	**Kharif**	**Rabi**	**Rice equivalent yield**
Rice-wheat				
Kalyani				
100% NPK	100% NPK	3904	3021	7131
50% NPK+ 50% N (CR)	100% NPK	4132	3469	7837
75% NPK+ 25%N (CR)	75% NPK	4051	3177	7444
Rice-rice				
Chiplima				
100% NPK	100% NPK	5335	5517	10852
75% NPK+ 25%N (CR)	75% NPK	5613	5534	11146
Bhubaneswar				
100% NPK	100% NPK	5358	5919	11277
50% NPK+ 50% N (CR)	100% NPK	5510	6215	11725
75% NPK+ 25%N (CR)	75% NPK	5537	6326	11863
Karmana				
100% NPK	100% NPK	4601	4432	9033
50% NPK+ 50% N (CR)	100% NPK	5048	4583	9631
75% NPK+ 25%N (CR)	75% NPK	4920	4511	9431
Rice-mustard				
Rudrur				
100% NPK	100% NPK	4869	810	6308
50% NPK+ 50% N (CR)	100% NPK	4922	887	6499
75% NPK+ 25%N (CR)	75% NPK	5100	806	6532
Peailmillet-wheat				
S. K. Nager				
100% NPK	100% NPK	1071	2305	2753
50% NPK+ 50% N (CR)	100% NPK	1345	2910	3470
75% NPK+ 25%N (CR)	75% NPK	1448	3015	3639
Junagadh				
100% NPK	100% NPK	1450	3325	3895
50% NPK+ 50% N (CR)	100% NPK	1548	3625	4219
75% NPK+ 25%N (CR)	75% NPK	1570	3356	4017

CR- Crop residue
Source : AICRP-IFS Reports

Table 8 : Effect of rice and wheat crop residues incorporation on productivity of the system and soil health

Treatment	Mean grain yield (t ha^{-1})			Soil fertility after 6 cycle		
	Rice	Wheat	Total	OC%	Av P (kg ha^{-1})	Av K (kg ha^{-1})
R. S. Pura (06 years)						
Rec. N, no N at CR addition	4.31	3.39	7.70	0.38	11.5	90
10 kg rec. N at CR addition	4.21	3.45	7.66	0.43	13.2	96
20 kg rec. N at CR addition	3.98	3.33	7.31	0.48	14.5	93
Rec. N+ 10 kg N at CR addition	4.61	3.73	8.34	0.48	15.2	99
Rec. N+ 20 kg N at CR addition	4.46	3.90	8.36	0.46	13.2	96
Kanpur (06 years)						
Rec. N, no N at CR addition	4.41	4.03	8.44	0.29	21.4	188
10 kg rec. N at CR addition	4.29	4.07	8.36	0.33	22.6	190
20 kg rec. N at CR addition	4.23	3.99	8.22	0.31	24.8	195
Rec. N+ 10 kg N at CR addition	4.56	4.36	8.92	0.36	25.8	200
Rec. N+ 20 kg N at CR addition	4.69	4.14	8.83	0.34	26.5	198

Initial values of OC, Available P and available K were 0.43 and 0.10%, 10.4 and 18.4 kg ha^{-1}, and 91.5 and 218 kg ha^{-1}, respectively at R.S. Pura and Kanpur.
Source: Yadav, 1997

Legume crops fix atmospheric N and enrich soil fertility, and could help to sustain the long-term productivity of cereal-based cropping systems. Depending on the soil and ecological stresses, rice-wheat cropping system can be diversified using legumes as a substitute crop (Yadav *et al.*, 1998; Gangwar *et al.*, 2007). In extensive field experiments conducted at different locations of AICRP_IFS has opined the possibilities of inclusion of legumes as a short duration grain crop, forage crop as substitute crops, or as break crop in different cereal based cropping system in order to conserve soil organic matter, improve nutrient use efficiency, and decrease NO_3-N leaching. Raising forage cowpea in summer increased N and P use efficiency in subsequent rice and wheat crops, besides minimizing movement of NO_3-N to deeper soil layers (Dwivedi *et al.*, 2003; Singh *et al.*, 2005). Inclusion of pigeon pea in place of rice in alluvial soils, and that of chickpea in place of wheat in *Tarai* soils increased over all system productivity, N use efficiency in wheat and improved soil organic matter (Singh *et al.*, 2001). The advantage of inclusion of pigeon pea can be increased by raising an extra short-duration (ESD) cv. ICPL 88039 and increasing its leaf residue recycling by foliar spray of 10% urea solution at maturity (Chauhan *et al.,* 2004). Since systems like rice and wheat are staple food grain crops, it is very difficult to substitute any one crop from the system. Under these circumstances, substitution of rice or wheat with a legume as break crop at 2-3 years intervals Say for cowpea in place of rice or berseem in place of wheat may become an alternate viable option for reducing weed infestation and sustaining system productivity (Singh and Mishra, 2010). The benefits of legumes in rotation are not solely due to biological nitrogen fixation but because of increased nutrient availability, improved soil structure, reduced disease incidence and increased mycorhyizal colonization also occur (Wani *et al.,* 1995). Soil organic carbon and available N, P and K increased markedly, when the wheat in the rice-wheat cropping system was substituted with a legume (Hegde and Dwivedi, 1992).

Benefit of bio fertilisers has been studies in detail in different crops with varying rate of success, depending on the type of microbial strains, crops, growing environment and soil fertility level, but cropping system based information is scanty. In multi-location trails carried out under Cropping Systems Research Project, *Rhizobium* inoculation of *kharif* legumes produced residual effect on succeeding wheat and *rabi* sorghum at only 4 out of 17 locations (AICRP-IFS Reports). Crop response to bio fertilizers in irrigated areas reviewed by Hegde and Dwivedi (1994) indicated that wheat responded significantly to *Azotobacter* inoculation in 342 out of 411 trials on cultivators' fields under Cropping Systems Research Project.

Under field conditions, responses to *Azatobacter* and Azospirillium were equivalent to 15-30 kg fertilizer N in cereals other than rice and sugarcane. In rice, *Azolla* and blue-green algae (BGA) contributed 20-30 kg N ha^{-1} (Singh, 1992). Most important characteristic common to bio fertilizers is the unpredictability of response, as the success rate with any biofertiliser in terms of significant impact on yield ranges from 30 to 65%. To harness potential benefits of biofertilizers in commercial agriculture, the consistency of their performance needs to be improved.

Experiments conducted on cultivators' field during 1990-91 to 1994-95 under Cropping Systems Research Network further revealed beneficial effect of integration of chemical fertilizers with green manuring or FYM, as the total productivity of the systems involving cereals, oilseeds and cotton increased by 7 to 45% over farmers' practice in different agro-ecological zones (Table 9).

Table 9 : Influence of integrated nutrient supply on system productivity at cultivators' fields

State/Location/Sequence	Treatment	Yield (kg ha^{-1})			% increase
		Kharif	Rabi	Total	over FP
Orissa/North Central Plateau	T_1	3853	1727	5580	1.0
Rice- groundnut	T_2	4590	2144	6734	21.9
	T_3	4291	2030	6321	14.5
	F.P.	3681	1842	5523	-
Rajasthan/Udaipur	T_1	2908	3760	6668	18.5
Maize-wheat	T_2	2789	3912	6701	19.1
	T_3	2705	3744	6449	14.6
	F.P.	2088	3540	5628	-
U.P./ Ghaziabad	T_1	2631	5193	7824	10.9
Maize-wheat	T_2	2718	5399	8117	15.1
	T_3	2830	5474	7665	8.7
	F.P.	2220	4835	7055	-
U.P./ Aligarh	T_1	2088	4472	6560	9.1
Maize-wheat	T_2	1906	4612	6518	8.4
	F.P.	1671	4342	6013	-
Manipur	T_1	4844	483	5327	9.2
Rice-mustard	T_2	5129	524	5653	15.8
	T_3	5456	540	5996	22.9
	F.P.	4438	442	4480	-
	T_1	1819	3522	5341	19.0
Haryana/ Hisar	T_2	1949	3762	5711	27.3
Cotton-wheat	T_3	1850	3638	5488	22.3
	F.P.	1374	3113	4487	-
	T_1	3409	3723	7132	3.5
West Bengal/ Kalyani	T_2	3243	4046	7289	5.8
Rice-wheat	T_3	4049	4249	8298	20.5
	F.P.	3174	3715	6889	-
	T_1	4646	3887	8533	42.9
U.P. Ghazipur	T_2	4124	4107	8231	37.8
Rice-wheat	T_3	4265	3959	8224	37.7
	F.P.	3060	2912	5972	-
	T_1	5293	4135	9428	20.1
Haryana/ Zind	T_2	5391	4221	9612	22.5
Rice-wheat	T_3	5353	4163	9516	21.3
	F.P.	4537	3311	7848	-

Contd...

M.P. Jabalpur	T_1	1935	1801	3736	31.3
Rice-wheat	T_2	1874	1725	3599	26.4
	T_3	1758	1598	3356	17.9
	F.P.	1467	1380	2847	-
Punjab/Kapurthala & Jalandhar	T_1	6151	4759	10910	1.3
Rice-wheat	T_2	6413	4825	11238	4.3
	T_3	6106	4918	11024	2.3
	F.P.	6611	5095	11706	-

F.P.=Farmers' Practice and T_1, T_2 and T_3 stand for recommended NPK, NPK+FYM and NPK+Green manure, respectively.

Source : AICRP-IFS Reports

b) INM under pulse based systems

Pulse crops are energy rich and remove sizeable quantity of nutrients from the soil. For producing 1 tonne of biomass, pulse crop generally remove 30 to 50 kg N, 2-7 kg P, 12-30 kg K, 3-10 kg Ca, 1-5 kg Mg, 1-3 kg S, 200-500 g Mn, 5 g B, 1g Cu and 0.5 g Mo (AICRP-DA, 2003). On the other hand, most of the pulses in the country are grown on degraded or marginal lands. Also the nutrient prescriptions for pulses are mainly confined up to N and P. Therefore, water and nutrient stress is a major cause of low productivity of these crops. Hence, there is a need to integrated organic manures with inorganic to improve the water holding capacity and nutrient use in pulse based production systems. The results of AICRP-DA in INM project indicated that use of 50% of organic in combination with 50% of inorganic not only enhanced the productivity of pulses in different cropping systems but also improve the fertility of the rain fed soils. Biomass production as a part of an organic source of fertilizer can be enhanced by growing bushes on field boundaries for lopping/ litter fall recycling, use of organic waste materials for vermin compost preparation, production of legumes as preceding crop (Summer or *Kharif* crop) and their incorporation such as urdbean or mungbean after picking of pods (ACIAR, 2002). Long-term studies conducted at PDFSR, Modipuram indicated that growing summer mung bean and its incorporation after picking of pods under rice-wheat system not only enhanced the system productivity in terms of rice equivalent yield but also mitigated sub-soil surface compaction induced by puddling and improved the soil health (Singh and Mishra, 2010; Gangwar *et al.*, 2007). The advantage of legume to succeeding crops due to its BNF ability are also reported by Dwivedi *et al.*, (2003), Singh *et al.*, (2001), Singh *et al.*, (2005), Singh and Dwivedi *et al.*, (2006); Gangwar and Gangwar (1996).

Table 10 : Effect of integrated nutrient management on yield and sustainability of pulses in intercropping systems

Center	Crops/ cropping systems	Nutrient management	Yield (kg ha^{-1})		SYI	
			Main crop	Inter crop	Main crop	Inter crop
Faizabad	Maize+ pigeonpea (5)	100% recommended N	1224	1060	0.38	0.31
		50% recommended N	948	718	0.25	0.16
Phulbani	Pigeonpea+rice (5)	45 kg N ha^{-1}(urea)	606	1576	0.17	0.52
		20 kg N (FYM) + 25 kg N ha^{-1} (urea)	785	1585	0.26	0.52
Ranchi	Rice+ urdbean (3)	40 kg N ha^{-1} (urea)	1944	820	0.74	0.73
		15 kg N (compost) + 20 kg N ha^{-1} (urea)	1691	916	0.62	0.82
	Rice+lentil (5)	100% RDF (inorganic)	2547	618	0.30	0.14
		50% N (urea)+ 50% N (FYM)	2701	903	0.33	0.31
		100% RDF (inorganic)	2285	692	0.24	0.22
		15 kg N (green leaf) +20 kg N ha^{-1} (urea)	2274	728	0.21	0.26
Bijapur	Rabi sorghum + chickpea (7)	100% RDF (inorganic)	1235	785	0.57	0.62
		15 kg N (compost) + 20 kg N/ha (urea)	1354	872	0.65	0.69
Bellary	Rabi sorghum + chickpea (7)	15 kg N (Green leaf)+ 20 kg N ha^{-1} (urea)	773	370	0.03	0.23
		15 kg N (Green leaf)+ 10 kg N ha^{-1} (urea)	719	353	0.01	0.21

Values under parenthesis indicates number of years of experimentation
Source: Ramakrishna *et al.*, 2007

With integrated use of organic and inorganic nutrients under different intercropping system not only sustainable productivity was noticed (Table 10) but also reduction in cost of production was ensured. Seed treatment along with *Rhizobium* inoculation added further advantages as it enhances

BNF, which met 80-90% N requirement of pulse crop also leaves 30-40 kg N for succeeding crop (Awonike *et al.*, 1990; Singh and Dwivedi, 2006). Studies conducted at Kanpur indicates that *Rhizobium* bio inoculants saved 50% of recommended doses of Fertilizers (RDF) like N, P and S under rain fed and irrigated conditions of clay and clay loam soils (TAR-IVLP-2005).

c) INM in oilseed based cropping systems

The nutrient requirement of oilseed crops in general is high for all the nutrients and need to be supplied in adequate quantities for higher yields (Hedge and Babu, 2001). Inadequate and/or imbalance use of fertilizers has been identified as one of the critical constraints limiting oilseed production. Subba Rao (1994) has reported the yield increase ranging from 26 to 300% with fertilizers application alone under rain fed areas which constitute about 75% of the total oilseeds acreage. Research experiences indicated that some of the cropping systems involving oilseeds crops may remove as much as 400 to 800 kg nutrients ha^{-1} year^{-1} under high productivity conditions (Table 11). On the other hand, wide spread nutrients deficiencies and responses to N, P, K, S and the micronutrients like Zn, Mn, Fe and B in one or the other oilseeds crops are very common. Replenishment of all these nutrients being removed by the crops is essential for maintaining soil fertility and sustainable productivity. For leguminous oilseeds like groundnuts and soybean, the main nutrient form fertilizer management point of view are P, S, Ca, and Zn while for rapeseed-mustard, application of N, P, S and in some case Zn and B plays key role to stepping up the yield. For safflowers and sunflower, the major requirements are of N, P and S in certain cases. In course textured soils at high yield levels and where leaching of K occurs, it application become critical. Iron assumes importance in alkaline-calcareous soils and Mo in very acidic soils, particularity in groundnut and soybean. An appropriate combination of fertilizer and organic sources like FYM, compost, green manures, legumes in system, crop residue, recyclable wastes and bio fertilizer can ensure many of these production related issues.

Fertilizer management on a system basis rather than an individual crop leads higher nutrient use efficiency and economics besides sustainability. The low level of utilization of nutrients supplied through fertilizers and manures can be overcome by choosing appropriate combination of sequence of crops to effectively utilize the nutrients. The specific attention needs to be given to harness the residual effects of fertilizers containing P and K that application of these two nutrients can be phased to get maximum benefit. In wheat-groundnut, maize-groundnut and *Raya*-groundnut cropping systems groundnut crop was observed to thrive on residual effect when the preceding crops received recommended fertilizers (Subba Rao, 1994). Long-term studies conducted at Directorate of Oilseed Research, Hyderabad indicated that 50% RDF saving in sunflower based system after third crop cycle when rotated with groundnut, which was comparable with that of 150% RDF in sunflower-sunflower sequence (Reddy and Sadhakar Babu, 1996). A sowing of 50% of N to sunflower was possible when preceding crops was green gram compared to sorghum or fallow (DOR, 1998).

Table 11 : Nutrient uptake in some cropping systems involving oilseeds

Cropping system (t ha^{-1})	Nutrient uptake (kg ha^{-1})			
	N	P_2O_5	K_2O	Total
Sorghum (3.6) - Groundnut (2.2)	216	74	184	474
Maize (3.0) - Mustard (2.0)	127	70	177	374
Cotton (2.0) - Sunflower (1.8)	203	90	376	669
Soybean (1.5) - Potato (26.0) - Wheat (4.0)	305	177	345	827
Sorghum (4.0) - Sunflower (1.5) - Groundnut (2.0)	301	121	385	807
Rice (4.8) - Rice (5.7) - Groundnut (2.0)	316	153	369	838
Rice (4.7) - Rice (5.6) - Sesame (0.7)	243	131	354	728
Sorghum (F) (18.9) - Potato (29.0) - Sunflower (1.8)	296	121	481	898
Maize (2.8) - Potato (27.0) - Sunflower (1.9)	296	110	479	885
Soybean (1.9) - Wheat (4.8)	252	100	267	619
Sunflower (1.6) - Groundnut (3.4)	294	97	304	700

Source: Hegde, 1992

In groundnut application of FYM @ 7.5 t ha^{-1} increased the pod yield by 60% over control and 27% over recommended NPK (Agasimani and Hosmani, 1989). The sesamum yield was improved by 29% with application 2.5 t ha^{-1} enriched FYM. AT Sardar Krishi Nagar, caster yield was similar when 25% N was substituted with mustard cake or 75% N through caster cake and 25 through urea N (DOR, 1992). The results obtained from Long-Term Fertilizer Experiments (LTFE) reveals that an integration of FYM over 100% recommended NPK had edge over 150% NPK application.

Groundnut and soybean being leguminous oilseed crops inclusion under non-legume crops may have saving of 30-40 kg ha^{-1} N to succeeding crops. Similarly, inclusion of non-leguminous oil seeds may have residual benefit of 20-30 N ha^{-1} to succeeding oilseed crop. In country like India, where fertilizer use is still very low and use of nutrient particularly in oilseed is very meagre, the residual fertility build-up due to legumes in obviously a major contribution, which must be exploited for enhancing oilseed based systems productivity.

Bio-fertilizers are a potential source of supply of nutrients at low cost. There is scope for exploiting the ability of leguminous oilseeds-groundnut (112-115 kg N and soybean (49-130 kg N ha^{-1} to fix atmospheric N to meet in part or fully N requirement (Wani and Lee, 1992). In groundnut new *Rhizobium* strain IGR-6 and IGR-50 were found tolerant to thiram and hence seed treatment with fungicide and inoculation of *Rhizobium* can go together. In groundnut, yield improvement of 5.5-17.1% was obtained due to use *Rhizobium* culture. Similarly *Azospririllum* seed treatment in sesame and *Azatobacter* in *Toria* and sunflower reduces up to 50% of the total N requirements of the crops.

Alike N, P requirement of oilseeds can be met through seed inoculation with P solubilises like *Pseudomonas straita* and *Paecilomyces fussispores* (Mehta *et al.*, 1995). Use of P solubilising bacteria in conjunction with *Neem* cake, caster cake or FYM had higher yield of several oilseed crops and its residual benefit to the systems was noticed by several workers (Dubey, 2001; Baldev Ram and Pareek, 2000; DOR, 2001; DOR, 2002).

Conclusions and future research priorities

Integrated nutrient management holds great promise in meeting the growing nutrient demands of different crops in intensive agriculture. It can also help in maintaining production sustainability without deterioration in the quality of plant's environment. Despite many constraints like use of cattle dung as fuel, increasing competitive value of crop residues as animal feed, extra cost and time required for raising green manure crop, poor and inconsistent responses to bio-fertilisers etc., efforts must be made to develop agro-ecoregion specific practical recommendations for different cropping systems utilising the already generated data on different components of INM. Many research gaps need to be filled and the following are some of the important future research priorities:

- There is a need to prepare an inventory of promising organic and biological sources of nutrients available under different agro-eco-regions involving diverse farming situations.
- For accruing maximum potential benefit of INM fertilizer recommendation packages for different cropping systems have to be refined by collating and synthesising the existing information's.
- The location based INM strategies for different cropping systems needs to be planned and efficiency as on its productivity, profitability and sustainability and eco-friendliness.

- Based on available information on LTEs on INM computer besed models has to be developed for various IPNS options available under different cropping systems. Such approach may be helpful for advising the site-specific IPNS packages for varying farming systems, farm size, input availability and socio-economic situations.
- There is a need to develop methodologies for quantifying the non-nutrient benefits of using organic sources of nutrients which are important for production sustainability of crop production.
- Different industrial wastes (solid and liquid) with potential manurial value need to be characterised and their long-term effects on soil-plant-human contineoum has to be studied.
- Studies on in-situ decomposition of crop residues including stubbles in relation to soil properties and crop productivity are needed on short, medium and long term basis.
- An elaborated programme for monitoring of long-term changes under soil chemical, physical, biological and environmental dynamics is a essential need for assessing the INM options on long-term basis.
- INM studies needs to be focussed on modern RCTS, which have potential to recycle the maximum in-situ crop residue and its impact on soil micro-flora and fauna has to be addressed.
- Studies on soil organic matter dynamics and carbon sequestration under already well established LTEs under AICRP_IFS or LTFE has to be made on regular basis and cropping system wise location specific information's needs to be generated. Such studies may become useful to establish the cause and effect relationship under various cropping systems associated with deterioration of soil health and reduction in soil organic C stock under varying nutrient management options.
- Since cereal-cereal system are prime contributors to national food production, it will be desirable to include a legume in the system. Therefore, inclusion of legume as break crop or catch crop or as summer crop or inter crop/mixed crop needs to be studied under different cropping systems.
- Biofertilizer technology needs to be refined to make it more acceptable to farmers. Microbial strains which can compete with indigenous ones and work efficiently over a wide range of soil edaphic and climatic conditions need to be isolated and multiplied.
- Agro-forestry and live-stock based integrated farming systems should be encouraged for enhancing the nutrient recycling under different cropping systems.

- Mass awareness on conservation of organic source and their recycling needs to be done and promotional literature should be prepared in local language and made available to the farmers and extension personnel in addition to some adaptive research, demonstrations and specific training programmes.

References

ACIAR (2002). Tools and indicators of sustainable soil management practices in semi-arid tracts of India. Final report jointly published by Central Research Institute for Dry land Agriculture in collaboration with ACIAR Team, Australia.

Agasimani, C.A. and Hosmani, M.M. (1989). Response of groundnut crop to farm yard manure, nitrogen and phosphorus in rice fallows in coastal sandy soils. *Journal of Oilseeds Research*, **6** (2): 360-363.

Agricultural Statistics at a Glance (2007). Directorate of Economics and Statistics, Department of Agriculture and Cooperation, Ministry of Agriculture, Govt. of India, New Delhi.

AICRPDA (2003). Annual progress report of All India Coordinated Research Project for Dryland Agriculture, Central Research Institute for Dryland Agriculture, Hyderabad.

AICRP-IFS Reports (2005-2010). Project Directorate for Farming Systems Research Modipuram, Meerut, India.

Awonaike, K.O., Kumarasinghe, K.S. and Danso, S.K.A. (1990). Nitrogen fixation and yield of cowpea as influenced by cultivar and *Bradyrhizobium* strain. *Field Crops Research*, **14** : 163-171.

Baldev Ram and Pareek, R.G. (1999). Effect of phosphorus, sulphur and PSB on growth and yield of mustard. *Agricultural Science Digest*, **19**: 203-206.

Buresh, R.J. and De Datta, S.K. (1991). Nitrogen dynamics and management in rice-legume cropping systems. *Advances in Agronomy* **45** : 1-59.

Chauhan, Y.S., Apphun, A., Singh, V.K. and Dwivedi, B.S. (2004). Foliar sprays of concentrated urea at maturity of pigeonpea to induce defoliation and increase its residual benefit to wheat. *Field Crops Research*. **89** : 17-25.

DOR (1992). AICORPO (Caster), Annual report, Directorate of Oilseed Research, Rajendranagar, Hyderabad.

DOR (1998). AICORPO (Safflower), Annual report, Directorate of Oilseed Research, Rajendranagar, Hyderabad.

DOR (2001). AICRP (Safflower), Annual report, Directorate of Oilseed Research, Rajendranagar, Hyderabad.

DOR (2002). AICRP (Safflower), Annual report, Directorate of Oilseed Research, Rajendranagar, Hyderabad.

Dubey, S.K., (2001). Balance sheet of N and P as influenced by phosphate solubilising bacteria and phosphate fertilisation in rain fed soybean grown Vertisols. *Annals of Agricultural Research*, **22** (10): 162-164.

Dwivedi, B. S., Singh, D., Chonkar, P.K., Sahoo, R.N. Sharma, S. K and Tiwari, K.N. (2006). Soil Fertility evaluation-a potential tool for balanced use of fertilizer, pp.1-66, IARI New Delhi-PPI-PPIC, India programme, Gurgaon, India.

Dwivedi, B.S., Shukla, A.K., Singh, V.K. and Yadav, R.L. (2001). Results of participatory diagnosis of constraints and opportunities (PDCO) based trials from the state of Uttar Pradesh. In: A. Subba Rao, S. Srivastava (Eds), *Development of farmers' resource-based integrated plant nutrient supply systems: experience of a FAO-ICAR-IFFCO collaborative project and AICRP on soil test crop response correlation* Bhopal: Indian Institute of Soil Science. pp. 50-75.

Dwivedi, B.S. Shukla, Arvind K., Singh, V.K. and Yadav, R.L. (2003). Improving nitrogen and phosphorus use efficiencies through inclusion of forage cowpea in the rice-wheat system in the Indo-Gangetic Plains of India. *Field Crops Research.* **80**: 167-193.

Dwivedi, B.S., Shukla, A.K., Singh, V.K. and Yadav, R.L. (2001). Sulphur fertilization for sustaining productivity of rice-wheat system in western Uttar Pradesh. PDCSR Bulletin No. 2001-1, 35 pp. PDCSR, Modipuram.

FAI (2010). Fertiliser Statistics, 2009-10. The Fertilizer Association of India, New Delhi.

Gangwar, B., Tripathi, S.C., Singh, J.P., Kumar, Ravi, Singh, R.M. and Samui, R.C. (2005). Diversification and resource management of rice-wheat system. Research Bulletin No- 05/1. PDCSR, Modipuram, Meerut, pp 1-68.

Gangwar, K.S. and Gangwar, B. (1996). Integrated nutrient management in cropping systems, A review. *Agricultural Review*, **17** (1): 57-73.

Gill, M.S. and Singh, V.K. (2009). Productivity enhancement of cereals through secondary and micronutrients application. *Indian Journal of Fertilizer* **5** (4): 59- 80 and 106.

Gill, M.S., Shukla, Arvind, K. and Pandey, P.S. (2008). *Indian Journal of Fertilizer* **4** (4) : 11-48.

Hegde, D.M. (1992). Cropping systems Research Highlights: Coordinator's Report 20th workshop of Project Directorate for Cropping Systems Research, 1-4 June , 1992. TNAU, Coimbatore, pp.39.

Hegde, D.M. and Dwivedi, B.S. (1992). Nutrient management in rice-wheat cropping system in India. *Fertilizer News* **37**(2) : 27-41.

Hegde, D.M. and Dwivedi, B.S. (1994). Crop response to bio fertilizers in irrigated areas. *Fertilizer News*, **39** : 19-26.

Hegde, D.M. and Sudhakarababu, S.N. (2001). Nutrient management strategies in agriculture: A future outlook. *Fertiliser News*, **46** (12) : 61-66 & 71-72.

Kanniayan, S. (1998). Integrated plant nutrient supply system. In: C.L. Acharya, K.P. Tomar, A. Suibba Rao, T.K. Ganguly, M.V. Singh, D.L.N. Rao, T.R. Rupa (Eds),

Integrated plant nutrient supply system for sustainable productivity : Indian Institute of Soil Science. pp. 10-13.

Katyal, V., Gangwar, B. and Gangwar, K.S. (1999). Cumulative effect of long- term fertiliser use on crop productivity in rice-wheat system. *Bioved*, **10** (1,2) : 25-30

Ladha, J.K., Dawe, D., Pathak, H. Padre, A.T., Yadav, R.L., Bijay-Singh, Yadvinder-Singh, Singh, Y., Singh, P., Kundu, A.L. Sakal, R., Ram, N., Regmi, A.P., Gami, S.K., Bhandari, A.L., Amin, R.C., Yadav, R. Bhattarai, S., Das, S., Aggarwal, H.P., Gupta, R.K. and Hobbs, P.R. (2003). How extensive are yield declines in long-term rice-wheat experiments in Asia? *Field Crops Research* **81** : 159–180.

Meelu, O.P., Palaniappam, S.P., Singh, Y and Singh, B. (1992). Integrated nutrient management in crops and cropping sequence for sustainable agriculture. *In: Proceedings, International Symposium "Nutrient management for sustained productivity"* pp. 104-114, PAU Ludhiana.

Mehta, A.C., Malavia, D.D., Kaneria, B.B. and Khanpara, V.B. (1995). Effect of phosphatic biofertilisers in conjunction with organic and inorganic fertilisers on growth and yield of groundnut (*Arachis hypogaea*). *Indian Journal of Agronomy*, **40** (4) : 709-710.

Rama Krishna, Y.S., Subba Reddy, G., Ravindra Chary, G. and Prasad, Y.G. (2009). Improving productivity of grainlegumes under moisturestrss conditions. In: Massod Ali, Sanjeev Gupta, P.S. basu and Naimuddin (Eds.). *Legumes for ecological sustaiunability*. Indian Society of Pulse research and development, IIPR, Kanpur, India, pp 92-117.

Reddy, B.N. and Sudhakarababu, S.N. (1996). Production potential, land utilization and economics of fertiliser management in summer sunflower (*Helianthus annuus* L.) based crop sequences. *Indian Journal of Agricultural Sciences*, **66** (1) : 16-19.

Sanyal, S.K. and Chatterjeee, S. (2007). Efficient use of soil, water and plant nutrients for food and environmental security. *Indian Journal of Fertilizer*, **3** (9) : 71-132.

Singh, B., Singh, Y. and Sekhon, G.S. (1995). Fertilizer N use efficiency and nitrate pollution of groundwater in developing countries. *Journal Contaminant Hydrology* **20** : 167-184.

Singh, M.V., 2001.Current status of micronutrients in different agro ecological zones of India. *Fertilizer News*. **46** (2) : 25-48.

Singh, P.K. (1992). Bio fertilizers for flooded rice ecosystem. *In: Fertilizers, Organic manures, recyclable waste and bio fertilizers* (Tandon, H.L.S. Eds), pp 113-131, FDCO, New Delhi.

Singh, R.P., Prasad, K. and Katyal, V. (1994). *Highlights of cropping systems research*, pp. 35, PDCSR, Modipuram.

Singh, V.K and Dwivedi, B.S. (2006). Yield and N use-efficiency in wheat, and soil fertility status as influenced by substitution of rice with pigeon pea in a rice-wheat cropping system. *Australian Journal of Experimental Agriculture*, **46** : 1185-1194.

Singh, V.K. and Gangwar, B. (2011). Sulphur management in crops and cropping systems for sustainable production. PDFSR Bulletin No. 1, pp. 70. Project Directorate for Farming Systems Research, Modipuram, Meerut, India.

Singh, V.K. and Mishra, R.P. (2010). Integrated nutrient management in transplanted rice-wheat system. PDCSR Annual Report 2007-2008, pp. 44-50, PDCSR, Modipuram, Meerut, India.

Singh, V.K., Dwivedi, B.S., Shukla, Arvind K., Chauhan, Y.S. and Yadav, R.L. (2005). Diversification of rice with pigeonpea in a rice-wheat cropping system on a Typic Ustochrept: effect on soil fertility, yield and nutrient use efficiency. *Field Crops Research,* **92** : 85-105.

Singh, V.K., Sharma, B.B. and Dwivedi, B.S. (2002). The impact of diversification of a rice-wheat cropping system on crop productivity and soil fertility. *Journal of Agricultural Science,* **139** : 405-412.

Singh, Y., Khind, C.S. and Singh, B. 1991 b, Efficient management of leguminous green manures in wet land rice. *Advances in Agronomy,* **45** : 135 –187.

Subba Rao, I.V. (1994). Integrated nutrient management for sustainable productivity of oilseeds in India. *In:* M.V.R. Prasad (Eds). *Sustainability in Oilseeds,* Indian Society of Oilseeds Research, DOR, Hyderabad, pp.264-270.

Tandon, H.L.S. (1997). Plant nutrients needs supply, efficiency and policy issues. NAAS, New Delhi, pp 15-28.

TAR-IVLP (2005). Technology assessment and refinement through Institute Village Linkage Programme (Completion Report, 1999-2004). Agro-Ecosystems Directorate (Rainfed) NATP, CRIDA, Hyderabad. pp 74-78.

Tiwari, K.N. (2002). Nutrient management for sustainable production. *Journal of the Indian Society of Soil Science,* **50** (4) : 374-397.

Wani, S.P., Rupela, O.P., and Lee, K.K. (1995). Sustainable agriculture in the semi-arid tropics through biological nitrogen fixation in grain legumes. *Plant and Soil,* **174** : 29-49.

Wani, S.P. and Lee (1992). Role of biofertilisers in upland crop production. *In:* H.L.S. Tandon (Ed). *Fertilisers Organic Manures, Recyclable Wastes and Biofertilisers,* Fertiliser Development and Consultation Organization, New Delhi. pp.91-112.

Yadav, R.L., Dwivedi, B.S., Gangwar, K.S. and Prasad, K. (1998). Over view and prospects for enhancing residual benefits of legumes in rice and wheat cropping systems in India. In: *Residual effects of legumes in rice and wheat cropping systems of the Indo-Gangetic Plain* (Eds J. V. D. K. Kumar Rao, C. Johansen and T. J. Rego). pp 207-225. Patancheru: International Crops Research Institute for the Semi-arid Tropics.

Yadav, R.L. (1997). Urea-N management in relation to crop residue recycling in rice-wheat system in north-western India. *Bioresource Technology,* **61** : 105-109.

Yadav, R.L., Dwivedi, B.S., Prasad, K., Tomar, O.K., Shurpali, N.J. and Pandey, P.S. (2000). Yield trends, and changes in soil organic-C and available NPK in a long-term rice-wheat system under irrigated use of manures and fertilizers. *Field Crops Research,* **68** : 219-246.

❑❑❑

System Based Integrated Nutrient Management, 2012
© B. Gangwar & V.K. Singh (eds.), pp. 29-41
New India Publishing Agency, New Delhi (India)
e-mail : info@nipabooks.com; website : www.nipabooks.com

CHAPTER 2

Crop Residue Management in Rice-Wheat System

YADVINDER SINGH

Rice-wheat (RW) is the major cropping system occupying about 10.5 mha in the IGP of India. It contributes more than 60% of the foodgrains to the central pool. With shrinking land resources and burgeoning population, the RW system will be under pressure to produce more grain output per unit area in the coming decades. Meeting this challenging task requires sustainable management of soil, water and plant nutrient resources. High yields of the irrigated RW system result in production of huge quantities of crop residues. Total production of crop residues in Punjab is about 47.2 mt which includes paddy straw, wheat straw, cotton sticks, sugarcane leaves, maize stalk, and other oilseeds and pulse residues (Beri and Gupta, 2003). Wheat and rice straw constitutes more than 70% of the crop residues produced in the country. Paddy straw alone constitutes more than 50%. Cotton and sunflower residues are generally used as fuel. Sugarcane trash (leaves) is generally burnt in-situ or used as fuel after collection from the field. Increasing constraints of labour and time have led to the adoption of mechanized farming in the highly intensive RW cropping system. Approximately 91% of the total rice area is mechanically harvested, while 82% of the total wheat area is mechanically harvested. Traditionally, wheat and paddy straws have been removed from the fields for use as cattle feed and for several other purposes such as livestock bedding, thatching material for houses, and fuel. While more than 80% of wheat residue is collected by the farmers after combine harvesting and often fed to animals, paddy straw is considered poor feed for animals due to its high silica content.

Substantial loss of plant nutrients (especially N and S) and organic carbon occurs during burning of crop residues, with important implications to soil health. About 40% of the N, 60-85% of the K, 30-35% of the P, and 40-50% of the S absorbed by rice remains in the vegetative parts at maturity. One tonne of paddy straw contains approximately 6-7 kg of N, 1.0-1.7 kg of P_2O_5 and 14-25 kg K_2O. Apart from huge loss of precious plant nutrients, burning of crop residues depletes soil health. Air pollution from stubble burning also impacts human and animal health both medically, and by traumatic road accidents due to restricted visibility. In view of the serious problems associated with the residue burning, new ways are being sought to efficiently utilize the huge amount of surplus residues produced in the country.

Management options for crop residues

There exist several options for managing crop residues. Residues can be removed from the field, left on the soil surface, incorporated into the soil, burnt in situ, composted or used as mulch for succeeding crops. Crop residues removed from the field can also be used as animal feed, bedding for animals, as a substrate for composting, biogas generation or mushroom culture or as a raw material for industry. Local conditions determine the disposal method. The complete removal of straw from the field is adopted in basmati type rice growing areas or where there is great demand of straw as fodder and bedding, or for industrial purposes, causing large nutrient export from paddy fields. Burning has to be discouraged in view of its harmful effects on the environment and soil productivity. In RW system, management of paddy straw, rather than wheat straw is a serious problem, because there is very little turn-around time between rice harvest and wheat sowing and due to the lack of proper technology for their recycling. There are few options for paddy straw because of poor quality for forage, bioconversion, and engineering applications. After combine-harvesting, paddy residues remain scattered in the field and are difficult to collect, and impede subsequent seedbed preparation. Rice growers are therefore seeking alternative options, including direct seeding of wheat into paddy residues.

In-situ incorporation of residues

Incorporating residues improves soil organic matter level and returns nutrients to the soil. However, incorporating rice residues before wheat planting is challenging for farmers because of the short interval between rice harvest and wheat planting. The incorporation of crop residues with high C-to-N ratio into soil typically results in microbial N immobilization and a temporary decrease in plant-available N. This initial period of several weeks of net N immobilization is followed by net N mineralization (Yadvinder-Singh *et al.,* 2005). The duration of net N immobilization and the net supply of N from crop residue to a subsequent crop depend upon decomposition rate, residue quality, and environmental conditions. Studies have shown no significant wheat yield increase upon rice residue incorporation when fertilizer N was applied at rates sufficient to meet the crop requirements for supplemental N (Yadvinder-Singh *et al.,* 2004; Bijay-Singh *et al.,* 2008). A 7-year study by Yadvinder-Singh *et al.* (2004) demonstrated that rice and wheat productivity were not adversely affected when rice residue was incorporated for at least 10 days and preferably 20 days prior to establishment of the succeeding crop (Table 1).

Table 1 : Effect of long-term (7 years) paddy straw management on wheat yield, N recovery efficiency and rice yield in subsequent crops (Yadvinder-Singh *et al.*, 2004)

Treatment to wheat	Wheat yield (t ha^{-1})	Recovery efficiency of N in wheat (%)	Rice yield (t ha^{-1})
Paddy straw removed	4.94a	52bc	6.19bc
Paddy straw Burned	5.10a	56ab	6.25bc
Paddy straw incorporated-40 days	5.17a	54ab	6.34bc
Paddy straw incorporated-20 days	5.22a	56ab	6.29bc
Paddy straw incorporated-10 days	4.95a	53ab	6.33bc
Paddy straw incorporated-20 days + 25% N at incorporation	4.97a	49c	6.29bc

This study showed that rice residue decomposition of about 25% during the pre-wheat fallow period was sufficient to avoid any detrimental effects on wheat yields. Under laboratory conditions, no immobilization of fertilizer N was observed when paddy residues were incorporated at 20 days or more before fertilizer application. The N was applied in two equal split doses; half at sowing and half at the time first irrigation as recommended for wheat with no residues in the field. Recovery efficiency of N was decreased by the application of starter-N. Thus, in general, paddy straw can be managed in situ successfully by allowing sufficient time between the incorporation and sowing of the wheat crop. Nitrogen release from rice residue ranged from 6 to 9 kg N ha^{-1} during the wheat season (Yadvinder-Singh *et al.,* 2004). Paddy straw incorporated in wheat did not show a residual effect on the succeeding rice crop.

Yadvinder-Singh *et al.* (2005) and Bijay-Singh *et al.* (2008) have concluded, based on a review of literature, that incorporation of rice residue has a small effect on subsequent wheat yields during the short term of 1 to 3 years. From a 4-year study, Gupta *et al.* (2007) reported that incorporation of paddy straw showed no effect on wheat yield during the first three years but wheat yield increased significantly in the 4th year compared with removal or burning of straw. Long-term incorporation of rice residue can increase readily mineralized organic soil N, suggesting potential after several years for reducing fertilizer N rates for optimal rice yield. The practice of straw incorporation has not been adopted by farmers because of high incorporation costs and higher energy and labour requirements. The technology for the incorporation of paddy straw involves use of straw chopper followed by rotavator and allowing residue to decompose for 15-20 days. This involves an additional cost of Rs. 2,500 ha^{-1}.

Crop residues as surface mulch

The use of crop residues as a mulching material under optimal conditions has been found beneficial as it reduces maximum soil temperature and conserves water. The effectiveness of mulch to reduce soil water evaporation depends upon the soil type, rainfall pattern and evaporative demand (Jalota *et al.*, 2001). The evaporation reduction is more and lasts longer in finer textured soils and under high evaporative demand. More favourable soil temperature and higher water content under mulched than un-mulched soil increases mineralization of soil and applied N. This has implications for N fertilization of crops. Straw mulching of crops has been found to be useful in both irrigated and rain-fed environments.

Straw mulching in wheat

Minimum and zero-till technologies for wheat have been demonstrated beneficial in terms of profitability, water savings and timeliness of sowing, in comparison with conventional tillage. However, there are problems with direct drilling of wheat into combine harvested rice fields. Loose straw accumulates in the seed drill furrow openers, the seed metering drive wheel loses traction due to the presence of loose straw and the depth of seed placement is non-uniform due to frequent lifting of the implement under heavy trash conditions. Until now, no suitable machine was available which can sow wheat directly into a heavy straw load without burning. New developments in machinery were needed to overcome the problems of direct sowing wheat into paddy residues. A new machine, known as the 'Happy Seeder', has been developed for this purpose by Punjab Agricultural University. Happy Seeder is capable of direct drilling wheat into heavy rice residue loads, without burning in a single operation. The machine achieves this by managing only that part of the straw load that comes just in front of furrow openers.

Nitrogen fertilizer and irrigation management in wheat sown into paddy residues

Nitrogen fertilizer management in wheat

To facilitate adoption of the Happy Seeder, farmers need clear guidelines for optimum fertiliser management in wheat. Adjustments in timing and rate of fertilizer N are necessary to optimally supply N to crops receiving residues. Reducing fertilizer N contact with the straw by drilling the fertilizer below the soil surface and/or delayed top dressings of fertilizer may reduce N losses and increase N use efficiency in wheat. A field experiment was conducted to study optimum dose, time and method of fertiliser N application for wheat sown with the HS. The study showed that N application in 3 splits doses (25 kg N ha^{-1} as DAP drilled at sowing and top dressing 48 kg N/ha each before first irrigation and second irrigation) resulted in significantly higher grain yield and N use efficiency than all other treatments including the recommended practice of two equal splits at sowing and with the 1st post-sowing irrigation (2).

Table 2 : Effect of method and time of N application on grain yield and N use efficiency of wheat sown with Happy Seeder

Treatment (N applied (kg ha^{-1}) at sowing-1st irrigation-2nd Irrigation)	Grain yield (t ha^{-1})	Recovery efficiency of N (%)
T1 -No N control	2.82	-
T2 - 25*D +35*B- 60-0	4.42	45.0
T3- 25D+35B-30-30	4.29	44.1
T4- 25D+65B-0-30	4.27	41.9
T5- 25D+95B-0-0	4.02	39.1
T6- 25D-48-48	4.79	56.7
T7- 25D+35*PSI-60-0	4.37	47.8
T8- 25D+35PSI-30-30	4.36	49.4
T9- 25D+65PSI-0-30	4.34	45.3
T10- 25D+95PSI-0-0	4.40	45.4
LSD (0.05)	0.31	2.8

D- drill, B-broadcast at sowing, PSI (Ed – need to define this acronym), before pre-sowing irrigation

Recycling of crop residues has often been promoted in rice-based systems for maintenance of soil fertility and savings in fertilizer use. Fifteen replicated single year on-farm experiments were conducted at three locations to study the response of wheat sown into paddy residues to fertilizer N application.

Nitrogen was applied in two equal split doses (one-half broadcast before sowing and remaining one-half top-dressed before first post-sowing irrigation). Wheat generally responded to the application of N up to 120 kg N ha^{-1} similar to that recommended for conventionally sown wheat. On the basis of these findings, there appears to be no need to apply higher doses of N to wheat sown into paddy residues than that recommended for conventional till wheat.

In contrast to the above results, in experiments conducted by Narang *et al.* (1999), wheat responded significantly to the application of 160 kg N ha^{-1} during the first two years of straw incorporation (both rice and wheat) as compared to recommended N rate of 120 kg N ha^{-1} when residues are removed. In the third year of the study, a significant response to fertilizer N was observed up to 120 kg N ha^{-1} in straw amended wheat plots. Results from this study suggested that it was beneficial to apply a further 25-30 kg ha^{-1} of N (compared to the rates recommended for straw removal fields) to wheat on straw amended fields during initial 1-2 years after residue incorporation. After this initial period, recommended N rates may be sufficient to achieve comparable or higher yields to paddocks where straw has not been recycled.

A three year study by Thuy *et al.* (2007) showed little or no net benefit of incorporating rice residue on the supply of N to wheat or in terms of wheat yield. It was found however, that fertilizer N required for rice to achieve the target agronomic efficiency of N (AEN) was typically reduced when rice residue was incorporated. Several years of residue incorporation in a RW system might be required before the N benefit of rice residue on the following rice is evident. In a long-term field study on rice-wheat system at PAU, Ludhiana (H.S. Kalsi, personal communication) it was found that the incorporation of rice straw in wheat for three years showed no wheat yield advantage. However, the yield of following rice crop, with the application of 90 kg N ha^{-1} was found to be similar to the application of 120 kg N ha^{-1} on straw removed plots (ie. a saving of 30 kg N ha^{-1}).

A 5-year study by Sidhu and Beri (2005) showed that drilling or broadcast application of 50% N (of the recommended 120 kg N ha^{-1}) at wheat sowing, and top dressing the remaining 50% N at first post-sowing irrigation produced the maximum wheat yield. Application of 50% of N at pre-sowing irrigation proved inferior to its application at wheat sowing. Proper management of N should reduce immobilization of N by crop residues and produce higher N-use efficiency. One obvious solution to the N immobilization problem would be to place the fertilizer below the C-enriched surface soil that results from surface placement of crop residues. In soils amended with crop residues, band placement of urea will immobilize significantly less fertilizer N than the mixed application of urea, possibly due to limited contact between fertilizer N and the decomposing microorganisms.

Irrigation management in wheat

Rice residue mulching in upland crops has a significant effect on soil water conservation in minimum tillage systems. The optimum strategy for irrigation scheduling may also be affected by mulching due to suppression of soil evaporation. Soil evaporation is considered to be a non-beneficial loss of water, aside from its effect in moderating vapour pressure deficit on crop water use. Studies showed that average daily soil evaporation, with and without mulch, were 0.62 and 0.89 mm day^{-1}, respectively (Balwinder-Singh, personal communication). Experiments in 2007-08 recorded average daily soil evaporation, with and without mulch, as 0.78 and 1.09 mm day^{-1}, respectively. The mulch lowered total soil evaporation by 42.4 and 48.0 mm each season, and much of this appeared to be partitioned in to transpiration which increased by 38.1 and 44 mm in 2006-07 and 2007-08, respectively. The additional water transpired by the mulched crops increased grain yield significantly in 2006-07 (by 610 kg ha^{-1}), but there was no effect on grain yield in 2007-08.

Straw mulch was also found to influence soil temperatures. Straw mulch in wheat lowered the maximum soil temperature by 1.5-2.0°C and increased the minimum soil temperature by 0.5-1.0°C during the first 21 days after sowing. Higher soil water content in the surface soil was observed under wheat with mulch compared with no mulch in the topsoil (0- 30 cm). This is due to lower evaporation losses under mulch. Studies suggest that in the presence of straw mulch, the first irrigation to wheat can be delayed by about 7 days compared to no mulch. Wheat can be sown with Happy Seeder immediately after rice harvest in the residual soil moisture and thus eliminating the need for pre-sowing irrigation.

Paddy straw decomposition and its effect on N mineralization-immobilization

Yadvinder-Singh et al. (2004) studied the in-situ decomposition and N release dynamics for incorporated rice residue using the litterbag decomposition technique. Time-of-incorporation had a significant effect on the decomposition of rice residue during the fallow phase after the rice harvest. At the time of wheat seeding, the mass loss of rice residue was 51% for the 40-day, 35% for the 20-day and 25% for 10-day decomposition treatment. The equations describing the amount of residue remaining under various decomposition periods can be used to predict mass loss. The N concentration of the rice residue increased continuously indicating N immobilization in the residue by microorganisms. Net release of N from paddy straw during wheat growing season was quite low (5-6 kg N ha^{-1}). Another study using litter bag technique showed that about 20% of buried rice residue and 50% of surface-placed residue remained undecomposed over the wheat growing cycle.

The incorporation of cereal residues often results in a net N immobilization phase followed by a net re-mineralization phase (Yadvinder-Singh et al., 2005). Soil mineral N (NH_4+ NO_3) 10 days after incubation was significantly lower in treatments in which paddy straw was incorporated at zero and 10 days before the application of fertilizer than in no straw treatment (Fig. 1). This suggests immobilization of fertilizer N with paddy straw (C/N ratio of 60:1) incorporation. The magnitude of immobilized N was influenced by the decomposition period of paddy straw prior to fertilizer application. Mineral N in soil was significantly higher under 20-day and 30-day pre-decomposition periods than under the no straw treatment up until 20 days after incubation. The total amount of N released, under different decomposition treatments during the life span of wheat crop (about 150 days), ranged from 6 to 9 kg N ha^{-1}. Thus with such a small amount of N released from incorporated residue, a significant effect on yield of crops can hardly be expected.

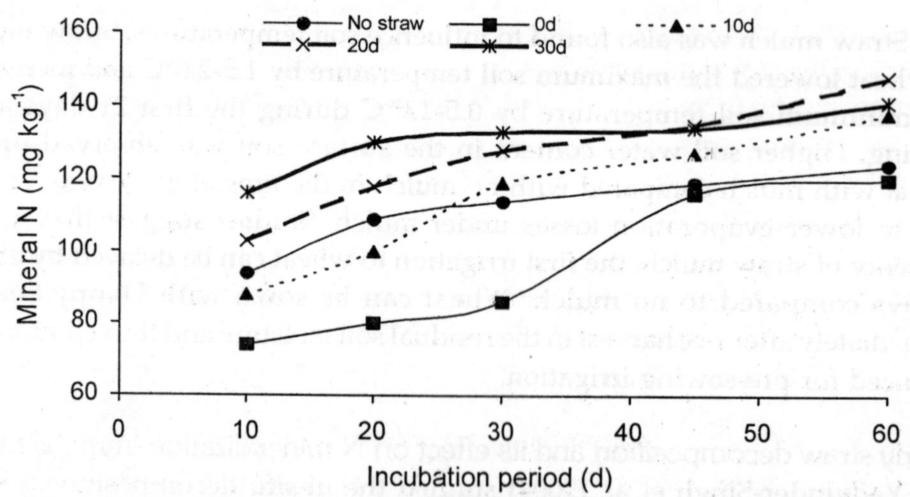

Fig. 1: Effect of pre-decomposition period of paddy straw on N mineralization in a sandy loam soil incubated at 35°C.

Effect of crop residue management on soil properties

Effect on soil fertility

Long-term straw retention will build soil organic matter and N reserves and also increase the availability of macro and micro-nutrients. The amount of soil organic matter in majority of soils of Punjab is low to medium due to fast rate of decomposition. The input of organic residues can result in an increased level of soil organic matter. The equilibrium level will depend upon the amount of organic materials added, management factors and the environment. Field experiments showed that incorporation of crop residues as compared to burning or removal of residues, increased organic carbon and total N content. Yadvinder-Singh *et al.* (2004) observed that organic C content in soil increased from 0.41 to 0.59 g kg^{-1} soil after 7 years of rice residue incorporation before sowing wheat. The per cent increase in organic carbon content is greater on sandy loams with lower initial organic carbon content than on silt loams (Yadvinder-Singh *et al.*, 2009). Thus recycling of straw can increase C accumulation in the soil, which can be advantageous in terms of both global warming and soil fertility. On the basis of soil C values and the amount of C applied, 12-15% of applied (Ed – applied or retained) paddy straw-C (incorporated into the soil) was sequestered by the soil after 4-7 years (Yadvinder-Singh *et al.*, 2004). In another study (Yadvinder-Singh et al., 2009), C sequestration in soil from straw mulch after 2.5 years was about 25% on both sandy loams and silt loams. The amount of C sequestration from the straw incorporation under conventional tillage was reduced to about 17%. The straw decomposition rates are generally slower for surface placed residues compared with incorporated residues. In an 11-year field experiment on a loamy sand soil, the incorporation of residues of both crops in the rice-wheat cropping system increased the total P, available P, and K contents in the soil over the removal of residues (Table 3). Total P, available P, and available S were in the order of residue incorporation > residue removal > residue burning.

Table 3 : Effect of crop residue management on soil fertility of a loamy sand soil over 11 years of the rice-wheat cropping system at Ludhiana (Beri *et al.*, 1995)

Soil property	Crop residue management		
	Burned	Removed	Incorporated
Organic C (%)	0.43	0.38	0.47
Total N (%)	0.55	0.51	0.56
Total P (mg kg^{-1})	390	420	612
Total K (g kg^{-1})	17.1	15.4	18.1
Olsen P (mg kg^{-1})	14.4	17.2	20.5
Available K (mg kg^{-1})	58	45	52
Available S (mg kg^{-1})	34	55	61

The concentration of P in cereal residues is usually low, resulting in their inability to provide sufficient P for crop growth upon incorporation. Long-term application of crop residues, however, increased availability of P and K in soil over straw burned (Yadvinder-Singh *et al.* (2009). Besides a direct supply of P, crop residues can lower the P sorption capacity and enhance nutrient availability. Both inorganic and organic P contents in soil increased with straw incorporation (Gupta *et al.*, 2007). Incorporation of both rice and wheat straw caused the maximum increase in inorganic P, followed by either rice or wheat straw removed or burned. The increase in inorganic P with the incorporation of rice and wheat straw was 27.8% higher than with straw removal. The P sorption decreased significantly with the incorporation of crop residues compared with removal of residues. Phosphorus release was significantly higher with straw incorporation than straw removal.

Crop residues, being very rich in K, with a large fraction in water soluble form, add a large amount of K to the soil. The release of K from paddy straw occurs at a fast rate and about 70% of total straw-K is released within 10 days after incorporation. Available soil K increased from 33 mg kg^{-1} with straw removal to 43-50 mg kg^{-1} in straw-amended plots after 4 years (Table 3). In another study, incorporation of rice residue caused small but a significant increase in available K content in soil over residue removal treatments (Yadvinder-Singh *et al.*, 2004). It is most likely that a part of straw K might have been fixed by clay minerals and some K might have been lost via leaching during rice cultivation on this permeable soil. The incorporation of crop residues on a long-term basis increased the DTPA-extractable Zn, Cu, Fe, and Mn content in the soil (Yadvinder-Singh *et al.*, 2000). Four years after practicing crop residue management options, available soil Zn, Mn and Fe contents increased when both rice and wheat straws were incorporated compared with straw removal (Gupta *et al.*, 2007). The effect of straw management on available soil Cu content was relatively small.

Effect on soil physical and hydraulic properties

Crop residues are important source of soil organic matter and upon incorporation may lead to improvement in soil physical parameters. Straw incorporation for 5 years on a sandy loam soil significantly increased mean weight diameter of aggregates, aggregate stability and total soil porosity than straw removal, and the straw burnt treatment was intermediate (Table 4). Bulk density and modulus of rupture were significantly smaller in the straw incorporation treatment compared with straw removal and straw burnt treatments (Table 4). Straw incorporation maintained a greater proportion of macro- (<0.1mm) and mesopores (0.1-0.01 mm) than in the straw removal and straw burnt treatments. It can be conclusively stated that in RW cropping system, use of crop residues restore the damaged soil structure (due to puddling in rice) by increasing its organic carbon content, size and stability of aggregates, water retention and infiltration; and decreasing bulk density, dispersion ratio and soil strength. The beneficial effects of crop residues confined to surface 15 cm soil layer.

Table 4 : Effect of paddy straw management on soil physical properties (Singh et al., 2005)

Straw management	Organic C (%)		Mean weight diameter (cm)		Aggregate stability (%)		Total porosity (%)	
	0-5 cm	5-10 cm	0-5 cm	5-10 cm	0-5 cm	5-10 cm	0-5 cm	5-10 cm
Removal	0.44	0.42	0.35	0.22	10.8	6.8	36.3	36.6
Burning	0.51	0.47	0.53	0.23	18.6	4.3	37.8	37.0
Incorporation	0.62	0.56	0.60	0.28	26.0	9.6	43.1	39.9
LSD(0.05)	0.02	0.07	2.4	1.2				

Straw management	Bulk density (Mg m-3)		Dispersion ratio		Modulus of rupture (10-4 kPa)		Infiltration rate (cm/h)
	0-5 cm	5-10 cm	0-5 cm	5-10 cm	0-5 cm	5-10 cm	
Removal	1.68	1.69	0.24	0.25	8.43	8.61	0.34
Burning	1.65	1.67	0.20	0.22	8.03	8.50	0.34
Incorporation	1.51	1.59	0.14	0.18	6.27	7.16	0.41
LSD (0.05)	0.03	0.03	0.02				

Soil biological properties

Crop residues provide energy for growth and activities of microbes and substrate for microbial biomass, and provide conditions for a source-sink of nutrients. Availability of nutrients like N, P, and S is particularly dependent upon microbial biomass and microbial activity, which in turn depends on the supply of organic substrates in soil. Sidhu and Beri (2005) observed that soil treated with crop residues inhabited 5-10 times more aerobic bacteria and 1.5 to 11 times more fungi than in soil where residues were either burned or removed. Crop residues also enhance dinitrogen fixation in soil by asymbiotic bacteria (*Azotobacter chroococcum* and *A. agilis*). Many of the changes in soil properties and nutrient cycling may become apparent only after many years (10 year or more) of residue management and long-term results may differ from those obtained over the short term. Thus, there is a strong need to establish long-term experiments on soils varying in soil texture and pH (irrigation water quality).

References

Beri, V., Sidhu B. S., Gupta A.P., Tiwari, R.C., Pareek, R.P., Rupela, O.P., Aluwalia, J.S. and Khera, R. (2003). Organic resources of a part of Indo-Gangetic plains and their utilization. Department of Soils, Punjab Agricultural University, Ludhiana, India. 93 pp.

Beri, V. and Gupta, A.P. (2003). Recycling of rural and urban wastes- A Review. Department of Soils, Punjab Agricultural University, Ludhiana, India. 145 pp.

Bijay-Singh, Shan, Y.H., Johnson-beeebout, S.E., Yadvinder-Singh and Buresh, R.J. (2008). Crop residue management for lowland rice-based cropping systems in Asia. *Advances in Agronomy*, **98** : 118-199.

Gupta, R.K.,Yadvinder-Singh, J.K. Ladha, Bijay-Singh, Jagmohan Singh, G. Singh and H. Pathak (2007). Yield and phosphorus transformations in a rice-wheat system with crop residue and phosphorus management. *Soil Science Society of America Journal*, **71** : 1500-1507.

Narang, R.S., Brar, S.S., and Kumar, S. (1999). Effect of crop-residue incorporation load on nitrogen requirement of succeeding crops and soil productivity in rice (Oryza sativa) - wheat (Triticum aestivum) system. *Indian Journal of Agronomy*, **44** : 8-11.

Sidhu, B.S. and V.Beri (2005). Experience with managing rice residues in intensive rice-wheat cropping system in Punjab. In: *Conservation Agriculture-Status and Prospects*, I.P. Abrol, R.K., Gupta, and R.K. Malik (Eds). CASA, New Delhi. pp. 55-63.

Sidhu, B.S., Beri, V., Jasbir Singh and Pannu R P S (2003).Crop residue and their utilization for crop production. Pp. 1-36. In: Recycling of rural and urban wastes - a review. Beri V and Gupta A P.(eds.) Department of Soils, PAU, Ludhiana, India.

Sidhu, B.S. and V.Beri (2005a). Experience with managing rice residues in intensive rice-wheat cropping system in Punjab. In: *Conservation Agriculture-Status and Prospects*, I.P. Abrol, R.K., Gupta, and R.K. Malik (Eds). CASA, New Delhi. pp. 55-63.

Sidhu, B. S., Goswami, K.P. and Pareek, R.P. (1994). Influence of paddy and wheat straw application on crop yields. *Journal of Research*, Punjab Agric. Univ. 31: 147-153.

Sidhu H.S Manpreet-Singh, Humphreys, E. Yadvinder-Singh, Balwinder-Singh, S.S. Dhillon, Blackwell, J. Bector, V. Malkeet-Singh and Sarbjeet-Singh (2007). The Happy Seeder enables direct drilling of wheat into rice stubble. *Australian Journal of Experimental Agriculture*, **47** : 844-854.

Singh, G., S.K. Jalota and B.S. Sidhu (2005). Soil physical and hydraulic properties in a rice-wheat cropping system in India: effects of rice-straw management. *Soil Use and Management*, **21** : 17-21.

Thuy, N.H., Shan Yuhua, Bijay-Singh, Kairong Wang, Zucong Cai, Yadvinder-Singh, Buresh R.J. (2008). Nitrogen supply in rice-based cropping systems as affected by crop residue management. *Soil Science Society of America Journal*, **72** : 514-523.

Yadvinder-Singh, Bijay-Singh, Meelu, O.P. and Khind, C.S. (2000). Long-term effects of organic manuring and crop residues on the productivity and sustainability of rice-wheat cropping system in Northwest India. pp. 149-162. In: *Long-term soil fertility experiments in rice-wheat cropping systems* (I. P. Abrol, K. F. Bronson, J. M. Duxbury and R.K. Gupta, eds.). Rice-wheat Consortium Paper Series 6. Rice-wheat Consortium for Indo-Gangetic Plains, New Delhi, India.

Yadvinder-Singh, Bijay-Singh, Ladha, J.K., C.S. Khind, R.K. Gupta, O.P. Meelu and Pasuquin, E. (2004a) Long-term effects of organic inputs on yield and soil fertility in the rice-wheat rotation. *Soil Science Society of America Journal*, **68** : 845-853.

Yadvinder-Singh, Bijay-Singh, Ladha, JK, Khind, C.S., Khera, T.S., and Bueno, C.S. (2004b) Effects of residue decomposition on productivity and soil fertility in rice-wheat rotation. *Soil Science Society of America Journal,* **68** : 854-864.

Yadvinder-Singh (2007). Final progress Report 2004-07 of the ICAR Ad-hoc project, "Managing rice straw in following wheat for enhanced soil and crop productivity in rice-wheat cropping system" Dept. of Soils, PAU, Ludhiana.

Yadvinder-Singh, Gupta R.K., Gurpreet-Singh, Jagmohan-Singh , Sidhu H. S. and Bijay-Singh (2009). Nitrogen and residue management effects on agronomic productivity and nitrogen use efficiency in rice-wheat system in Indian Punjab. *Nutrient Cycling in Agro-ecosystems,* **84** : 141-154.

□□□

System Based Integrated Nutrient Management, 2012
© B. Gangwar & V.K. Singh (eds.), pp. 43-53
New India Publishing Agency, New Delhi (India)
e-mail : info@nipabooks.com; website : www.nipabooks.com

Integrated Nutrient Management for Irrigated Ecosystems

B.S. DWIVEDI

All the states have some area under irrigation. Thus the term 'irrigated ecosystem' is rather loosely defined, and the states having distinctly large net irrigated area may constitute this ecosystem. Two such areas can be easily carved out i.e., Indo-Gangetic Plain region representing five states namely Punjab, Haryana, U.P., Bihar and West Bengal, and coastal areas of two southern states i.e., A.P. and Tamil Nadu. About 56% of the net irrigated area of the country exists in these states, and the cropping intensity is also much higher (144%) as against national average of 132%. With demographic viewpoint also, these areas are densely populated. Of late agriculture in this ecosystem has become unsustainable owing to over-exploitation of natural resources. Studies have conclusively established the superiority and worth of integrated nutrient management in restoration of soil health and sustenance of crop productivity and profits.

The basic concept underlying the principles of integrated nutrient management (INM) is the maintenance, and possibly improvement, of soil fertility for sustaining crop productivity on long-term basis. This may be achieved through combined use of all possible sources of nutrients and their scientific management for optimum growth, yield and quality of different crops and cropping systems. INM is not a new concept, but an age-old practice. Its importance was, however, not realized earlier due to low nutrient turn over in soil-plant system and almost all the nutrient needs were met through organic sources, which supplied secondary and micronutrients also besides major nutrients. INM has now assumed great significance mainly because of two reasons. First, the need for continued increase in agricultural production based on increase in per hectare yields requires growing application of nutrients, and the present level of fertiliser production in india is not enough to meet the total plant nutrient requirement. Second, the results of a large number of experiments on manures and fertilisers conducted in india and other countries reveal that neither the chemical fertilisers alone nor the organic sources exclusively can achieve the production sustainability under intensive cropping systems (Hegde and Dwivedi, 1993). Even the so called balanced use of chemical fertilisers would not be able to sustain high productivity due to emergence of the deficiency of one or more of the secondary and micro-nutrients (Swarup and Wanjari, 2000). The interactive advantages of combining organic and inorganic sources of nutrients in INM have proved superior to the use of its each component separately. The advantages of INM can be broadly enumerated as following:

- It helps to restore and sustain soil fertility and crop productivity.
- INM may also help to check the emerging deficiency of nutrients other than N, P and K. It brings economy in fertiliser use and improves the nutrient use-efficiency.
- INM favourably affects the physical, chemical and biological health of soils.

Fertilisers, organic manures, legumes, crop residues/wastes and biofertilisers are the main components of INM. At present, information on use and availability of components other than fertilisers is scanty.

Organic Manures

Organic manures like farmyard manure (FYM) and compost have been traditionally important inputs for maintaining soil fertility and ensuring yield stability. As these nutrient sources are bulky in nature with low nutrient content and short in supply, they lost their relative importance over time in crop production to readily available chemical fertilisers. However, escalating cost of fertilisers due to energy crisis and their limited supply made it necessary to search for alternative and renewable sources of plant nutrients leading to major interest in organic recycling.

The rural and urban compost potential in India is estimated at 600 and 16 million tonnes, respectively. Only a-half of this vast potential is actually produced with an average use of just 2 tonnes ha^{-1} $year^{-1}$. Less than 50% of the manurial potential of livestock is utilized at present, as a large proportion is lost as fuel and droppings in non-agricultural areas. Even out of the amount used for agricultural production, substantial proportion of nutrients is lost due to improper handling, storage and incorporation in soil. In this regard, there is need to learn a lesson from the meticulous performance of China in mobilizing all organic sources of plant nutrients including human excreta towards agricultural production.

There are other potential organic sources of nutrients such as urban compost, pressmud, non-edible oil-cakes and wastes from food processing industries. Nutrient composition of some important organic sources/manures is given in Table 1.

Table 1 : Nutrient content of some organic manures and recyclable wastes

Category	Source	Nutrient content (%)		
		N	P_2O_5	K_2O
FYM/ Composts	Farmyard manure	0.5-1.0	0.15-0.20	0.5-0.6
	Poultry manure	2.90	2.9	2.3
	Urban compost	1.5-2.0	1.0	1.5
	Rural compost	0.5-1.0	0.2	0.5
Animal wastes	Cattle dung	0.3-0.4	0.10-0.15	0.15-0.20
	Cattle urine	0.80	0.01-0.02	0.5-0.7
	Sheep & goat dung	0.65	0.50	0.03
	Night soil	1.2-1.5	0.80	0.5
Oil cakes	Castor	5.5-5.8	1.8	1.0
	Coconut	3.0-3.2	1.8	1.7
	Neem	5.2	1.0	1.4
Animal meals	Horn & hoof	13.0	0.3-0.5	-
	Fish	4-10	3-9	1.8
	Raw bone	3-4	20-25	-

These are all very valuable sources of nutrients and there is practically no information on the extent of their use in agriculture. Organic manures play a direct role in supplying macro and macronutrients and an indirect one by improving the physical, chemical and biological properties of the soils. These manures, besides supplying nutrients to the current crop, also leave substantial residual effect on succeeding crop in the system.

Long-term experiments (LTEs) being carried out under All India Coordinated Research Project on Integrated Farming Systems (AICRP-IFS, earlier AICRP-CS) have indicated that FYM can substitute a part of fertiliser N needs of monsoon crops without any adverse effect on the total productivity of major wheat-based cropping systems (Table 2). It was further noticed that fertiliser needs of the winter could be reduced by 25% by substituting 25% N needs of preceding monsoon crop through FYM at some locations. The integrated use of chemical fertilisers and FYM resulted in markedly higher productivity than that obtained with the application of chemical fertilisers alone. Similarly, LTEs with maize-wheat and soybeen-wheat systems on different soils continuing under aegis of AICRP-LTFF invariably showed the superiority of IPNS options involving FYM over sole fertilisers (Table 3).

Table 2 : Effect of integrated nutrient supply through fertilisers and FYM on the grain yield of crops in different wheat-based cropping systems under AICRP-CS experiments

Treatments		Grain yield (kg ha^{-1})		
Monsoon	Winter	Monsoon	Winter	Total
Parbhani (sorghum-wheat) av of 7 yrs				
100% NPK	100% NPK	2.97	2.64	5.62
50% NPK + 50% N (FYM)	100% NPK	2.85	2.78	6.53
Ranchi (maize-wheat) av of 8 yrs				
100% NPK	100% NPK	2.92	2.71	5.63
75% NPK + 25% N (FYM)	75% NPK	3.30	2.40	5.70

100% NPK : recommended fertiliser NPK dose.
Source : AICRP - CS Annual Reports

Besides FYM and compost, there are several other organic sources with good nutrient potential. However, there is no peoper evaluation of these nutrients-carriers to establish their fertiliser equivalent value. There is need to integrate in appropriate crops and cropping systems all these sources at locations where these are available. The by-products of factories like spent- wash from distillery, molasses, pressmud etc. from sugar industry, and wastes from other food processing industries have good manurial value. Sulphitation pressmud (SPM) has a great potential to supply nutrients in addition to favourable effects on soil properties. During the last three decades, SPM has assumed great importance as a nutrient supplement in sugarcane-ratoon-wheat and other intensive cropping systems of the sugarcane dominated areas.

Legumes

Legumes have a long-standing history of being soil fertility restorers due to their ability to obtain N from the atomosphere in symbiosis with Rhizobia.

Legumes can form an important component of INM when grown for grain or fodder in a cropping system, or when introduced as a green manuring crop in a cropping system. Legumes grown as green manure, forage or grain crops improve the productivity of rice-wheat cropping system and regenerate soil fertility (Singh *et al.* 1991; Mann and Garrity, 1994; Ahlawat *et al.*, 1998).

Table 3 : Effect of IPNS on the long-term sustainability of crop yields in wheat based cropping systems

Location	Crops	Grain yield (t ha^{-1})			
		Control	100% NPK	150% NPK	100% NPK+ FYM
Inceptisols					
Delhi (1994-99)	Maize	1.2	2.1	2.4	2.4
	Wheat	2.5	4.6	4.9	5.0
Ludhiana (1971-99)	Maize	0.4	2.6	2.5	3.2
	Wheat	1.0	4.8	4.9	5.0
Vertisols					
Jabalpur (1972-99)	Soybean	0.9	2.1	2.1	2.2
	Wheat	1.1	4.2	4.4	4.6
Alfisols					
Palampur (1973-98)	Maize	0.3	3.2	4.0	4.6
	Wheat	0.3	2.5	3.0	3.3
Ranchi (1973-99)	Soybean	0.7	1.6	1.5	1.9
	Wheat	0.8	2.6	2.8	3.1

Source: Swarup and Wanjari (2000)

(i) Legumes grown in rotation

Different legumes have the capacity to leave behind different amounts of N for use by the succeeding crop. Further, fodder legumes contribute more N than grain legumes. A number of leguminous crops have been evaluated for the contribution which they make in meeting the N requirement of the succeeding crop, and it has been found that as much as 50 to 60 kg N ha^{-1} may be made available. The carry over of N for succeeding cereal crop may be 60-120 kg in berseem, 75 kg in Indian clover, 35-60 kg in fodder cowpea, 55 kg in blackgram, 60 kg in groundnut, 68 kg in gram and 50 kg in lathyrus. Grain yield of succeeding crop increased markedly when legumes preceded them, compared with that when cereals preceded. In general, pulse crops do leave a residual amount of as much as 30 to 50 kg N ha^{-1} for utilization by the succeeding crop (Table 4). Pulse crops can be made more effective by inoculation with proper species of *Rhizobia*. In a country where the average consumption of plant nutrients from chemical fertilisers on national basis is very low, the residual fertility build-up due to legumes is obviously a major contribution, which must be fully exploited.

Table 4 : Effect of pulse crops on the N economy in succeeding wheat

Preceding winter crop	Fertiliser N (kg ha^{-1})				
	0	**40**	**80**	**120**	**Mean**
	Grain yield (t ha^{-1}) of wheat				
Gram	2.01	2.35	2.62	2.80	2.44
Lentil	1.97	2.39	2.53	2.66	2.41
Peas	2.13	2.56	2.63	2.73	2.03
80 kg N ha^{-1}	1.55	1.97	2.21	2.41	2.03
120 kg N ha^{-1}	1.61	1.92	2.44	2.61	2.14

Source: Tauro (1983)

The effect of preceding crops, rice or legume, was evaluated on wheat under field conditions, wherein the N use efficiency indices, viz. AE_N, AR_N and PFP_N were significantly greater in wheat following a legume (cowpea) than wheat following rice (Yadav *et al.*, 2003). Inclusion of forage legumes in rice-wheat system also increased nutrient use efficiency (Table 5).

Table 5 : Recovery efficiency of N and P fertilizers in rice-wheat system as influenced by inclusion of summer forage cowpea

Fertilizer NP rate (kg ha^{-1})	Rice		Wheat	
	Summer Fallow	**Summer Cowpea**	**Summer Fallow**	**Summer Cowpea**
Recovery efficiency of N (%)				
$N_{120}P_0$	34.8	35.3	42.3	38.3
$N_{120}P_{60}$	36.4	41.2	54.5	61.7
Recovery efficiency of P (%)				
N_0P_{26}	11.6	15.6	11.2	12.6
$N_{120}P_{26}$	22.7	25.0	27.9	30.4

Source: Dwivedi *et al.* (2003)

(ii) Legumes as green manures

Before the advent of mineral fertilisers, green manuring was considered as an indispensable practice. However, with the easy availability of fertilisers and adoption of intensive cropping systems, the practice of green manuring was almost given up. In recent years, with an indication of declining trend in productivity due to indiscriminate use of only chemical fertilisers (NPK), there has been revival of interest in green manuring. Green manuring with legumes enriches soil N due to fixation of atmospheric N. The decomposing green manure has a solubilising effect on N, P, K and micro-nutrients in the soil. It also reduces leaching and gaseous losses of N. Apart from this, green manures also improve the physical, chemical and biological properties of soil. Sunnhemp (*Crotolaria juncea*) and *dhaincha* (*Sesbania aculeata*) are the most important green manure crops, while clusterbean, berseem, Indian clover etc. are also used occasionally. Long-term studies on integrated use of fertilisers and green manures have indicated that it is possible to substitute a part of fertiliser N needs of monsoon crop by green manure in different cropping systems with or without some gain in the total productivity of the system (Table 6).

Table 6 : Effect of integrated nutrient supply through green manures on the grain yield of crops in different cropping systems

Treatments		**Grain yield (kg ha^{-1})**		
Monsoon	**Winter**	**Monsoon**	**Winter**	**Total**
Rahuri (Sorghum-Wheat) av. of 7 yrs				
100% NPK	100% NPK	4.74	3.39	8.13
50% NPK + 50% N (GM)	100% NPK	4.66	3.47	8.13
Hanumangarh (Pearlmillet-Wheat) av. 5 yrs				
100% NPK	100% NPK	2.72	3.61	6.33
50% NPK + 50 % N (GM)	100% NPK	2.86	3.96	6.82
75% NPK + 25% N (GM)	75% NPK	2.76	3.57	6.32

GM: Green manure; 100% NPK : Recommended fertiliser NPK dose.
Source : AICRP - CS Reports

With intensification of cropping, it is becoming difficult to adjust green manure crops within intensive cropping systems, and farmers do not wish to set apart 6-8 weeks exclusively for growing green manure crop with no direct cash benefit. Nonetheless, in regions where double cropping is practiced and water availbility is not a constraint, summer green manuring can be introduced provided it does not compete with the main-crop. However, in other cases, raising a green gram crop in summer and incorporation of above ground green biomass after pod picking may also serve as green manuring (Dwivedi *et al.*, 2002). Green gram residue incorporation could easily substitute 50% NPK needs of rice amounting to 60 kg N, 30 kg P_2O_5 and 15 kg K_2O ha^{-1} in rice-wheat system without any adverse effect on total productivity (Bhardwaj and Prasad, 1981). This practice also helped in mobilizing the availability of N and micronutrients like Zn, Fe, Mn and Cu.

Where framers are not willing to spare their scarce land resources and inputs for growing a green manure crop, fresh loppings of perennial leguminous trees like *Glyricidia* grown in hedge-rows and field-bunds may be used for incorporation into soil as a source of N. This practice of green leaf manuring is common in southern parts of India.

Crop residues

India has a vast potential of crop residues. However, many crop residues having feed value are needed to support large livestock population, and may not be available for using as complementary resources to chemical fertilisers. But in the regions like those of rice-wheat growing areas of north-west India where mechanical harvesting is practiced, a sizeable quantity of residue is left in the field, which can form a part of nutrient supply. There are large amounts of residues of other crops like potato, sugarcane, vegetables etc., which are practically wasted in most cases.

Although cereal crop residues are valuable cattle-feed, they can be supplemented to the chemical fertilisers wherever they are in excess of the local needs and ready market is not available. Studies being carried out at several locations under AICRP-CS have revealed the possibility of substituting a part of fertiliser N needs of monsoon crops by cereal crop residues in intensive cereal-cereal production systems.

The pace of immobilization-mineralization cycle, particularly that of N, is determined by a number of factors. Inconsistent crop responses to incorporation of crop residues under different situations may, therefore, be expected. Stubbles left in the field even in traditional harvesting methods range from 0.5 to 1.5 t ha^{-1} in different crops. When mechanical harvesting is done, this amount is much greater. Stubbles of coarse cereals such as sorghum, maize, pearlmillet etc., which are difficult to decompose are normally collected and burnt during land preparation causing significant loss of plant nutrients. There is a need to evolve appropriate management practices to make use of these stubbles either by applying a part of chemical N recommended for the succeeding crop during land preparation or by adding cellulose decomposing microbial cultures.

Biofertilisers

Materials containing agriculturally beneficial microorganisms are called microbial inoculants or biofertilisers. Rhizobium is the most widely used biofertiliser, which colonizes the roots of specific legumes to form root nodules.

These nodules act as factories of ammonia production. The Rhizobium-legume association can fix up to 100-300 kg N ha^{-1} in one crop season and in certain situations leave substantial N for the following crop. This symbiosis can meet more than 80% of the N needs of the legume crops.

Another group of bacteria, which are free-living N-fixer, is Azotobacter. The mechanism by which the plants inoculated with Azotobacter derive positive benefits in terms of increased grain yield and N uptake are attributed to small increase in N input from Biological Nitrogen fixation (BNF), development and branching of roots, production of plant growth hormones, enhancement in uptake of NO^-_3, NH^+_4, $H_2PO^-_4$, K^+ and Fe^{++}, improved water status of the plants, increased nitrate-reductase activity and anti-fungal compounds. In irrigated wheat, significant response to inoculation was recorded in 342 out of 411 on-farm trials, and the response ranged from 34 to 247 kg ha^{-1} (Table 7). Based on a large number of experiments in cereal crops, Azotobacter has been found to contribute, in general, 20-25 kg N ha^{-1}.

Table 7 : Response of irrigated wheat to Azotobacter inoculation on farmers' fields

District	No. of on-farm trials	Response (kg ha^{-1}) to inoculation
Vidisha	63	34
Dungarpur	6	68
Dhar	60	104
Narsinghpur	80	163
Bilaspur	47	247
Hoshiarpur	28	150
Patiala	89	142
Nizamabad	51	212
Shivpuri	59	92

Source: AICRP-CS (Now AICRP-IFS) Annual Reports

Azospirillum colonises the root mass and fixes N in loose association with plants. It has shown positive interaction with applied N in several field crops with an average response equivalent to 15-20 kg ha^{-1} of applied N.

Rock phosphate holds potential in acid soils, but there is little possibility for its use in neutral and alkaline soils. In recent years, several strains of P solubilizing bacteria and fungi have been isolated. Inoculation with P-solubilizing microbial cultures is known to increase the dissolution of insoluble and sparingly soluble P in the soil (Whitelaw, 2000). Integrated use of the microbial cultures along with low-grade rock phosphate might add about 30-35 kg P_2O_5 ha^{-1}. Soil inoculation with P solubilising organism *Pseudomonas striata*, besides increasing grain yield of wheat, showed residual effect by increasing grain yield of succeeding maize crop in alluvial soil at Delhi (Hegde and Dwivedi, 1993). In a 3-year field study on sandy loam alluvial soils of Modipuram, however, PSM (P-solubilizing microbial culture containing *Aspergillus awamori*) inoculation of wheat seed did increase crop response to P, and soil available P content over the non-inoculated treatments, but the magnitude of increase in these parameters was generally too small to attain statistical significance (Dwivedi *et al.*, 2004).

Epilogue

Restoration and improvement of soil quality is crucial to sustain future agricultural productivity. Integrated nutrient supply and management would be the most practically viable technique which holds the key to sustain crop yield and quality of crops without adversely affecting the environment. Multi-locational long-term fertiliser experiments established that the crop productivity under intensive cropping systems can not be further increased with incremental use of chemical fertiliser alone, but the addition of organic sources could increase the yield through increased soil productivity and nutrient use efficiency. Although constraints like use of cattle dung as fuel, increasing competitive value of crop residues as cattle feed, extra-cost and time required for raising green manure crop, poor and inconsistent response to biofertilisers *etc* limit their large scale adoption, yet the consciousness about IPNS is increasing. Site-specific and cropping system based IPNS recommendation that support the maintenance or even increase in organic matter content need to be developed in consonance with other management practices responsible for conservation of soil organic matter. Quantification of fertiliser equivalents of organic resources, flows and fluxes of nutrients under IPNS, *in situ* composting of crop residues and nutrient input potential of dual-purpose legumes are the other important areas demanding greater research thrust.

References

Ahlawat, I.P.S., Ali, M., Yadav, R.L., Kumar Rao, J.V.D.K., Rego, T.J. and Singh, R.P. (1998). Biological nitrogen fixation and residual effect of summer and rainy season grain legumes in rice and wheat cropping systems of the Indo-Gangetic Plain. In 'Residual Effects of Legumes in Rice and Wheat Cropping Systems of the Indo-Gangetic Plain' (Eds J.V.D.K. Kumar Rao, C. Johansen and T.J. Rego), pp 31-54, International Crops Research Institute for the Semi-arid Tropics, Patancheru, India.

Bhardwaj, S.P. and Prasad, S.N. (1981). Response of rice-wheat rotation to zinc under irrigated conditions in Doon valley. *Journal of the Indian Society of Soil Science*, **29** : 220-224.

Dwivedi, B.S., Shukla, Arvind K. and Singh, V.K. (2002). Integrated nutrient management in transplanted rice-wheat system. PDCSR Annual Report (2001-02), PDCSR, Modipuram.

Dwivedi, B.S., Shukla, Arvind K., Singh, V.K. and Yadav, R.L. 2003. Improving nitrogen and phosphorus use efficiencies through inclusion of forage cowpea in the rice-wheat system in the Indo-Gangetic Plains of India. *Field Crops Research*, **84** : 399-418.

Dwivedi, B.S., Singh, V.K. and Dwivedi, V. (2004). Application of phosphate rock, with or without *Aspergillus awamori* inoculation, to meet P demands of rice-wheat systems in the Indo-Gangetic Plains of India. *Australian Journal of Experimental Agriculture*, **44** : 1041-1050.

Hegde, D.M. and Dwivedi, B.S. (1993). Integrated nutrient supply and management as a strategy to meet nutrient demand. *Fertilizer News*, **38** (12) : 49-59.

Hegde, D.M. and Dwivedi, B.S. (1994). Crop response to biofertilisers in irrigated areas. *Fertilizer News*, **39** (4) : 19-26.

Mann, R.A. and Garrity, D.P. (1994). Green manures in rice-wheat cropping systems in Asia. In 'Green Manure Production Systems for Asian Ricelands' (Eds J. K. Ladha and D.P. Garrity), pp. 27-42, International Rice Research Institute, Los Banos, Philippines.

Swarup, A. and Wanjari, R.H. (2000). Three decades of All India Co-ordinated Research Project on Long-term Fertiliser Experiments to study changes in soil quality, crop productivity and sustainability, 59 pp, Indian Institute of Soil Science, Bhopal.

Tauro, p. (1983). Proc. FAI Seminar on Systems Approach to Fertiliser Industry. AGS ii-4/ 1-8.

Whitelaw, M.A. (2000) Growth promotion of plants inoculated with phosphate solubilizing fungi. *Advances in Agronomy*, **69** : 100-151.

Yadav, R.L., Singh, V.K., Dwivedi, B.S. and Shukla, Arvind K. (2003). Wheat productivity and N use-efficiency as influenced by inclusion of cowpea as a grain legume in a rice-wheat system. *Journal of Agricultural Science*, **141** : 213-220.

Yadvinder-Singh, Khind, C.S. and Bijay-Singh (1991). Efficient management of leguminous green manures in wetland rice. *Advances in Agronomy*, **45** : 135-189.

□□□

System Based Integrated Nutrient Management, 2012
© B. Gangwar & V.K. Singh (eds.), pp. 55-66
New India Publishing Agency, New Delhi (India)
e-mail : info@nipabooks.com; website : www.nipabooks.com

CHAPTER **4**

Growth and Productivity in Relation to Nutrient Management

D. SINGH

Limited availability of additional land for crop production, along with declining yield growth for major food crops, have heightened concerns about agriculture's ability to feed a world population expected to exceed 7.5 billion by the year 2020. Decreasing soil fertility has also raised concerns about the sustainability of agricultural production at current levels. Future strategies for increasing agricultural productivity will have to focus on using available nutrient resources more efficiently, effectively, and sustainably than in the past. Management of the nutrients needed for proper plant growth, together with effective crop, water, soil, and land management, will be critical for sustaining agriculture over the long term. Plants, like all other living things, need food for their growth and development. Growth is the function of various factors like light, carbon dioxide gas, water and mineral nutrients. Plants require 16 essential elements. Carbon, hydrogen, and oxygen are derived from the atmosphere and soil water. The remaining 13 essential elements (nitrogen, phosphorus, potassium, calcium, magnesium, sulfur, iron, zinc, manganese, copper, boron, molybdenum, and chlorine) are supplied either from soil minerals and soil organic matter or by organic or inorganic fertilizers. For plants to utilize these nutrients efficiently, light, heat, and water must be adequately supplied. Cultural practices and control of diseases and insects also play important roles in crop production. Each type of plant is unique and has an optimum nutrient range as well as a minimum requirement level. Below this minimum level, plants start to show nutrient deficiency symptoms. Excessive nutrient uptake can also cause poor growth because of toxicity. Therefore, the proper amount of application and the placement of nutrients are important.

Nutrient management strategies

Plant nutrition is only one of more than fifty factors which directly affect both crop yield and quality. The availability of required nutrients, together with the degree of interaction between these nutrients and the soil, play a vital role in crop development. A deficiency in any one required nutrient or, a soil condition that limits or prevents a metabolic function from occurring can limit plant growth. A soil nutrient management plan should include analyzing soil deficiencies to determine the type, application rate, application interval, and the placement of any nutrients required to optimize short and long term productivity. There is a significant difference between an induced deficiency and a real soil deficiency. For example, certain crops require the addition of molybdenum at a specific rate for optimum growth. This is a real deficiency. In other crops zinc or iron deficiencies, caused by high levels of phosphorus and active calcium, can result in reduced yield. This is an induced deficiency. Typically, when deficiencies occur, the tendency is to foliar or soil apply copious amounts of product and hope for a favorable result. This ad hoc approach seldom achieves the expected result and is certainly not cost effective. The simple fact is diagnosis is the first step in determining an appropriate corrective action which many include (1) a combination of treatments or (2) a program that incorporates several applications of different products at different application rates and intervals. When soil is depleted there are two methods for restoring its fertility (1) it can be left idle for several years allowing it to rebuild naturally or (2) organic matter, in the form of crop residue, together with microbial based inoculants can be applied from an external source. In the latter case, the rebuilding process is accelerated and optimal conditions for soil biological activity and long term soil fertility are maintained.

Soil organic matter is vital in rebuilding depleted soil as it ensures a continuous energy source for soil biomass. Soil biomass, consisting of microbes, fungi, algae, protozoa etc. (1) transform organic molecules into mineral elements that are readily available to plants and (2) help maintain good soil structure by transforming organic matter into humus and producing compounds that cement small soil particles together, promoting both increased drainage and moisture retention.

Soil nutrient management involves not only the physical properties and mineral structure of the soil, but also the balance between soil pathogens and beneficial microbes. Beneficial microbes increase nutrient availability, reduce disease, reduce nutrient losses, and help degrade toxic compounds. Plants thrive or suffer, depending on the type of microbes in the rhizoshere (the area around the roots.) In a healthy rhizoshere, dominated by beneficial microbes, plant life and soil life work together to produce healthy plants. Conversely, in unhealthy soil, dominated by pathogenic microbes, optimum plant growth is unattainable.

Law of the Minimum

The Law of the Minimum is an important concept in any nutrient management strategy. It states that crop yield will be limited by any essential component of plant growth which is insufficiently present. Since plant nutrients are essential to plant growth, any deficiencies in any one nutrient can limit crop yield. For example, even if nitrogen is sufficiently supplied, a phosphorous deficiency will cause yields to be much less than if both were adequately present (Fig. 1).

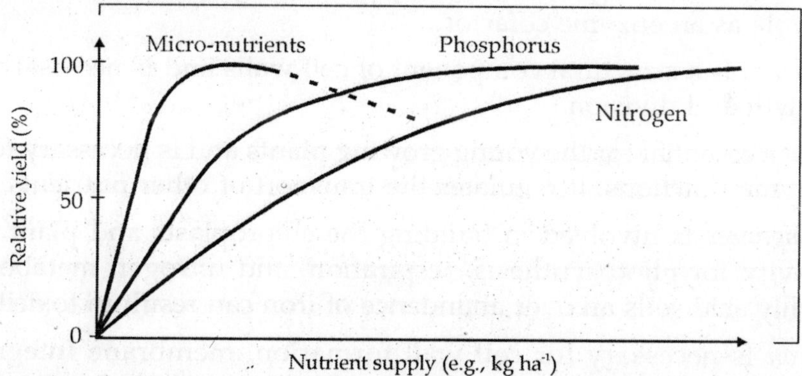

Fig. 1 : Yield response curve for nitrogen, phosphorus and micronutrients (Marschner, 1997)

Critical Nutrient Levels

There are 13 essential elements which plants obtain from the soil that are commonly managed by growers. In addition, plants require carbon, hydrogen, and oxygen to grow.

What makes an element essential to plant growth?

- An element is essential if the plant cannot complete its life cycle without the element.
- It is essential if the element is directly or indirectly involved in the metabolic processes of the plant (i.e. photosynthesis or respiration).

- A deficiency in an essential nutrient will result in the development of a characteristic, visual symptom.

Role of Nutrients on Plant Growth and Development

- *Nitrogen* An over abundance of nitrogen can delay flowering while a deficiency can reduce growth and yield. N is essential component of all proteins.
- *Phosphorus* is necessary for almost all aspects of plant growth and is essential for flower and fruit formation and it is important in plant bioenergetics i.e. assimilation of energy.
- *Potassium* is necessary for the formation of sugars, starches, carbohydrates, protein synthesis and cell division in roots and other parts of the plant. It regulates opening and closing of stomata in leaf.
- *Sulfur* is a structural component of some amino acids & vitamins and is required for the production of chlorophyll.
- *Magnesium* is necessary for seed germination and the production of carbohydrates, sugars and chlorophyll through production of ATP and its role as an enzyme cofactor.
- *Calcium* is a structural component of cell walls and is necessary for cell growth and division.
- *Iron* is essential for the young growing plants and is necessary for many enzyme functions. It regulates the transport of other nutrients.
- *Manganese* is involved in building the chloroplasts and plant enzyme activity for photosynthesis, respiration, and nitrogen metabolism In highly acid soils an over abundance of iron can result in toxicity.
- *Boron* is necessary for cell wall formation, membrane integrity and calcium uptake. The functions affected by boron include flowering, pollen germination, fruiting, cell division, water relationships and the movement of sugars and hormones.
- *Zinc* is essential to carbohydrate metabolism, protein synthesis and stem growth. Zinc deficiency may lead to iron deficiency. Lowering soil pH can result in an over abundance of zinc.
- *Copper* is concentrated in roots of plants and plays a part in nitrogen metabolism and photosynthesis. An over abundance of copper can cause toxicity.
- *Molybdenum* is a structural component of the enzyme that reduces nitrates to ammonia. Without it, the synthesis of proteins is blocked and plant growth ceases.

- *Chlorine* is involved in osmosis, the ionic balance necessary for plants to take up mineral elements and in photosynthesis.
- *Nickel* is an essential trace element for plants. It is required for urease activity to break down urea to liberate the nitrogen into a usable form for plants. Nickel is required for iron absorption. Seeds need nickel in order to germinate.
- *Sodium* is involved in water movement and ionic balance in plants.
- *Cobalt* is required for nitrogen fixation in legumes and in the root nodules of non-legumes. The demand for cobalt is much higher for nitrogen fixation than for ammonium nutrition.
- *Silicon* is found as a component of cell walls. Plants with supplies of soluble silicon produce stronger and tougher cell walls.

Growth rate and nutrient supply

Plant nutrients both applied and available are known to regulate the foliage and reproductive growth of the plants. Shoot elongation, leaf growth and biomass productivity are directly associated with nitrogen and other essential nutrients. Nutrients also play a decisive role in controlling the metabolic activities related to source and sink developments in plants. The development of plants during the vegetative growth phase (source) control both biological and economic yields (sink). The important crop physiological activities influenced by nutrition and their efficient management are discussed here :

1. Leaf area index and net photosynthesis

Vegetative growth consists mainly the growth and formation of new leaves, stems and roots. As the meristematic tissues have a very active protein metabolism, photosynthates transported to these sites are used predominantly in the sysnthesis of nucleic acid and proteins. It is for this reason that during the vegetative stage, the N nutrition of the plants to a large extent controls the growth rate of the plant. A high rate of growth only occurs when abundant N is available. Under low N supply, photosynthates can be utilized to a limited extent in the sysnthesis of organic N compounds and the remainder is stored in the form of starch and polyfrutosans (Table 1). As vegetative growth proceeds the carbohydrates levels of annual plants generally increase. The effect of N nutrition on carbohydrate accumulation is therefore often less marked towards the end of the vegetative period and the onset of flowering. Under low N nutrition the life cycle of the plant is shortened and the plant matures early resulting in poor economic yield. Furthermore, under N deficiency conditions early senescence occurs not due to lack of N for protein sysnthesis but it is induced by a depression in cytokinin synthesis (Michael and Beringer, 1980). Amino acids are required for the synthesis of cytokinin so that cytokinin metabolism is very closely associated with N- nutrition.

Table 1 : Effect of N supply on the dry matter yield and organic constituents in ryegrass. (Hehl and Mengel, 1972)

	N supply (g N pot^{-1})	
	0.5	2.0
Dry matter (g/pot)	14.9	26.0
Crude protein (% DM)	12.3	26.4
Sucrose (% DM)	7.7	6.3
Polyfructosans (% DM)	10.0	1.0
Starch (% DM)	6.1	1.4
Cellulose (% DM)	14.4	17.6

The density of crop population is expressed in terms of leaf area index (LAI), which is defined as the leaf area of plants per unit area of the soil. The yield of crop increases until the optimum value reached in the range of 3-6, the exact value depending on many factors like plant species, leaf shape, leaf angle, light intensity and nutrient availability etc. When the nutrient supply is suboptimal, the leaf growth rate, and thus the LAI, can be limited by low rate of photosynthesis or insufficient cell expansion or both these factors. This is particularly evident with suboptimal supply of nitrogen and phosphorus. In plants suffering from N deficiency, elongation rate of leaves may decline before there is any reduction in net photosynthesis (Chapin *et al.*, 1988). The effect of nitrogen deficiency on leaf expansion differs between plant species and is associated with morphological differences and transpiration rate (Table 2). In cereals (monocotyledons), eg. wheat, barley, maize and sorghum, cell expansion occurs at the base of leaf blade, which is protected form the atmosphere by the sheath of the preceding leaf, so that little transpiration occurs from this zone of elongation. In dicotyledons (sunflower, cotton soybean and radish etc), cell expansion occurs in the leaf blades which are exposed to the atmosphere and therefore experience high rate of transpiration during the day. In contrast to leaf expansion, net photosynthesis per unit leaf area is depressed to a similar extent in both groups of plants by nitrogen deficiency. Similarly, phosphorus deficiency severely inhibited leaf growth rate in cotton (Radin and Eidenbock, 1984). A mineral nutrient deficiency can also depress net photosynthesis by influencing CO_2 fixation reaction and entry of CO_2 through stomata. Finally, starch synthesis in the chloroplast and transport of sugars across the chloroplast envelope into the cytoplasm are directly controlled by the concentration of inorganic phosphate (Marschner, 1997). Manganese deficiency in the leaves drastically reduced photosynthesis per unit chlorophyll which could be restored within two days after foliar application of manganese, indicating a direct effect on Photosystem II (Fig. 2). Mineral nutrients like magnesium, manganese, sulphur and iron etc, are directly involved in the light reaction of photosynthesis controlling various oxidation/reduction processes in electron transport chain during light energy assimilation. Magnesium play important role in the absorption of photon by chlorophyll in PS-I and PS-II and thereby initiating the electron flow. Manganese clusters act as device for storing energy prier to oxidation of water molecules. It presumably acts as the binding site for water molecules which are oxidized (Rutherford, 1989). Electron transfer between PS-I and PS-II is mediated via iron-sulphur atoms and plastocyanin – a copper containing protein. Further, ferredoxin, which is a 9 kDa iron-sulphur protein, regulate electron to NADP which gets reduced to NADPH by ferredoxin-$NADP^+$ - oxidoreductase enzyme. Mineral nutrients like iron and sulphur also regulate reduction of nitrite (NO_2^-) and sulphite (SO_2^-) in the plants via light reaction of photosynthesis. These are also associated with the formation of organic compounds like amino acids in nitrogen metabolism. Similar to nitrogen, phosphorus deficiency also affect photosynthesis due to poor light utilization, however, carbohydrates accumulation in leaves and roots have been noticed in such plants. The mineral nutrient deficiency not only decrease the photosynthesis and LAI but also *leaf area duration* (LAD), that is the length of time in which the source leaves supply the photosynthates to sink sites.

In many situations it is observed that for optimum growth of the plants there must be a balance between the rate of photosynthesis and the rate of N assimilation. Under conditions when high photosynthesis occur (under high light intensity, optimum temperature and no water stress), the level of N nutrition must also be high and *vice-versa*. Therefore, under growth conditions, the plant with high CO_2 fixation rate (C_4 species), the N demand is considerably high.

The level of N nutrition required for optimum growth phase must be balanced by the presence of other nutrients in adequate amount. The synthesis of organic N compounds depends on a number of organic ions, including Mg^{2+} for the formation of chlorophyll, and phosphate for the synthesis of nucleic acids. Both the uptake of nitrate, and especially its assimilation into protein, is also considerably influenced by the K^+ status. Potassium is important for growth and elongation probably in its function as an osmoticum and may react synergistically with IAA and GA (Table 2).

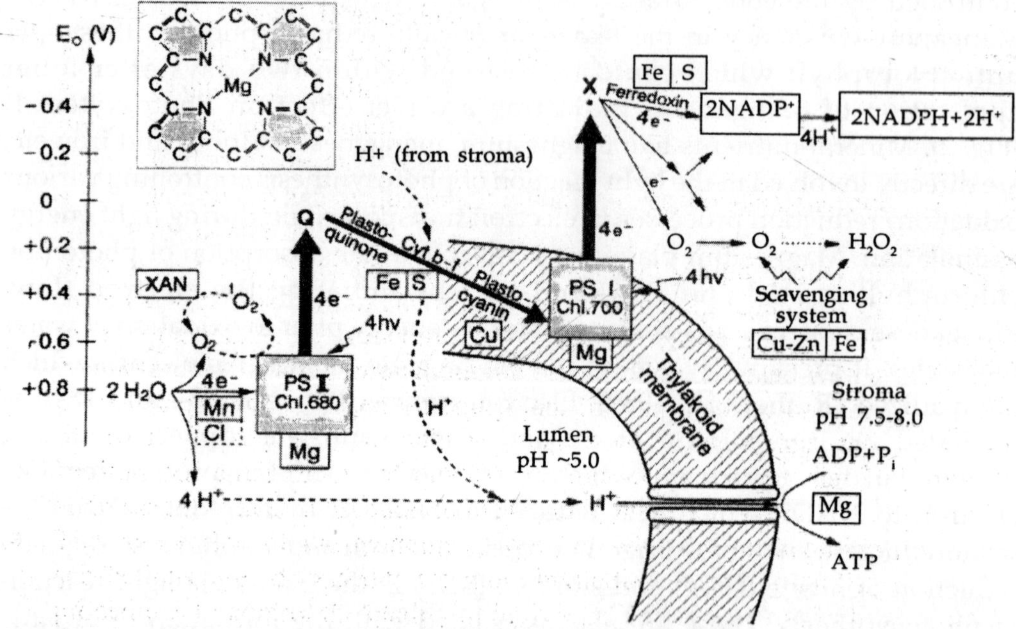

Fig. 2 : Nutrients regulated flow of electron in photophosphorylation (Marschner, 1997)

Table 2 : Effect of K^+ and gibberellic acid on growth rate (internodes elongation) in sunflower (Guardia and Benlloch, 1980)

Conc. of KCL (mM)	GA_3 application (µg)	Growth rate (mm)
0.0	0	3.8
0.5	0	6.0
5.0	0	19.0
0.0	100	29.2
0.5	100	41.4
5.0	100	56.6

In general, it may be concluded that mineral nutrient requirement during the vegetative growth period is primarily determined by the rate of CO_2 assimilation. If the rate of photosynthate production is high, the amount of inorganic nutrients must also be correspondingly high in order to convert the photosynthates into numerous metabolites needed for vegetative growth.

2. Mineral nutrients and sink development

In crop plants the response of mineral nutrients is often a reflection of sink limitations imposed by either a deficiency or an excessive supply of nutrients during certain critical period of plant development, including flower induction, pollination and tuber initiation. These effects can be either direct (as in case of nutrient deficiency) or in direct (eg. effects on the level of photosynthates or phytohormones). Flower formation in many plants is regulated by N, P and K via formation of phytohormone in general and cytokinin in particular. This is also true for the beneficial effect of N fertilization before anthesis in increasing grain number per ear in wheat (Herzog, 1981) or seed number per plant in sunflower (Steer, *et al.*, 1984). Mineral nutrients may also affect flower initiation and seed set by increasing the supply of photosynthates during critical period of reproductive growth (Marschner, 1997). The deficiency of copper, zink, manganese and boron affect reproductive development in cereals (wheat and maize) more than vegetative growth. These are known to regulate pollination and fertilization via pollen germination and pollen tube growth in low land rice spikelet sterility may be considerably reduced with high potassium supply. Increase in potassium content in dry matter (from 0.61 to 2.36) decreased spikelet sterility from 75% to 11%. It could perhaps be due to imbalance nutrients owing to high nitrogen availability (Haque, 1981). N & P deficiecy often leads to flower and fruit drops in many crop plants eg. soybean and cotton and there by decrease in yields (Marschner, 1997). Though the flower and fruit drop is controlled by various factors, however, nutrients controlled phytohormones level (cytokinin and ABA) is associated with this phenomenon.

3. Source and sink relationship

A high nitrogen supply is important for rapid leaf and shoots growth for obtaining an optimum leaf area index (LAI) 3-5, necessary for higher productivity in many crop plants. Unlike cereals, the sink development in tuber crops (storage sink) often exhibit a distinct sink competition between vegetative shoot growth and storage tissue growth (eg. tuber in potato) with the application of mineral nutrients (eg. nitrogen). A large and continuous supply of nitrogen to the roots of potato delays or even prevents tuberization. The tuber growth rate is also drastically reduced by high N application, while the growth rate of vegetative shoot is enhanced.

The sink competition between vegetative shoot and tuber can be readily manipulated by interruption and resupply of N. This is related with N induced changes in the phytohormones balance both in vegetative shoot and the tuber. Thus, a lower but more continuous supply of N allows an early tuberization and continuous root growth and cytokinin production which is more effective on LAD (lead area duration) than on LAI. This situation is more conducive for higher tuber yield than faster development of high LAI by high N supply during early growth stage. Competition for N rather than carbohydrates supplied from source leaves is the main limiting factor for seed yield in crop plants. Developing seed and leaves compete for nitrogen. The seed growth and final yield are primarily determined by the size of N pool in vegetative parts. For example in mustard, application of nitrogen at onset of flowering reduced dry matter in source (leaf) but increased the dry matter of flower/ pods/ seeds (sink). This is known as source and sinks manipulation. Removal of source (leaf) reduced yield by 50 %, largely due to limitation of N and other nutrients to the sink. On the other hand, shading of source (leaf) imposes limitation of photosynthates supply to sink and reduced yield only by 20%, as stem and pods also provide photosynthates to the developing seeds (sink). Young growing leaf needs lot of nutrients for growth and expansion and acts as a strong sink during vegetative phase. After fully developed the same leaf become the source of nutrients and assimilates to the growing reproductive sink (Fig. 3).

In mature cereal plants, 80% of the total amount of N or P is located in the grains as compared with less than 20% of the total K. Thus, in cereals plants where there is a suboptimal supply of these 3 mineral nutrients during vegetative phase, source limitation during grain filling may be likely induced by N or P and not by K. In moderately P deficient plants, source limitation may also become dominant factor for grain yield. In wheat these relationship are prominent for flag leaf area which may deliver 51-89% of P found in the grain at harvest (Batten et al, 1986). All these examples indicate the role of mineral nutrients as yield limiting factors when fruits, seeds or other organs are dominant sink sites and mineral nutrients uptake by the roots (source) is declining.

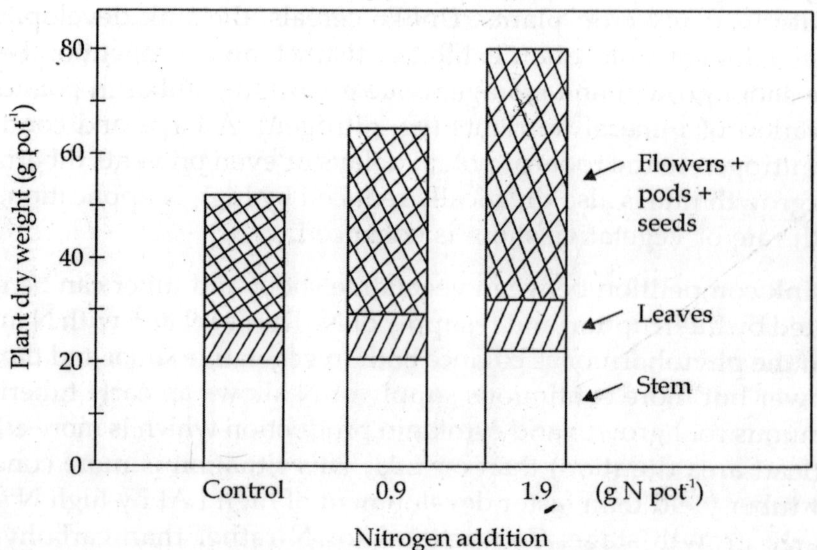

Fig. 3 : Effect of nitrogen application at flowering on dry matter production and distribution in mustard (Based on Trobisch and Schilling, 1970)

4. *Foliar application of fertilizers*

Foliar applied fertilizers provide a quicker response and are more effective than soil applied fertilizers. A multi-purpose adjuvant and compatibility agent can increase the efficacy of foliar sprays i.e. fertilizers, herbicides, pesticides, fungicides etc. by improving wetting, sticking, and absorption of foliar sprays. Foliar feeding is an effective method for correcting soil deficiencies and overcoming the soil's inability to transfer nutrients to the plant. Tests have shown that foliar feeding can be 8 to 10 times more effective than soil feeding and up to 90 percent of a foliar fed nutrient solution can be found in the smallest root of a plant within 60 minutes of application. The effectiveness of foliar applied nutrients is determined by (1) The condition of the leaf surface, in particular the waxy cuticle. The cuticle is only partially permeable to water and dissolved nutrients and, as a result, it can limit nutrient uptake. (2) The length of time the nutrient remains dissolved in the solution on the leaf's surface. (3) Diffusion, the movement of elements from a high concentration to a low concentration. For diffusion to occur, the nutrient must dissolve, and (4) The type of formulation. Water-soluble formulations generally work better for foliar applications as they are more easily absorbed when compared to insoluble solutions.

5. *Site-specific nutrient management (SSNM)*

The SSNM approach was developed to increase mineral fertilizer use efficiency and achieve balanced plant nutrition (Witt and Dobermann, 2002). Field- and season-specific fertilizer rates were calculated after taking into account indigenous soil nutrient supplies, plant nutrient demand (based on yield targets), and interactions among nitrogen (N), phosphorus (P), and potassium (K). This SSNM concept is valid for modern, high yielding varieties with a harvest index (ratio of grain to dry matter) of about 0.5.

A recommendation is provided for the total NPK fertilizer requirement depending on cropping season, crop establishment method, and inputs from other nutrient sources such as straw or manure. To improve the match between plant N requirements and fertilizer N supply, the SSNM strategy provides guidelines for splitting and timing fertilizer N applications according to crop growth stage. Nitrogen applications are fine-tuned during the season using a chlorophyll meter (SPAD) or leaf color chart (LCC) and when growing conditions in the season differ from the assumptions used in the N fertilizer recommendation model. Field-specific fertilizer N, P, and K recommendations are calculated in five key steps.

Step 1: Yield goal selection

Step 2: Estimation of crop nutrient requirements

Step 3: Estimation of indigenous nutrient supplies

Step 4: Calculation of fertilizer rates

Step 5: Dynamic adjustment of fertilizer N applications

Inadequate and imbalanced plant nutrition leads to limit the yield. SSNM strategies that include site and season-specific knowledge of crop nutrient requirements and indigenous nutrient supplies are required to increase productivity, yields, and nutrient use efficiency of crops. SSNM with potassium, sulphur and micro-nutrients help to sustain high yields of rice-wheat cropping system.

References

Batten C.D., Wardlaw, I.F. and Aston, M.J. (1986) Growth and the distribution of phosphorous in wheat developed under various P and temperature regimes. *Australian Journal of Agricultural Sciences*, **37** : 459-69.

Chapin, F.S., Walter C.H.S. and Clarson, D.T. (1988) Growth response of barley and tomato in nitrogen stress and its control by ABA. water relation and photosynthesis. *Planta*, **173** : 352-66.

Haque, M.Z. (1988) Effect of NPK on spikelets' sterility induced by low temperature at reproductive stage of rice. *Plant and Soil*, **109** : 31-36.

Krauss, A. (1980) Influence of N nutrition on tuber initiation of potato. In : Physiological aspects of crop productivity, *International Potash Institute*, Bern, pp. 175-84.

Michael, G. and Beringer, H. (1980) Role of hormones in yield formation. In : Proceeding of 15th Colloq. Int. Potash Instt., Bern, pp 85-116.

Marschner, H.(1997) Mineral nutrition of higher plants; Academic Press, London, New York, pp. 689.

Mengle, K. and Kerby, E.A. (1996) Principal of Plant Nutrition, Panima Publishing Corporation, New Delhi, pp. 687.

Radin, J.W. and Eidenbock, M.P. (1984) Hydraulic conductance as a factor limiting leaf expansion of P deficient cotton plant. *Plant Physiology*, **75** : 372-77.

Rutherford, A.W. (1989) Photosystem II, the water splitting enzyme. *Trends in Biochemical Science*, **14** : 227-32.

Witt, C.and Dobermann, A. (2002) A Site-Specific Nutrient Management Approach for Irrigated, Lowland Rice in Asia. *Better Crops International*, **16** (1) : 20-24.

□□□

System Based Integrated Nutrient Management, 2012
© *B. Gangwar & V.K. Singh (eds.), pp. 67-78*
New India Publishing Agency, New Delhi (India)
e-mail : info@nipabooks.com; website : www.nipabooks.com

CHAPTER **5**

Soil Testing for Higher Productivity and Soil Health Rejuvenation

B.S. DWIVEDI

Sustainable agriculture warrants a wise management of natural resources and inputs to satisfy human needs in the fast changing scenario while maintaining, if not improving, the quality of environment. Soil as a natural resource is facing more and more serious degradation threats under modern agriculture. Incidence and expansion of multi-nutrient deficiencies in Indian soils owing to inadequate and unbalanced nutrient input through fertilizers is considered one of the major reasons for decline in factor productivity of crops. The problem of soil fertility depletion is spectacular across the soil types and agro-climatic zones, but it is more acute in intensively cropped areas, where annual nutrient removal by the crops often far exceeds replenishments. Recent diagnostic surveys indicate that in several high productivity areas of irrigated ecosystems like, Trans-and Upper Gangetic Plain zones, farmers are resorting to even excessive use of fertilizers, especially N, to maintain the yields at levels attained previously with lower fertilizer application rates. With an unabated deterioration in soil health in general, and soil fertility in particular, and a mounting pressure on finite land resources due to increase in population, the interest in soil fertility and soil health has revived due to several reasons:

- Stagnation in growth rates of production and productivity of staple foodgrains - threat to food security

- Decline in factor productivity, thus increasing cost of production
- Nutrient withdrawals in excess of replenishment
- Low nutritional value of food and feed grown on nutrient-starved soils – threat to nutritional security
- Poor quality of produce fetching low price – threat to economic security

Public interest in soil health (that necessarily include soil fertility) is increasing throughout the world, as humankind recognizes the fragility of earth's soil, water and air resources, and the need to protect them to sustain the civilization. As research evidences amply suggest that a single intervention of soil fertility improvement may account for 50-70% difference in crop yields (Swaminathan, 2006), various aspects related to soil fertility evaluation and management have to be discussed and debated in detail to understand the soil fertility evaluation protocols and bring about need-based refinement so as to prevent any further degradation in soil fertility and also meet farmers' high productivity aspirations.

Some important definitions

Soil fertility and soil productivity

Soil fertility is the inherent capacity of soil to supply plant nutrients in right amount and right proportion. Thus it is more related to nutrient availability to the plants. On the other hand, soil productivity is a broader term that indicates capacity of soil to produce biomass depending on soil fertility and other factors like soil texture, soil-born diseases, waterlogging (drainage), soil depth, erodibility *etc.*

Soil health and soil quality

Soil health could be defined as 'the capacity of soil to function within ecosystem boundaries, to sustain biological productivity, maintain environmental quality and promote plant and animal health' (Doran *et al.*, 1996). Many a times, the state of soil is also explained loosely in terms of soil health, soil productivity, soil environment *etc.* On the other hand, soil quality connotes soil's usefulness for a particular purpose over a long time scale (Goswami 2006). Nonetheless, scientific literature as well as popular press often uses the terms 'soil health' and 'soil quality' interchangeably, with farmers preferring the former and scientists the later.

Soil testing for soil fertility evaluation

In Indian context, diagnosis of soil health is often considered synonymous to soil testing, carried out for assessment of soil fertility status or appraisal of salinity/alkalinity problems (Gajri *et al.*, 2006). While not a "perfect" system, soil testing is the only science-based tool for nutrient recommendations. To date, the results are impressive - soil testing does change the fertilizer rates applied, and increases yield, and/or quality. Making it work in a region, however, requires a regional effort to make it appropriate and efficient. As the current infrastructure for soil health monitoring is not adequate to support the analysis of even minimum soil fertility parameters required for formulation of balanced fertiliser schedules, measurement of biological and physical characteristics of soils by soil testing laboratories would be unlikely in near future. Research laboratories are, however, studying the impact of input use, cropping systems, climate *etc*, on different soil health parameters.

Soil testing as a means of judicious fertilizer use works on the principle of probability, meaning that if all other factors of productivity are at optimum and none of them is limiting, there is all probability to obtain more profitable response to applied nutrients based on soil testing than to those applied on ***ad-hoc*** basis. In case of soil fertility evaluation, that part of the total nutrient content which is accessible and present in the plant utilizable form is determined. Thus, it is the index of availability of nutrients that serves as a guide for fertilizer use. Standard analytical procedures are employed for this. A suitable procedure has to satisfy three criteria, namely: (i) rapidity *i.e.* less time taking nature, (ii) reproducibility of results, and (iii) predictability *i.e.* ability to predict the crop response obtainable from applied nutrients based on soil test values.

Objectives of soil testing

- To assess the soil fertility status and recommend suitable and economic nutrient doses for different crops and cropping systems
- To identify the type and degree of degradation problems/abnormalities like soil acidity, salinity and sodicity *etc*, and to suggest effective remedial measures
- To generate data for compilation of soil fertility maps
- To study soil pollution-related aspects and to advise preventive as well as remedial measures for a safe food-chain
- To make continued efforts for improving the science of wholesome analysis and interpretation of the test data for more meaningful use of this tool for soil care and crop production

Soil-test parameters

As already indicated in the objectives, soil testing includes analysis of soils for detection of soil problems also along with available nutrient status. Major parameters testes are:

Problem-related parameters: Salinity, acidity, alkalinity (sodicity)

Nutrient availability: Available N, P, K, S, Zn, Fe, Cu, Mn, B

Auxiliary tests: Calcium carbonate, texture

Special tests: Need-based only such as heavy metals, BOD, COD *etc.*

Soil fertility ratings

Majority of soil testing laboratories (STLs) follow a 3-category soil fertility rating system (Table 1). These rating proposed in 1950s have become redundant in several farming situations owing to changed nutrient uptake pattern of modern cultivars and also due to ever-increasing multi-nutrient deficiencies. At IARI and in some of the states, however, the fertility ratings are revised within broad framework of the earlier categories, and the revised ones are used for fertility evaluation.

Table 1 : Soil fertility ratings adopted in the STLs

Soil Parameters	Low	Medium	High
Organic Carbon (%)	<0.5	0.5-0.75	>0.75
Available Nitrogen (N, kg ha^{-1})	<280	280-560	>560
Available Phosphorus (P, kg ha^{-1})	<10	10-25	>25
Available Potassium (K, kg ha^{-1})	<110	110-280	>280

For secondary and micronutrients, critical limits are used to differentiate deficient soils from non-deficient ones. A deficiency critical limit is a value beyond which sufficiency of the nutrient in question sets-up, with lesser predictability of crop response to applied nutrients. It is, thus, apparent that the critical limits would vary according to soils and crops, and also with productivity levels. Nonetheless, on the basis of analysis of large number of soil samples as well as crop response studies, some critical levels are suggested (Table 2) for use in generalized fertilizer prescriptions and for soil fertility mapping.

Fertility status of Indian soils

Soil organic matter content

Soil organic matter (SOM) is known to serve as soil conditioner, nutrient source, substrate for microbial activity, preserver of environment and major determinant for sustaining or increasing agricultural productivity. Soil fertility aspects controlled by SOM include direct contribution of nutrients, influence on cation exchange capacity and binding of heavy metals and pesticides. Tropical Indian soils, majority of which belong to arid and semiarid climate, rarely exhibit organic C levels exceeding 0.6% (Virmani *et al.*, 1982), whereas those in temperate environment have organic C levels ranging between 1.2 to 2.5%. Lal and Kang (1982) reviewing SOM management in tropical soils, observed the necessity for at least 4 times higher organic inputs in tropical than in temperate environments to maintain the native SOM levels.

Table 2 : Critical limits of secondary and micronutrients

Nutrient	Extractant	Critical limit (mg kg^{-1} of soil)
S	0.15% $CaCl_2$	10
Zn	DTPA	0.6
Fe	DTPA	4.5
Cu	DTPA	0.2
Mn	DTPA	2.0
B	Hot water	0.5

NPK fertility maps

The soil test data obtained from soil testing laboratories have been compiled from time to time for soil fertility appraisal and development of fertility maps. Nutrient index (NI) values were used to indicate fertility status of a district as a mapping unit. The scenario of soil fertility as revealed through these maps, prepared by Ramamurthy and Bajaj (1969), Ghosh and Hasan (1976, 1979, 1980) and Motsara (2002) on the basis of analysis of 1.3, 9.2 and 3.65 million soil samples, respectively, is presented in Table 3. Some typical and unexplainable trends emerged out of these compilations, especially with respect to temporal changes in K fertility status. The percentage of districts belonging to low or medium K fertility category decreased, whereas those falling in high K category increased with the passage of time, thus leading to an erroneous conclusion regarding K status of soils. On the other hand, fertiliser consumption statistics, experimental results on crop response, apparent K balance sheets and soil K status as measured in soil fertility experiments support a reverse trend. Any build-up in soil K status over the years is unlikely because of continuously low fertilizer K consumption and an increase in crop response to fertiliser K.

Similarly, a build up in N and P fertility status over the years, as revealed by an increase in percentage of districts belonging to high N or P soil fertility class, is also difficult to explain. Apparent reasons for such significant discrepancies in fertility appraisal are: lack of sampling norms *i.e.* minimum number of samples per unit area for mapping purposes, lack of stratification in sampling, poor recording of information on sampling sites, and use of variable number of samples for preparation of these maps. It is not understood as to what extent these maps are really useful to agricultural administrators, policy makers or other stakeholders.

Table 3 : NPK fertility status of Indian soils as evaluated on the basis of STL database

Nutrient	% Districts in different fertility classes		
	Low	Medium	High
Ramamoorthy and Bajaj (1969): 1.3 million samples from 184 to 226 districts			
Nitrogen	52	43	5
Phosphorus	47	49	4
Potassium	20	53	27
Ghosh and Hasan (1976, 1979, 1980): 9.2 million samples from 310 to 365 districts			
Nitrogen	62	33	5
Phosphorus	46	52	2
Potassium	20	42	38
Motsara (2002): 3.65 million samples from 450 districts			
Nitrogen	63	26	11
Phosphorus	42	38	20
Potassium	13	37	50

Secondary and micronutrients

Past decades of intensive cropping brought a general depletion in the availability status of S and micronutrient (especially Zn, B and Fe) status of the soils, which led to increased incidence of their deficiencies in plants and also crop responses to their application (Dwivedi et al., 2006 ; TSI, 1994). Since currently used high analysis fertilisers do not allow periodic addition and accretion of S in soils, the crops largely meet their S requirement through native S reserves (Tiwari, 1997). As a consequence, S deficiencies are no longer confined to coarse-textured soils and oilseed growing regions, but expended to almost all soils and all major crops/cropping systems. Extensive field experimentation under recently concluded TSI/FAI/IFA collaborative project on 'Sulphur in Balanced Fertilisation' revealed an enhanced mining of soil S due to neglect of S application and also a build-up of soil S consequent to inclusion of S in fertiliser schedules (Dwivedi *et al.*, 2001).

Among micronutrients, Zn deficiency is the most common soil disorder accounting for nearly 48% of the soil samples analyzed under ICAR's AICRP on Secondary and Micronutrients and Pollutant Elements (Singh, 1998). Next to Zn are B (38%) and Fe (12%) deficiencies. Since these estimates exclude any data on micronutrient status generated outside the AICRP, the actual magnitude of micronutrient deficiencies may be still greater. Further, the estimates on B deficiencies are based on a limited number of soil samples, because B determinations are not as frequent as those of DTPA extractable Zn, Fe, Cu and Mn. Recent studies, however, revealed occurrence of B deficiencies even in the soils of arid and semiarid regions, which are otherwise considered adequate in B supply. It is also true that farmers' awareness on Zn deficiency and application of Zn fertilisers by them increased in recent years, but there are no verifiable evidences regarding a decrease in the extent of Zn deficiency.

Multi-nutrient deficiencies

Emergence of multi-nutrient deficiencies in soils is repeatedly referred to in the literature, but no quantitative assessment of the nature and extent of this complex soil fertility problem exists. Databases generated under different projects and soil fertility maps developed so far deal with individual nutrients in isolation. It is not known if a soil assessed deficient in a nutrient X is simultaneously deficient in nutrient Y or Z also. Pilot-scale studies on evaluation of multi-nutrient deficiencies recently taken-up at IARI indicated widespread deficiencies of at least 6 nutrients *viz.* N, P, K, S, Zn and B in soils representing different agro-ecological regions, with NK, NPKB, PKZn, NKS and NPB as some of the prominent multi-nutrient deficiency combinations (Dwivedi *et al.*, 2006).

Soil testing service in India

Recognizing the multi-facet significance of soil testing, a modest beginning of soil testing service was made in India way back in 1955-56 with the establishment of 16 STLs under the 'Indo-US Operational Agreement for Determination of Soil Fertility and Fertiliser Use'. The network of STLs has been expanding continuously, and at present a total of 686 STLs are functioning with an annual analyzing capacity of 7.21 million soil samples (Table 4). Out of these, 647 STLs (94%) belong to State Govts, and remaining 39 (6%) STLs are with fertiliser industry. The capacity utilization varies from 53% in east zone to 82% in south zone. The capacity utilization in states like Jharkhand, W.B. and Nagaland is below 30% whereas the same is above 100% in A.P., Gujarat and Maharashtra. There is an urgent need to expand the analyzing capacity as well as its efficient utilization.

Table 4 : Soil testing laboratories (STLs) and their capacity utilization during 2007-08

Zone	Number of STLs							Annual analyzing capacity (000')
	State Govts.		Fertiliser industry		State Govt. + Industry			
	Static	Mobile	Static	Mobile	Static	Mobile	Total	
South	136	33	6	9	142	42	184	2015 (82)*
West	126	17	10	7	136	24	160	1514 (75)
North	188	37	4	3	192	40	232	2946 (80)
East	90	20	-	-	90	20	110	735 (53)
All India	540	107	20	19	560	126	686	7210 (75)

*Per cent capacity utilization

In addition to the STLs owned by State Governments and fertiliser industry, soil and water testing facilities have also been created with 300 Krishi Vigyan Kendras (KVKs). SAUs and some ICAR institutes also offer soil testing service to the farmers on a limited scale. Surprisingly, these STLs are not listed along with the STL network. Also, data generated by them is neither included in soil test databases nor utilized at state/national level for soil fertility mapping and delineation purposes.

Major weaknesses of soil testing service

Despite constant expansion of STL network since its inception, the soil testing service could not become saleable. Listed hereunder are some specific reasons responsible foı poor demand of soil testing:

Sub-optimal and incomplete fertiliser prescription

This is possibly the first and foremost reason for detraction of farmers' faith. The so called 'soil test-based' prescriptions advocated by STLs often fail to meet high yield or high profit aspirations of the farmers. Since the fertiliser NPK prescriptions overlook crop productivity level, and any possible deficiency of secondary or micronutrients, these actually prove equally unbalanced and incomplete as the farmers' fertiliser practice, leaving a poor or negative impact of soil testing on the farmers.

Lack of norms for soil sampling

Soil samples brought to a STL are mostly collected by the extension personnel. In absence of well-defined norms for sampling density, time or frequency, the utility of soil test data is actually compromised. These soil test data could neither be used meaningfully for generation of area-specific soil fertility maps, nor for monitoring of management-induced temporal variability in soil fertility parameters.

Inadequate analytical facilities and scarcity of trained human resources

The STLs analyze soil samples for organic C (as an index of N availability) and available P and K status, and prescribed recommendations for NPK only. The STLs, barring a few, are either not equipped for micronutrient analysis or the staff is not trained enough to handle sophisticated instruments like atomic absorption spectrophotometer (AAS). Even the STLs having AAS may not provide quality soil test data on micronutrients in the absence of trained technical staff, analytical grade reagents, glass distilled water and utmost care in collection, processing and storage of soil samples.

Quality of analysis?

Significant variations in training and subject matter expertise of the soil testing staff, efficiency of lab equipments with respect to reproducibility of results, quality of chemicals and quality of water in different STLs warrants for monitoring of the quality of soil analysis. Unless the soil test data are reliable and accurate, fertiliser prescriptions are unlikely to perform in the field whatsoever is the criterion of recommendations.

Poor linkage with research institutions

Soil testing service operates in almost isolation from other concerned professional organizations, and any linkages with SAUs/ICAR institutes are practically absent. Hence, transmission of any new knowledge or developments in science to the STLs remains restricted, though the same may be significant and crucial for maintaining efficiency and robustness of the service.

Suggestions to strengthen soil fertility evaluation and monitoring system

Action plan for researchers

Revisit fertiliser prescription approaches: Maximum economic yields can hardly be achieved without inclusion of those nutrients in fertiliser schedule, which are otherwise considered non-limiting for relatively low yield goals. The targeted yield approach advocated by AICRP-STCR, though scientifically sound than other contemporary approaches, is often criticized for not making progress beyond what was proposed by Ramamoorthy *et al.* (1967). Possibilities need to be explored to expand the boundaries of this novel approach beyond NPK, for the arena of balanced nutrition has already expanded. The tools like leaf color chart (LCC) and chlorophyll meter (SPAD) need to be evaluated for crops other than rice and wheat also and threshold levels worked out.

Ready-reckoners on soil test-based fertiliser recommendations: AICRP-STCR should be entrusted with a time-bound mandate of developing ready-reckoners on fertiliser recommendations for at least two yield targets of major field crops grown on different soils. The network centers could be made responsible for this exercise for the agro-ecological regions (AERs) wherein they are located. This would of course require strengthening of the AICRP in terms of scientific/ technical personnel and budget. For the AERs not covered under AICRP network, the project should have provision to outsource the services to SAUs or ICAR institutes located in those AERs.

Revision of soil fertility ratings: The soil fertility ratings (low, medium, high) proposed during 1950s on the basis of magnitude of crop response to nutrient input remained almost unchanged although the entire spectrum of agriculture has transformed since then, particularly with respect to nutrient removal and response pattern of exhaustive crop varieties. When crop responses to nutrient application are similar for both 'medium' and 'low' fertility soil as indicated by on-farm experiments (Table 5), a fertiliser prescription formulated for such a 'medium' fertility soil would be essentially sub-optimal. Hence, these ratings need to be revised in the light of current crop responses to applied nutrients on different soils. Only revised ratings should be used for interpretation of soil test data.

Table 5 : Average response of wheat to 60 kg P_2O_5 ha^{-1} or K_2O in on-farm trials on the soils of low, medium and high fertility status

Nutrients	Fertility rating	Districts	Trials	Response (kg ha^{-1})
P_2O_5	Low	21	2,140	680
P_2O_5	Medium	17	2,446	669
P_2O_5	High	1	147	486
K_2O	Low	6	749	245
K_2O	Medium	22	2,167	250
K_2O	High	12	1,834	192

Source: Tiwari, (2006)

Upward revision of fertiliser recommendations : The *ad hoc* fertiliser recommendations for different crops, as published in the state package of practices, need to be revised in the light of changing crop responses to nutrient inputs. Adoption of any sub-optimal nutrient recommendations brings deterioration in soil fertility, with low productivity, low produce quality and low farm income as ultimate results.

Development and refinement of soil test procedures: The search of an efficient soil test that works across the soils of varying characteristics is unending. In this context, the view point of Nelson and Anderson (1977) expressed some three decades ago is worth citing, "*much of the research in soil testing has dealt with developing chemical procedures for evaluating the levels of available nutrients in the soil. The value of soil testing for advisory purposes lies in improvement in the accuracy of crop response relationship and development of soil test-based fertiliser recommendations"*. Nonetheless, specific problems associated with soil test procedures as encountered in different determinations, have to be addressed through problem-solving and focused research.

Soil health monitoring in long-term experiments (LTEs) : The LTEs are precious assets to assess the impact of crop management on soil health and also to forewarn against management-induced soil problems. Data collection in the LTEs continuing under different AICRPs is, however, hopeless. What is needed is to identify a minimum data set for soil physical, biological and fertility parameters to be essentially collected by all LTEs across the AICRPs. Information generated under different LTEs should also be converged at some point - may be through joint workshop kind of events.

Strengthening soil testing service

For revamping of soil testing, some suggestions are offered hereunder:

- Establishment of new STLs
- Modernization of existing STLs
- Transforming STLs to soil-health clinics
- Developing rational soil sampling norms
- Implementing sample exchange programme amongst STLs
- Timely delivery of recommendations
- Human resource development
- Establishing feedback mechanism

Epilogue

A nutrient-starved and physically/biologically deteriorated soil cannot support high crop productivity. Any yield gain achieved on such soil owing to varietal or crop management intervention other than judicious nutrient input would not only be temporary but also encourage mining, leaving the soil further depleted of its native nutrients reserves. The physical and biological environment of a low fertility soil, that supports low yields and obviously restricted root system, is often unhealthy because of lesser recycling of belowground root mass. It is, therefore, inevitable to precisely monitor the existing as well as emerging soil fertility problems under different agro-ecologies, generate pragmatic prescriptions and convey them in-time to the farmers. Equally important is to enhance awareness of farmers regarding the threats of extractive farming practices, as also regarding long-term benefits of investing in soil health improvement. Soil fertility evaluation needs to be more knowledge-intensive so as to develop optimal nutrient prescriptions to address the current complex fertility problems like multi-nutrient deficiencies.

References

Doran, J.W., Sarrantonio, M. and Liebig, M.A. (1996). *Adv. Agron.* **56**: 1-54.

Dwivedi, B.S., Dhyan Singh, Chhonkar, P.K., Sahoo, R.N., Sharma, P.K. and Tiwari, K.N. (2006). Soil Fertility Evaluation - a potential tool for balanced use of fertilisers, IARI, New Delhi/PPIC-India Programme, Gurgaon, pp. 1-60.

Dwivedi, B.S., Shukla, A.K., Singh, V.K. and Yadav, R.L. (2001). Sulphur fertilisation for sustaining productivity of rice-wheat system in western Uttar Pradesh, Tech. Bull. 2001-1, PDCSR, Modipuram, pp. 1-35.

Gajri, P.R., Rao, D.L.N. and Singh, M.V. (2006). *In:* Proc. Int. Conf. Soil, Water Environ. Quality - Issues & Strategies, ISSS, New Delhi, pp. 74-83.

Ghosh, A.B. and Hasan, R. (1976). In : *ISSS Bulletin No.10*, ISSS, New Delhi, pp. 1-5.

Ghosh, A.B. and Hasan, R. (1979). In : *ISSS Bulletin No.12*, ISSS, New Delhi, pp. 1-8.

Ghosh, A.B. and Hasan, R. (1980). *Fertilizer News*, **25** (11): 19-24.

Goswami, N.N. (2006). In : *Proc. Int. Cong. Soil, Water Environ. Quality - Issues & Strategies*, ISSS, New Delhi, pp. 43-58.

Lal, R. and Kang, B.T. (1982). In : Non-symbiotic N Fixation and Organic Matter in the Tropics. *Trans. 12th Int. Congr. Soil Sci.*, New Delhi, pp. 152-178.

Motsara, M.R. (2002). *Fert. News* **47** (8): 15-21.

Nelson, L.A. and Anderson, R.L. (1977). In : *ASA Special Publication No. 29*, Madison WI, pp. 1-18.

Ramamoorthy, B. and Bajaj, J.C. (1969). *Fert. News*, **14** (8): 25-36.

Ramamoorthy, B., Narsimhan, R.L. and Dinesh, R.S. (1967). *Indian Fmg.* **17** (5): 43-45.

Singh, M.V. (1998). *In:* Singh G.B. and Sharma, B.R. (Eds) 50 Years of Natural Resource Management Research, ICAR, New Delhi, pp. 177-198.

Swaminathan, M.S. (2006). In : *Proc. Int. Conf. Soil, Water Environ. Quality - Issues & Strategies*, ISSS, New Delhi, pp. 6-8.

Tiwari, K.N. (2006). *Indian J. Fert.* **2** (9): 73-98.

Tiwari, K.N. (1997). *Proc. TSI-FAI-IFA Symp. Sulphur in Balanced Fetilisation*, New Delhi, pp. SI/1/1-5.

TSI. (1994). Sulphur: its role and place in balanced fertiliser use in Indian agriculture, Bulletin 16, TSI, Washington DC.

Virmani, S.M., Sahrawat, K.L. and Burford, J.R. (1982). In : Vertisols and Rice Soils of the Tropics. *Trans 12th Int. Cong. Soil Sci.*, New Delhi, pp. 80-93.

□□□

System Based Integrated Nutrient Management, 2012
© B. Gangwar & V.K. Singh (eds.), pp. 79-91
New India Publishing Agency, New Delhi (India)
e-mail : info@nipabooks.com; website : www.nipabooks.com

Nutrients Economy through Integrated Farming System Approach

B. GANGWAR AND J.P. SINGH

Post independence era brought about a radical change in agricultural scenario which improved economy and living standard of Indian farmers to a great extent. Green revolution (field crops) during early sixties followed by white revolution (milk) and thereafter yellow (oilseeds) and blue (fish) revolutions made the country not only self sufficient in most of the food commodities but also exporter in many. Several calamities including droughts, floods, and ill effects of national and international conflicts faced by the people of the country during post independence era could not affect the economy and stability of the country much because of rich agricultural heritage. As usual, the rapid and rather un-planned development creates several problems and this was true in our conditions too. Loss of indigenous land races, depletion of soil nutrients and water resources, creation of salinity and water logging, resurgence of pest and diseases, declining factor productivity and farm profits and increased environmental (air, water & soil) pollution are among such problems have great concern to agricultural scientists and planners in coming future. The situation further aggravated because of industry oriented government policies and globalization which supported growth of industries and trades rather than agriculture and hence the growth rate in these sectors was considerably high. This fact is more visible by the figures on relative contribution of different sectors in the national GDP during last sixty years

(Table 1). At the time of independence of the country more than 75% people were depend on agriculture related occupations which came down to mere 67.4% in 2009-10. Similar is the case of total contribution of agriculture in national GDP which decreased from 57% in 1951-52 to as low as 14.6% in 2009-10. Besides government policies, major factor identified responsible for degradation of resources, stability in production and decreased factor productivity is chemical based farming and that too in imbalanced and improper methods of application practiced by the farmers for getting more and more production and profits per unit area per unit time. The solution of this lies in the increased use of organic sources of nutrients in proper ratio along with chemical fertilizers – IPNS (Integrated Plant Nutrient System). Livelihood of marginal and small land holders who represent a large section (more than 86%) of farming community and remained deprived of advancements in the field of agriculture has been the main agenda of the Indian Government in XIth, five year plan and to continue in XIIth plan too. To ensure livelihood improvement with focus on small and marginal farmers Integrated Farming System (IFS) has been proved to be an effective approach. Accordingly, the Project Directorate of Cropping Systems Research renamed as Project Directorate for Farming Systems Research with AICRP on Integrated Farming Systems Research and Network Project on Organic Farming mandated with farming systems perspectives. Since, integrated farming system is a holistic approach wherein all the wastes and residues are recycled within the system and nothing go waste. Output of one enterprise is used as input for the others. This way, farming becomes more economic and environmentally safe.

Table 1 : Relative contribution of agriculture in national GDP

Sector	Percent contribution in national GDP			Dependency of Indian people on a particular sector during 2009-10 (%)
	1950-51	2008-09	2009-10	
Agriculture	57	18	14.6	67.4
Industries	14	26	56.9	22.4
Services	29	56	28.5	10.6

Source : Daily News Paper "Hindustan" Hindi Edition, Dated, the 26th, August, 2010 pp: 15

Nutrient management in cropping system perspective

Nutrient management on a cropping system basis often involves discounting for the residual effects if any and contributions from *in situ* residue recycling. Differences between direct and residual responses are likely to be wider when a particular cropping system has been adopted but may diminish with time as rotation goes through a number of cycles and a crop cumulatively benefits from all previous nutrient additions and any other improvement in soil productivity.

Nitrogen, phosphorus and potash are the three most researched plant nutrients. Experiments using ^{15}N have shown that a considerable portion of applied N may be left in the soil, however it's availability to the succeeding crop is less than 2%. Oza and Subbiah (1973) while working on wheat-moong-maize cropping system, observed that the utilization of applied N by wheat was up to 48% but for succeeding crops moong and maize, the value was below 2% in all cases.

The behaviour of applied P, however, is different to that of N and it allows the distribution of P among the component crops when grown in a system according to their responsiveness to P and aims at taking residual effect in to account while making recommendations to the farmers. Agronomic differences between crops and seasons decline as the P fertility is built up and the residual P is accessible to all crops in the system, though not to the same degree. As evidenced from the literature, hardly one fifth of the applied P is used by the first crop, while the rest gradually become available to succeeding crops. Legumes incorporated in to a cropping system act as a catalyst to augment availability of native and fixed P in the soil. Sinha *et al.* (1994) observed that the highest utilization of P occurred when P was banded near each legume row. In another study where the main crop as well as intercrop was grain legume, it was observed that sole crop pigeonpea or groundnut and their intercropping systems responded up to 90 kg $P_2O_5ha^{-1}$. Phosphate fertilization of legumes during rainy season increased the available P status of soil and gave higher yield of succeeding crops. P fertilization also enhance N fixation in legumes (Soni, P.N. and Mukharjee, A.K. 1982).

The importance of K management in cropping system approach will certainly increase with increasing cropping intensity and system productivity. Depending upon the level of inputs use in relation to crop demand, each successive cropping cycle may have a more depleted soil or a more fertile soil. Veeranna *et al.* (1996) reported that the K application is essential to sustain the productivity of sorghum-wheat system in long run even on K rich soils. Among the short duration crops, potato is perhaps the only one which receives moderate to optimum level of potassium fertilizer in addition to organic manures. Potato is grown in rotations with a variety of important crops viz; rice, wheat, sorghum, maize in different crop sequences and has also attracted the attention of research workers with regard to K management in a cropping system. In sugarcane intercropped with different crops, exchangeable potassium in the soil after 12 years was the highest in sugarcane + potato system and the lowest under continuous sugarcane (Yadav, R.L. and Singh K. 1986). Also exchangeable K status of the soil was higher in green manured plots than under continuous sugarcane without any green manuring.

Besides NPK, secondary nutrient sulphur (S) and micro nutrients Zn and Fe are considered important plant nutrients in last two decades or so. Dwivedi *et al.* (2001) made an extensive survey in Meerut and Jyotiba Phulenagar , the representative districts of western plain zone of Uttar Pradesh. On an average the magnitude of deficiency of sulphur was higher in the soils of Hasanpur (36.1%) and Gajraula (30.8%) blocks of J.P.Nagar districts compared to Daurala ((26.4%) and Hastinapur (19.1%) blocks of Meerut. They also reached to the conclusion that sulphur deficiency was associated with soil organic matter content and nutrient management practices in more intensively cropped areas. Continued use of high analysis fertilizers such as DAP, and Urea having no S and reduced use of cowdung and green manuring are the major factor responsible for such wide spread S and micronutrients deficiencies in soils. Further, importance of Zn was well recognized long back when a well known scientist Y.L.Nene (1966) identified *Khaira disease* in rice which was mainly because of Zn deficiency. Since then a lot of research work has been done and a blanket recommendation is made to apply 25 kg ha^{-1} of Zinc Sulphate in rice-wheat and also many other crop sequences.

Due to increasing fertilizer cost and associated soil and environmental problems, integrated nutrient management (INM), later known as Integrated Plant Nutrient System (IPNS) including organic manures, crop residues, green manure crops, dual purpose grain legumes such as mungbean, urdbean and cowpea and bio-fertilizers have aroused considerable interests to the mind of researchers as well as planners. In Brazil, more than 58% farmers use bacterial culture in soybean. In Indian conditions use of bacterial cultures in legume crops have been found to save 50 kg N per season with a range of 50-300 kg ha^{-1} and increase in yield ranging from 7-35% in situ and 2.4-16.3% in subsequent crops. The effects of different organic sources of nutrients on the productivity of different crops have also been reported. Application of BGA in rice has found to save N up to 30 kg ha^{-1} and increases yield by 10-12%. P solubilising microorganism PSB including yeasts (*Schwanniomyces occidentalis*), fungi (*Aspergillus awamori* and *Penecillium degitatum*) and bacteria (*Pseudomonas sp.bacillus*) inoculates with and without P application have been found to solubilize the inherent as well as fixed P in various soils. Gaur (1988) reported sizable yield increase of paddy, wheat, lentil, chickpea, maize and potato by use of PSB. Similarly use of PSB increases availability of fixed P and save 20 kg N ha^{-1}. N fixation and addition of litter fall in legume crops add 20-50 kg of N in to soil. Farm yard compost which is the major source of organic nutrition add 25-50 kg of N (based on nutrient content in the compost used) in to the soil and increase yield varying from 155 to 100% in some cases. The results of AICRP on long-term fertilizer experiments brought out the importance of INM in Indian agriculture. Mahaptra *et al.* (1987) reported that rice yield with 30 kg N/ha along with Azolla inoculation were same as with 90 kg N/ha added as prilled Urea. Singh *et al.* (1990) reported that green manuring, BGA and Neem Cake coated urea technologies can be simultaneous in the rice fields. In upland fields beneficial effects of Azotobactor and Azospirillum have also been reported by Rao (1993) and Wani and Lee (1997).

Nutrient management in farming systems perspective – Case studies (India)

Looking in to the present scenario of nutrient supply system in field and plantation crops and apparent negative impact on productivity, profitability and sustainability of agriculture as a whole, the only options left behind is the increased use of organics in agriculture either by inclusion of dual purpose legumes as main crops and or intercrops in different cropping systems, use of bio-fertilizers and bio agents directly in soil or as seed inoculants and last but not least important, the conservation of all the sources of organics available and it's rational use through proper collection, composting and recycling in to the soil which in turn will add nutrients and improve the health of the soil (physical, chemical and biological). This, however, is possible in an Integrated Farming System Approach essentially on small farm holders.

The farming system approach makes the system so holistic that fulfil a major part of the total requirements of major nutrients NPK and also other secondary and micro nutrients by recycling all the farm wastes, crop and other residues, cow dung and urine, and green manuring etc. Korikanthimath and Manjunath (2004) working in Goa conditions gave a very comprehensive account (Table 2) of recyclable manurial resources from rice based farming.

The data presented in the Table 2, indicate the importance of integration of crops related leguminous species and diversification of crop enterprise with subsidiary enterprises such as mushroom, poultry etc. wherein total recyclable organic biomass could be enhanced manifold by adopting a IFS approach rather than traditional farming of rice alone.

Similar studies (2004-05 to 2009-10) were also conducted at PDFSR, Modipuram, Meerut, Uttar Pradesh and more than 36% of total annual NPK requirements of field and plantation crops could be met by use of available cow dung (FYM, Vermicompost), urine, recycling of crop residues and intercropping of dual purpose legumes etc. within the system itself (Table 3). In addition to this the silt of fish pond was also excavated and mixed in to soil once in three years. A total amount of 18.56 kg N, 6.21 kg P and 74.24 kg K was added by excavation of 15 cm deep ground soil surface of 800 m^2 pond area saving an amount of about rupees nine hundred fifty. The OC% of the soil was as high as 1.20 with an average value of 0.95. Average total NPK need of the system tried in the IFS model at PDFSR, Modipuram were 285.5, 116.3 and 109.9 kg/ha, respectively.

Table 2 : Quantification of recyclable manorial resources from rice based farming systems

Farm enterprises	Pooled values (1999-2000 and 2000-01)		
	Recyclable Manorial resource	Potential quantity for recycling (t ha^{-1} year^{-1})	
		1999-2000	2000-01
Rice alone	Paddy straw	4.69	2.34
Rice-groundnut	Paddystraw+Groundnut haulm	7.12	4.78
Rice-cowpea	Paddy straw + cowpea haulm	7.29	4.94
Rice-brinjal	Paddy straw + brinjal stalks	6.37	4.03
Rice-sunhemp	Paddy straw + sunnhemp GM	16.45	14.11
Mushroom	Mushroom spent substrate	0.70	0.70
Poultry rearing	Poultry manure	1.40	1.40
Dairy (2 cows)	FYM	4.71	4.71
Rice-groundnut + mushroom + poultry	As above	9.22	6.88
Rice-cowpea + mushroom poultry	As above	9.39	7.04
Rice-brinjal + mushroom poultry	As above	8.47	6.13
Rice-sunnhemp + mushroom poultry	As above	18.55	16.21
Rice-alone +mushroom poultry	As above	6.79	4.44

Source: Korikanthimath and Manjunath ; 2004

This nutrient budgeting clearly advocates the self sustainability of the system which not only reduces the dependency on the external inputs but will also saves money to be spent on costly chemical fertilizers. In this way IFS approach met the challenges of organic farming to a great extent and in subsequent years the whole system can be converted in to organic farming through IFS activities.

Table 3 : Nutrient budgeting under Integrated Farming System at PDFSR, Modipuram

Source of nutrients and per cent nutrient content (N:P:K) on dry wt.basis	Available quantity (kg)	Released quantity of nutrients (kg)		
		N	P	K
		Green manure		
Sesbania spp. (1.29:0.36:1.64)	8800	18.9	5.3	24.0
Cowpea (1.29:0.36:1.64)	8500	18.3	5.1	23.2
		Crop residues (dry wt.)		
Sugarcane leaves (0.4:0.18:1.28)	900	3.6	1.6	11.5
Arhar leaves (1.29:0.36:1.64)	232	3.0	0.8	3.8
Potato leaves (0.52:0.21:1.06)	1450	7.5	3.0	15.4
		Animal excrete		
Cow dung (dry wt.) (0.4: 1.2 : 1.9)	17600	70.4	211.0	334.0
Total	-	121.7	226.8	411.9
Considering 30% use efficiency of released NPK through organic sources	-	36.51	68.0	123.6
Fish pond surface soil (once in every third year-15 cm deep soil from 800 sq.m. pond area		18.56	6.21	74.24
Nutrient requirement/year (field + plantation crops)	-	285.3	116.3	109.9

Source: Singh, J.P. and Gill, M.S.; 2010

Multiple land use through integration of crops with livestock can give the best and optimum production per unit land area (Misra and Mandal 2010). The integrated livestock provides food, fibre, skin, traction, fertiliser and fuel and also act as "bank on the hooves" providing flexible financial reserve in times of emergency and serve as "insurance" against crop failure for survival. The potency for amount of nutrients that can be recycled by cattle and other domestic species although depends upon age, breed, type of animals, quality of feed offered etc., however, amount of manure (dung + urine) produced by adult unit that can be recycled for crop husbandry and nutrient content of manure of different animal species are given below (Table 4).

Table 4 : Manure (Dung + Urine) produced by adult cattle and other livestock species.

Species	(%) of total live body weight / day	Fresh manure/ animal/ year (kg)	Dry matter / animal/year (kg)	Moisture Content (%)	OM (%)	Nitrogen	Phosphorus (P_2O_5) (%)
Dairy cattle	9.4	6000	1260	79	17	0.5	0.1
Sheep/goat	3.6	800	290	64	-	1.1	0.3
Pig	5.1	3000	-	-	-	-	-
Chicken	6.6	25	6-11	56	26	1.6	1.5
Duck	3.9	55-75	24-32	57	26	1.0	1.4

Source : Misra and Mandal; 2010

Five different IFS models were examined at ARS, Kovilpatti in Tamilnadu during 2003-2005. The animal waste from cow @ 20kg dung day^{-1} animal^{-1}, sheep and goat @ 400g/animal/day and poultry litter @ 30 kg/batch of 20 birds along with the unutilized feed were collected, composted and applied to the field of the respective IFS models. The soil fertility improvement because of effective recycling of organic residues and animal wastes from different IFS modules in the study are given in Table 5. Maximum fertility improvement recorded was under the IFS model having crop + goat + sheep + dairy wherein the soil N improved from 120 to 134 kg/ha, soil P from 7.4 to 8.5 kg ha^{-1}, soil K from 314 to 378 kg/ha and OC from 0.27 to 0.35%, during a period of 3 years.

Table 5 : Percent increase in available soil N, P and K (kg/ha) and organic carbon (%) due to effective management of organic sources in different IFS models.

IFS Model	% increase over initial			
	N	P	K	OC
Crop alone	2.6	5.2	10.3	10.7
Crop + goat	8.8	10.8	12.2	32.0
Crop + goat + dairy	10.0	13.7	17.1	30.8
Crop + goat + sheep	8.5	12.0	12.6	22.2
Crop + goat + sheep + dairy	10.7	14.9	20.4	29.6
Mean	8.1	11.3	14.5	25.1
SD (±)	3.2	6.2	4.1	8.9
CV(%)	39.6	14.5	28.4	35.4

Source: Solaiappan *et al.*, 2007

Under rain- fed vertisols of Tamilnadu, wherein integration of goat was made with conventional cropping, recycling of goat manure to substitute 50 percent of the recommended nitrogen resulted in gain of N content in the soil as compared to the sole sorghum with inorganic fertiliser alone (Radhamani and Chinusamy 2002). Among the systems, the loss of P was more with *C. pantandra* than with other tree species. Gain in K content was higher in *E. officinalis* followed by *A. excelsa* in integrated farming systems than sole sorghum. Apart from getting additional yield and improving soil fertility, goat manure can effectively be utilised for better productivity of cropping which reduced the cost of cultivation (Table-6).

Table 6 : Nutrient balance through manure recycling in IFS under rainfed vertisols

Farming systems	Nitrogen (kg ha^{-1})		Phosphorus (kg ha^{-1})		Potassium (kg ha^{-1})	
	1999-00	2000-01	1999-00	2000-01	1999-00	2000-01
FS1-Conventional cropping (1.0 ha)	- 15.9	-6.3	- 12.1	-11.6	13.2	12.0
FS2-*Ailanthus excesla* +crop*+goat(0.99ha)	7.59	9.57	- 1.85	-3.73	17.16	15.35
CS1	8.09	5.78	- 2.01	-3.14	30.20	28.88
CS2	1.65	2.64	-2.01	1.75	9.41	14.19
CS3						
FS3-*Ceiba pentandra* +crop* +goat(0.99ha)	7.92	4.79	-2.84	-3.66	10.73	14.19
CS1	4.46	2.64	- 4.06	-3.33	24.09	30.01
CS2	0.50	1.82	- 2.24	-2.28	5.12	8.09
CS3						
FS4-*Emblica officinalis* + crop* +goat (0.99)	8.25	13.37	-1.65	-3.33	23.76	19.64
CS1	9.74	10.40	-1.85	-2.48	30.03	31.52
CS2	1.16	2.81	-1.88	1.85	8.25	12.71
CS3						

* Crop CSI – Grain sorghum+cowpea (0.33 ha); CS2- Fodder sorghum + cowpea (0.33 ha)
CS3 – Cenchrus glaucus (0.33 ha)
Source: Radhamani and Chinusamy; 2002

Studies on N management in low land (wetlands) under rice-fish-Azolla integrated farming system at ARS Bhavanisagar in Tamilnadu revealed that quantum of organic residue addition and N added through recycling were higher in rice-rice-Azolla + fish farming with *Sesbania rostrata incorporation* (Table 7). The unutilized fish feed, decayed Azolla and excreta settled at the fish trench bottom had a higher nutrient value, which can be recycled to enrich the soil (Balasamy 1996).

Table 7 : Nitrogen added through recycling of organic residue (kg ha^{-1}) under different IFS modules.

IFS modules	N applied through residue recycling					
	Fish trench	I crop	II crop	Azolla	Green Leaf Manure (GLM)	Total
Rice-rice	0	9.2	9.9	0	0	19.1
Rice-rice +fish + 100% N	12.9	8.7	9.6	0	0	31.2
Rice-rice +Azolla+fish + 100% N	16.9	9.0	10.5	102.2	0	138.6
Rice-rice+fish + GLM +100%N	13.4	9.2	11.1	0	19.4	53.1
Rice-rice +Azolla + fish +GLM +100% N	17.3	9.2	11.3	107.5	19.4	164.7
Rice-rice+fish +75%N	12.4	6.8	7.5	0	0	26.7
Rice-rice+Azolla+fish+75%N	16.5	8.0	11.1	0	19.4	52.1
Rice-rice+Azolla +fish +GLM +75%N	17.0	9.1	11.2	106.3	19.4	163.0

Source: Balasamy;1996.

On-farm Integrated Farming Systems models developed and studied at CARI, Port Blair in four different micro-farming situations (MFS-I : Hilly; MFS-II: Sloping Hilly; MFS- III: Medium upland valley; MFS -IV: Low lying valley) of Bay Islands, revealed that integration of livestock, poultry and fish component in to prevailing system of cropping alone, gave additional income of Rs. 64300/-, Rs. 66970/-,Rs. 99818/-, Rs. 99893/- and Rs. 107200/- in IFS-I,IFS-II,IFS-III and IFS-IV, respectively . Study on bacterial load in different samples in different seasons of fish + poultry + duck system revealed that bacterial load has been found low before monsoon (Table 8) due to various management practices in the pond like removing of excess silt and application of lime just before rain. However, during monsoon season, bacterial load especially Salmonella sp. In the pond gets increased due to the factors like poultry drop gets accumulated uniformly in the pond water. Bacterial load sharply gets increased during summer mainly low level of water, higher droppings and increase in temperature which favours the growth of bacteria (Ravishankar *et al.,* 2007). Hence, the pond water of fish cum poultry cum duck should not be used for house hold purposes.

Table 8 : Microbial load -bacterial colonies (no's CFU^{-1} 0.1 ml) in different MFS

Samples	Pre monsoon	Monsoon	Summer
MFS-I	32 x 10^{-4}	45 x 10^{-5}	78 x 10^{-5}
MFS-II	32 x 10^{-4}	38 x 10^{-5}	94 x 10^{-5}
MFS-III	36 x 10^{-4}	44 x 10^{-5}	30 x 10^{-5}
MFS-IV	36 x 10^{-4}	48 x 10^{-5}	67 x 10^{-5}

Source: Ravisankar *et al.,* 2007

In fact huge quantities of organic resources are available for agriculture in India and tangible quantities of nutrients are tapable. The estimates on the availability of organic resources for agriculture in India and amount of tapable nutrients during 2000–2025 revealed that about 600-700 m.t. of agricultural wastes can be converted into biogas, nutrients and fuel. In addition to this 1800 mt animal dung–leaving 1/3 used for fuel, 2/3 of it used for biogas can produce 440 mt manure equivalent to 2.90 mt N + 2.75 mt P_2O_5 and 1.89 mt K_2O. Total potential of organic and biological resources estimated was 9.85 N + 2.66 P and 4.35 mt K. Total legume area in India is about 34.28 m.ha (18.24% of total cultivated area) and annual atmospheric N–fixation is equivalent to 2.48 mt. According to Tiwari *et al.* (2005) these values were equivalent to 5.05, 6.24 and 7.75 mt of NPK in 2000, 2010 and 2025, respectively.

To make farming productive, profitable and sustainable along with ensured livelihood at small farm holders, integrated farming system approach is considered to be the most powerful tool in coming future. Integrated farming system approach which ensures addition of lot of nutrients in the soil through recycling all the farm residues, wastes, excretes of the animals etc., generate additional employment, ensure livelihood, provides nutritional security and safe environment as well. Several studies on IFS approach conducted in the country and discussed in the paper confirmed that the IFS approach besides ensured livelihood of small land holders, generate/save a sizable amount ranging from Rs. 50,000 to Rs. 75,000 per hectare per year for meeting other liabilities including education, health and social obligations to a great extent. This saving is exclusive of household food and fodder needs as well as cost of production. Hence, there is urgent need to fine tune of research and developmental strategies in farming system perspective.

References

Balasamy (1996). Studies onN management in low land rice-fish-Azolla Integrated farming System- PhD Thesis, TNAU, Coimbatore.

Dwivedi, B.S., Shukla, A.K., Singh, V.K. and Yadav, R.L. (2001). Sulphur fertilization for sustaining productivity of rice-wheat system in western Uttar Pradesh.PDCSR Technical Bulletin No.2001-1.

Gaur, A.C. (1988). Phosphate solubilising bio-fertilizers in crop productivity and their interaction with VAM. In Mycorrhiza. Round table proceedings of a national workshop. IDRC-CRID-CIID, New Delhi, India, pp: 505-529.

Korikanthimath, V. S and Manjunath B.L. (2004). Resource use efficiency in Integrated farming systems. In: Proceedings, National Symposium on Alternative farming Systems- Enhanced income and employment generation options for small and marginal farmers held at PDFSR, Modipuram, 16-18 Sept., 2004, pp: 109-118.

Mahaptra, B.S., Sharma, K.C. and Sharma, C.L. (1987). *IRRI Newsletter,* **12** (1):32

Misra, A.K. and Mandal , D.K.. (2010). Interaction between crop and livestock in a cattle based farming system. Paper presented in the Trainning Programme on IFS methodologies for KVKs' SMSs and Programme Coordinators of Zone-I, held at PDFSR, Modipuram during 22-24 July,2010.

Nambiar, K.K.M. (1994) . Soil fertility and crop productivity under long term fertilizer use in India. A Publication of ICAR, New Delhi.

Nene,Y.L. . (1966). Symptoms, cause and control of Khaira disease of paddy. *Bulletin. India Phytopathol Society,* **3** : 97-191.

Oza, A.M. and Subbiah, B.V. .(1973). *ISNA News Letter,* **2** : 55.

Radhamani, S. and Chinusamy, C. (2002) . Integrated farming systems for the dry land vertisol areas of western zone of Tamilnadu. *Journal of Farming Systems Research and Development,* **8** (1&2) : 1-9.

Ravisankar, N., Pramanik, S.C., Jeyakumar, S., Singh, D.R., Nabisat Bibi, Nawaj, S., and Biswas, T.K. (2007). Study on Integrated Farming System (IFS) under different resources conditions of Island Ecosystem. *Journal of Farming Systems Research and Development,* **13** (1): 1-9.

Singh, S., Prasad, R. Singh, B.V.,Goyal, S.K. and Sharma, S.N. (1990). Effect of green manuring, blue green algae and neem cake coated urea on wet land rice. *Biology and Fertility of Soil,* **95** : 235-238.

Singh, J.P. and Gill, M.S. (2010). Livelihood security and resource conservation through IFS. *Indian Farming,* **60** (6) : 3-7.

Sinha, M.N., Hampiah, R. and Rai, R.K. (1994). Effect of phosphorus on grain and green fodder of kharif legume using ^{32}P as tracer. *Journal of Nuclear and Agricultural Biology,* **23** : 102-106.

Solaiappan, U., Subramanian, V. and Maruthi Sankar, G.R. (2007). Selection of suitable IFS model for rainfed semi arid vertic inceptisols in tamil Nadu. *Indian Journal of Agronomy,* **52** (3) : 194-197.

Soni, P.N. and Mukharjee,A.K. (1982). Phosphorus fertilization in a crop sequence. *Fertilizer News,* **27** (3) : 41-43.

Subba Rao, N.S. (1993). Bio-fertilizers in agriculture and forestry . 3rd' Edn. Oxford &IBH, New Delhi, pp. 242.

Tiwari, S.P., Ravi, R.,Nandeha, K.L., Vardia, H.K., Sharma, R.B. and Rajgopal, S. (1999). Augmentation of economic status of Bastar tribals through integrated (Crop, livestock, fishry, duck and poultry) farming system. *Indian Journal of Animal Sciences,* **69** (6): 448-52.

Veeranna,V.S., Panchaksharaiah, S., Vageesh, T.S.Shetty, R.A.and Sharma, K.M.DS. (1996). Potassium application for sustained productivity of sorghum-wheat sequence under long term fertilizer management. *Journal of Potassium Research,* **12** : 39-47.

Wani, S.P. and Lee, K.K. (1997). Role of bio-fertilizers in upland crop production. In: Fertilizers Organic manures, Recycled wastes and bio-fertilizers. (HLS Tandan, ed.), FDCO, New Delhi, pp: 91-112.

Yadav, R.L. and Singh, K. (1986). Long term experiments with sugarcane under intensive cropping system and variation in soil fertility. *Indian Journal of Agronomy*, **31** : 322-325.

❑❑❑

System Based Integrated Nutrient Management, 2012
© B. Gangwar & V.K. Singh (eds.), pp. 93-110
New India Publishing Agency, New Delhi (India)
e-mail : info@nipabooks.com; website : www.nipabooks.com

CHAPTER 7

Legume : A Panacea for Sustained Productivity and Soil Health

N.K. JAT AND V.K. SINGH

Food security, nutritional security, sustainability and profitability are the main foci of present agricultural scenario in India. In the absence of additional land for bringing under plough vertical growth in agriculture through intensive cropping and enhanced productivity becomes inevitable. Intensification of agriculture to meet the growing needs of the burgeoning population has brought with it numerous productivity related constraints. It has triggered multiple nutrient deficiencies in soil and this has raised doubts on the soils capacity to sustain anticipated production levels. One of the key inputs in enhancing and maintaining the productivity of intensified agricultural systems is chemical fertilizer which provides nutrients in readily available form. But long term experiments on various cropping systems at various agroecological regions and soil types revealed that continuous use of chemical fertilizers in unbalanced and indiscriminate manner deteriorates the soil health and leads to deleterious effect on long term soil fertility and yield sustainability. Apart from the soil productivity issue, the use of chemical fertilizer is also becoming more and more costly due to scarcity during peak demand. The excessive use of chemical fertilizers is not only costly but they pollute the production environment (Singh *et al.*, 2005; Dwivedi *et al.*, 2003).

In the coming decades, a major issue in designing sustainable agriculture system will be the management and rational use of organic inputs such as animal manure, green manure, crop residues and biofertilizers etc. (Powlson,

1994). Long term studies indicated that current fertilizer recommendations are inadequate for maintaining yields (Bhandari *et al.*, 2000; Singh and Mishra, 2010; Tiwari *et al.*, 2006). It is, therefore, to sustain high crop yield without deterioration of soil fertility it is important to apply required amount of plant nutrients through fertilizers in combination with organic sources, crop residues, biofertilizers etc.

Soil health defined

Conceptually, the intrinsic health (or quality) of a soil can be viewed simply as 'its capacity to function' (SSSA, 1995). Generally, the soil health can be defined as a state of dynamic equilibrium between flora and fauna and their surrounding soil environment in which all the metabolic activities of the former proceed optimally without any hindrance, stress or impedance from the latter. It is the state of the soil at a particular time, equivalent to the dynamic soil properties that change in short term (Carter *et al.*, 1997). Many a times, the state of soil is also explained loosely in terms of soil health, soil productivity, soil environment *etc.* The terms 'soil health' and 'soil quality' are often used interchangeably (Harris and Bezdicek, 1994). The basic soil health indicators in addition to meeting other criteria should (i) integrate soil physical, chemical and biological properties and processes, (ii) be sensitive to variations in management and climate, and (iii) be measurable or accessible by as many people as possible.

Soil health parameters

The major soil health parameters are broadly classified into three groups, namely physical, biological and chemical (particularly soil fertility, in terms of nutrient content and availability). Although individually they are considered as independent ones for ease but their behaviour and relationship with each other as well as with biomass production is quite complex. These are briefly discussed below:

Physical parameters: Physical parameters like soil texture, structure, bulk density, aggregate stability, water holding capacity, water infiltration, hydraulic conductivity, aeration *etc.* not only influenced by of nutrients and microbial activities in soil, but also exert profound effect on seed germination, seedling emergence and root growth (Prihar, 2000). Crop management practices like excessive tillage, use of heavy machinery, excessive puddling for rice, removal of crop residues and low turnover of organic matter deteriorate soil physical health (Dwivedi *et al.*, 2003). On the other hand, conservation agriculture envisages minimum disturbance of soil and retention of crop residues (Lal and Kimble, 2000), use of organics and balanced fertilization (Swarup and Wanjari, 2000), and crop diversification (Singh *et al.*, 2005) improve it.

Biological parameters: Biological parameters reveals about the organims that form the soil food web, which are responsible for decomposition of organic matter and nutrient cycling. These include measures of active organic matter pools, such as microbial biomass carbon, diversity of various soil organism populations such as earthworms, nematodes and microarthropods and biological activity such a enzyme activity, potentionally mineralizable N or respiration (CO_2 production). Due to a delayed realization of their significance in soil health sustenance, soil biological parameters have not been studied as frequently as physical or chemical indices. Recent studies with long-term experiments, however, revealed superiority of integrated nutrient supply (fertilizer NPK+organic manures) over fertilizer NPK alone in improving microbial biomass carbon and microbial biomass nitrogen contents and activities of enzymes namely dehydrogenase, urease and alkaline phosphatase.

Chemical parameters: Chemical parameters can give information about the equilibrium between soil solution (soil, water and nutrients) and exchange sites (clay particles, organic matter), nutritional requirement of plant and the availability and uptake of nutrient by plant. Different parameters include soil organic matter, pH, electrical conductivity and soil available N, P, K and micronutrients. Intensive cropping during last three decades has led to a general depletion of S and micronutrients, especially Zn, B and Fe. Consequently, the incidence of their deficiencies in plants and also crop responses to their application increased (Dwivedi *et al., 2006*).

Soil health care - necessity and current scenario

The current status of agricultural production in the country is a matter of great concern for both scientists and the farmers. Stagnation of crop yields at low to moderate levels despite the recently developed adoptable technologies has to be monitored seriously in view of the ever increasing population coupled with the fast shrinking per capita availability of land. Misuse and abuse of the technologies, mostly due to unawareness and occasionally due to greed, however, led to overexploitation of natural resources (soil, water and biodiversity). As a result, a constant decline in factor productivity has been noticed in recent past, which is in fact an indicator of non-sustainability of agricultural production system. In this situation, it may not be possible to achieve the target of 300 Mt food grains by the year 2025 unless the associated issues are addressed through a focused research and development and policy support. Diversion of arable lands towards non-agricultural purposes poses another serious challenge. Any future addition to the net cultivated area has to come from less-productive marginal lands requiring very high cost of restoration/amelioration or from deforestation which will be disastrous. Therefore, various aspects related to soil health (including soil fertility management) have to be discussed and debated in detail in order to arrive at sound management strategies for sustaining high productivity. Soil being the natural dynamic resource base for agricultural activities maintaining its health attains added significance because research evidences amply suggest that a single intervention of soil fertility improvement may account for 50-70 % difference in crop yields (Swaminathan, 2006).

Soil degradation of various types is also a severe problem in India, rendering vast areas unproductive or barren. These include soil erosion, salinity, alkalinity, waterlogging, and depletion of nutrients and organic matter. Mining of soil nutrient reserves due to suboptimal and unbalanced use of plant nutrients has been a common phenomenon in Indian agriculture. The pace of soil fertility depletion due to excessive nutrient mining is naturally more acute and obvious in high productivity zones, for greater withdrawals of nutrients through intensive cropping systems, and continuous neglect of their replenishment through fertilizers and manures. As a result, large scale multi-nutrient deficiencies in the soils have become spectacular, necessitating for enhanced fertilizer application rates and also for inclusion of need-based secondary and micronutrients in fertilizer schedules (Dwivedi *et al.*, 2006; Tiwari *et al*, 2006).

Although the fertilizer application plays tremendous roleduring post-green revolution period but a decline in the growth rate of food grain production in respect of productivity, fertilizer responses and factor productivity has been observed recently. The overall fertilizer response ratio in irrigated areas of the country has decreased nearly three times from 13.4 kg grain kg^{-1} nutrient in 1970 to 3.7 kg grain/kg nutrient in 2005. The individual nutrient ratios in respect of N, P and K also followed the same trend. While only 54 kg ha^{-1} NPK was required to produce around 2 t ha^{-1} in 1970, around 218 kg ha^{-1} are being added presently to sustain the same yield (Sharma, 2008). This undesirable situation has occurred due to ever increasing multi-nutrient deficiencies under different soil types, crops and cropping systems.

Expansion of multi-nutrient deficiencies and nutrient balance sheet

On the basis of 3.65 million soil samples analyzed for available N, P and K status during 1997-99 by different soil testing laboratories it was observed that at the national level aggregate N status can be categorized as low having nutrient index value of 1.48, P as medium being 1.78 and K as high being 2.37. For available P in Indian soils, 42 per cent samples fall in low, 38% in medium and 20 percent in high category (Motsara, 2002). Similarly, for N and K 63.0 and 13.0%, 26.0 and 37.0%, and 11.0 and 50.0% soils have been reported under low, medium and high categories, respectively. Based on the work done under the auspices of the All India Co-ordinated Scheme of Micronutrients in Soils and Plants (ICAR), analysis of 2.52 lakh soil samples drawn from 20 states of the country indicated 49, 12, 33 and 13 percent deficiency of Zn, Fe, B and Mn, respectively (Singh, 1998) (Table 1). Recent studies conducted by *Dwivedi et. al., 2006* clearly shows that nutrient deficiencies are not only restricted to N,P,K but also it has expanded to number of secondary and micro-nutrients also (Fig. 1). It indicated that Indian agriculture is now well into the era of multiple nutrient deficiencies, which is likely to be increased in the future due to mining of nutrients through intensive cropping to meet the growing demand of food, fiber, fuel and fodder due to population explosion.

Table 1 : Extent of micronutrient deficiency in India

State/Union Territory	Percentage deficit samples			
	Zn	Fe	B	Mo
Rajasthan	21	0	0	0
M.P	43.9	7	22	18
Uttar Pradesh	45.7	6	24	0
Gujarat	23.8	8	2	10
Haryana	60.5	20	0	28
Punjab	48.1	14	13	0
All India	48.5	12	33	13

Source: Singh (1998)

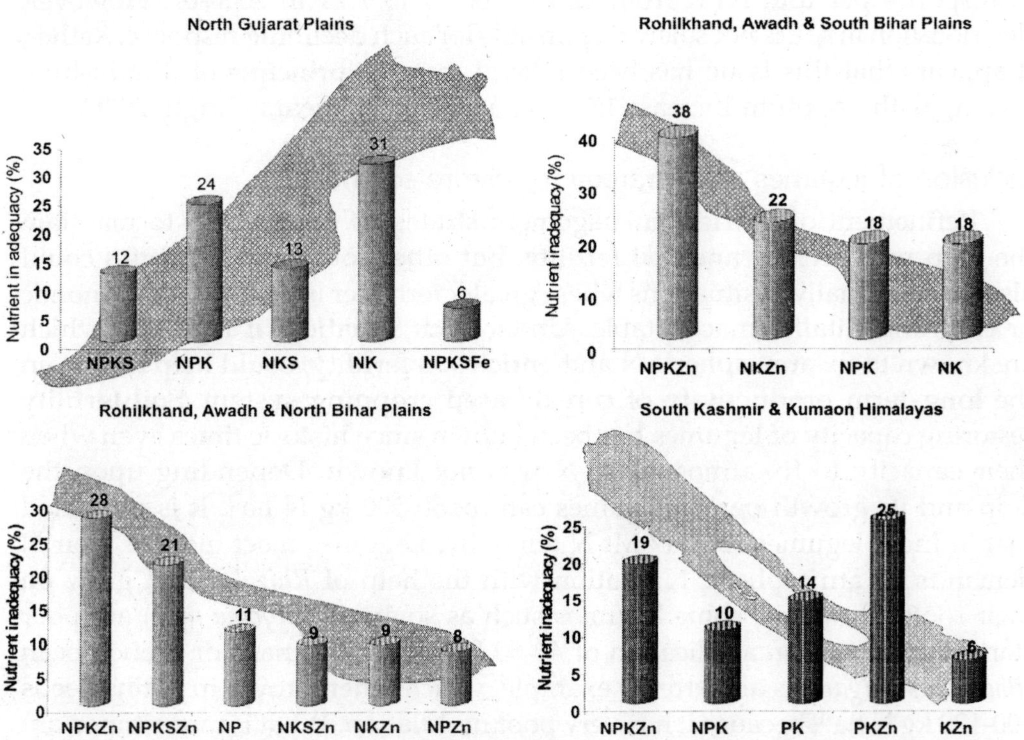

Fig. 1 : Multi-nutrient inadequacy in different agro-ecological sub-regions.
(Source : Dwivedi *et al.*, 2006)

For the present level of production, there is net negative balance of about 8-10 Mt of NPK in Indian agriculture. It is difficult to prepare an accurate balance sheet of nutrient in Indian conditions in the absence of reliable data on the various sources of addition and removal of nutrients. For instance, the estimated addition of S through various sources is 0.8 Mt annum^{-1}, while the removal is estimated at 1.3 Mt annum^{-1} indicating a net negative balance of 0.50 Mt (Biswas *et al.*, 2004). Deficiency of S in Indian soils is on increase due to extensive use of S free fertilizers coupled with intensive cultivation of high S demanding crops (Singh and Gangwar, 2011). S deficiency has been found in 41 percent soils of the country (Singh, 2001).

This undesirable situation has occurred due to inadequate and imbalanced application of fertilizer nutrients. The imbalanced fertilizer use in terms of NPK is evidenced by their wider consumption ratios for quite some time against ideal one of 4:2:1. Besides, the use efficiency of applied N, P, K, Zn, Fe and Cu in Indian soils is restricted in the range of 30-50, 15-20, 70-80, 2-5, 1-2 and 12%, respectively. Thus, problems of nutrient deficiencies are aggravated further because of low use efficiency of applied nutrient, in general for P and micro nutrient in particular. Concerns have been expressed over the decline in response per unit NPK from 12 in 1960-69 to 7.79 in 2000-07. However, deterioration in soil is not solely responsible for such declining response. Rather, it appears that this issue has been related more to principle of diminishing returns with quantum increase in NPK use over the years (Singh, 2008).

Inclusion of legumes as a strategy to restore soil health

Refinement of nutrient management strategies would help to maintain the crop productivity and soil fertility, but other rotational strategies could also help especially in situations where greater fertilizer use may be uneconomic or environmentally unacceptable. Under such situation, if legumes, which are known to fix atmospheric N and enrich soil fertility, could help to sustain the long-term productivity of cereal based cropping system. Soil fertility restoring capacity of legumes has been known since historic times even when their capacity to fix atmospheric N was not known. Depending upon the crop and its growth period legumes can fix 50-500 kg N ha^{-1}. It is estimated that in India legumes fix 2.47 Mt N annually. Legumes meet most of their N demands by atmospheric N fixation with the help of *Rhizobia* that grow on their roots. However, some legumes such as soybean (*Glycine max*) are poor starters and need an application of 40-60 kg N ha^{-1}. Rajmash or French bean (*Phaseolus vulgare*) is an extreme example, which when grown in plains needs 100-120 kg N ha^{-1}, because it has very poor nodulation. What is more important from the IPNS viewpoint is the amount of N that legumes leave for the succeeding crops. These amounts vary from 25-100 kg N ha^{-1}. Depending on the soil and ecological stress, the cereal based cropping systems can be diversified using legume as a inter-crop, substitute crop, catch crop, break crop in the form of grain, fodder, or green manure crop.

Legumes as intercrop

Intercropping is a traditional practice to reduce the risk of complete crop failure during aberrant weather. The scope of legume intercrop under rice-wheat system further increases under water scares conditions or aberrant weather situation, as it not only increase total productivity of the system but also play an important role in economizing the use of resources, particularity N fertilizer. Although the potential of legumes as intercrops in rice-wheat cropping system has not yet been fully exploited, but with advent of modern tools like bed planting, multi-crop seed drills, it may emerge one of the promising options for rice-wheat system sustainability. Co- culture of *Sesbania* along with direct seeded/ bed planted rice reveals further scope for inclusion of legumes as intercrop in rice- wheat cropping system. Legumes such as pigeon pea, black gram, mungbean, and soybean are ideal intercrops in upland direct seeded rice (Lal and Singh, 1990).

Under rainfed conditions growing pearl millet in association with cluster bean was better in normal rainfall years while in sub-normal rainfall years cowpea was a better associate (Misra, 1971). In a two years study (Panjab Singh and Joshi, 1980) double row of dew gram, cluster bean and green gram planted in the interspaces of paired rows of pearl millet (30/70 cm) yielded 381, 381 and 458 kg ha^{-1} additional grain without adversely effecting the yield of pearl millet have studies In the years of good rainfall various intercropping systems at Jodhpur recorded higher total productivity in case of intercropping of cowpea or mungbean with pearl millet (Daulay *et al.*, 1986). Maize also recorded higher yield of in intercropping systems with mungbean and the total productivity of different intercropping systems was 3 to 5.7 q ha^{-1} higher than the sole maize (Dhingra *et al.*, 1991).

Legume as catch crop

The most feasible way of including legumes in rice-wheat cropping system without decreasing land area of the cereal crops is to grow legume as catch crop. In rice-wheat cropping system medium duration rice varieties followed by wheat varieties suited for normal sowing or late sowing in North-eastern plain did not left scope to grow legume between rice harvest and wheat sowing. However, the period between wheat harvesting and rice planting can be utilized for growing short duration (60-70 days) summer legumes. Experiment at Masodha, Varanasi and Ludhiana indicated that inclusion of mung bean in rice-wheat cropping system enhance the total productivity and net returns as compared to rice-wheat cropping system alone (AICRP_CS report). In other studies, the yield of rice in rice-wheat system generally remained higher when preceded by mungbean (Singh and Sharma, 2001). Incorporation of summer mung bean biomass to soil after picking the pods not only improves succeeding rice yields but also curtail the fertilizer N dose by 25% (Singh and Mishra, 2010).

Legume as substitute crop

The substitution of rice or wheat largely depends on nature of stress increased in different agro-ecologies. For instance, in Trans-Gangetic Plain (TGP), where water table depletion is serious concern, there is scope for substituting rice with pigeon pea. Similarity in eastern part of India, where wheat productivity is generally low because of climatic constraints particularity higher thermal regimes, wheat need to be substituted with chickpea, lentil, pea or groundnut. Extensive studies on comparative performance of rice-wheat vs. rice–legume or legume-wheat at different location of AICRP_CS reveals that growing legume in system with recommended rate of NPK helped to improve system productivity and profitability than non-legume based system. In *Tarai* region, the annual productivity in terms of rice equivalent yields, energy production and net return increased tremendously in rice-wheat system, when a green manure or maize+cowpea fodder crop was introduced during post-wheat summer season (Table 2). Diversifying wheat with chickpea was further advantageous. In double crop sequences involving cereal crops only, diversifying *kharif* or *rabi* cereal with a grain legume has shown promising results.

Table 2: Effect of legume inclusion on system productivity, energy production and economics at Pantnagar, India

Crop sequence	Rice equivalent yield (t ha^{-1})	Production efficiency (kg/ha/day)	Chemical energy equivalent (k cal 10^{-4} ha^{-1})	Economics (Rs ha^{-1})	
				Gross return	Return over variable cost
Wheat-rice	10.3	28.28	3533	48302	32188
Chickpea-rice	10.8	29.79	3132	49203	35467
Wheat-*Sesbania* (green manure)-rice	11.2	30.75	3851	52715	34942
Wheat-maize+ cowpea (fodder)-rice	13.4	36.74	5051	61682	40793
Chickpea-maize+ cowpea (fodder)-rice	13.9	38.09	4865	62503	43992
CD 5%	0.5	6.64	333	3803	2955

Source: Singh and Sharma (2002)

The beneficial effects of substituting wheat or rice with a legume on soil organic carbon, mineral-N and Olsen-P content was also reported by Yadav *et al.*, 2003; Singh *et al.*, 2005. The substitution of wheat with pea at Ludhiana, lentil at Pantnagar, chickpea at Kanpur, blackgram at Bhubneswar, and groundnut at chiplima resulted marked increase in organic carbon and available NPK, when compared with rice-wheat cropping system (Hegde and Dewivedi, 1992). Raising a legume crop in rice-wheat system can also minimize the adverse effect of continuous puddling on the soil compaction. Data illustrated in Fig. 2 clearly show a rise in soil bulk density in sub-surface layers with rice-wheat cropping for three years. Taking a sort duration pigeon pea, on the other hand, mitigated such an effect of puddling. In other words, inclusion of legume may help a better crop establishment and root growth of wheat following rice, by way of reducing soil compaction.

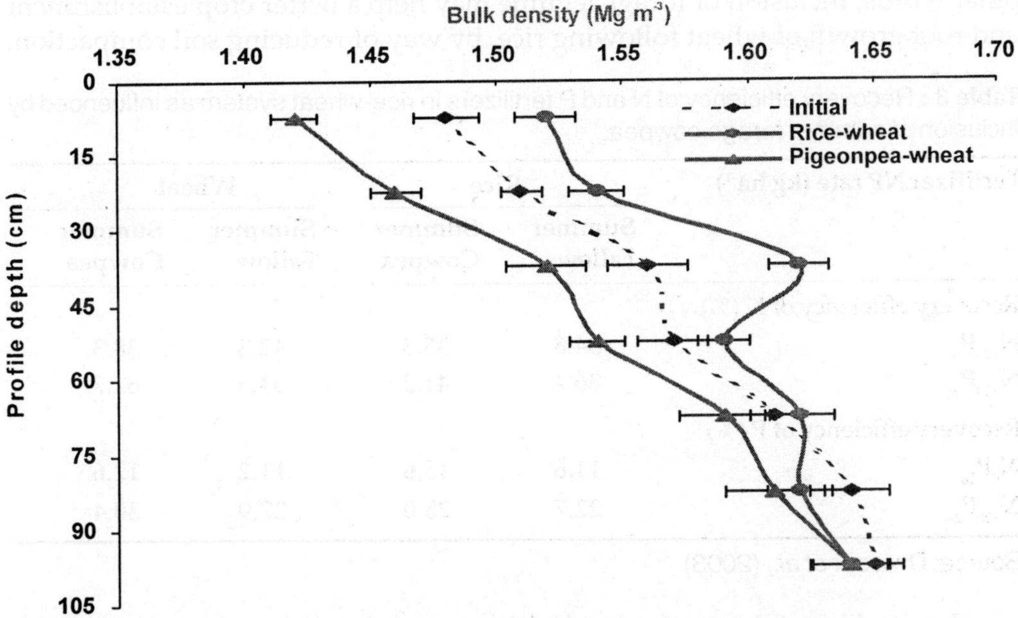

Fig. 2 : Changes in bulk density of soil at different profile-depths after three crop cycles as influenced by substitution of rice with pigeon pea in RWCS (Singh *et al.*, 2005).

Legume as fodder crop

Inclusion of fast growing forage legumes is another promising agro-technique, though the advantage of forage legumes in terms of nutrient recycling, and improvement in soil physical and chemical properties are relatively less explored. Recent work at PDCSR, Modipuram revealed an appreciable increase in the use-efficiency of N and P fertilizers in rice-wheat system with the inclusion of summer cowpea (forage) (Table 3). Study conducted on NO_3-N leaching pattern at Modipuram showed that the inclusion of cowpea as fodder crop during summer under rice-wheat system not only helped in trapping the NO_3-N from deeper soil layers, but also made it availability in upper profile layer for succeeding crops.

Inclusion of forage legumes as summer crop in rice-wheat system also increased N and P use efficiency on coarse-textured soils of Upper Gangetic Plains (Dwivedi *et al,* 2003). The problem of sub-soil compaction arising due to puddling in rice-wheat system could also be addressed to a great extent with the inclusion of legumes. Raising a legume crop in rice-wheat system could minimize the adverse effect of continuous puddling on the soil compaction. Experiments revealed a rise in soil bulk density in sub-surface layers with continuous rice-wheat cropping, whereas taking summer forage crop of cowpea mitigated such an effect of puddling (Dwivedi *et al,* 2003). In other words, inclusion of forage legume may help a better crop establishment and root growth of wheat following rice, by way of reducing soil compaction.

Table 3 : Recovery efficiency of N and P fertilizers in rice-wheat system as influenced by inclusion of summer forage cowpea.

Fertilizer NP rate (kg ha^{-1})	Rice		Wheat	
	Summer Fallow	Summer Cowpea	Summer Fallow	Summer Cowpea
Recovery efficiency of N (%)				
$N_{120}P_{0}$	34.8	35.3	42.3	38.3
$N_{120}P_{60}$	36.4	41.2	54.5	61.7
Recovery efficiency of P (%)				
$N_{0}P_{26}$	11.6	15.6	11.2	12.6
$N_{120}P_{26}$	22.7	25.0	27.9	30.4

Source: Dwivedi *et al.*, (2003)

A number of leguminous fodder crops have been evaluated for contribution which they make in meeting N demand of succeeding cereals and it has been found that as much as 35-120 kg N ha^{-1} can be made available. The carryover effect of N for succeeding cereal is about 120 kg ha^{-1} in berseem, 75 kg ha^{-1} in lucern and 35-60 kg ha^{-1} in fodder cowpea. These fodder legumes yielding 40-90 t ha^{-1} green fodder offers ample scope for inclusion as break crop.

Legume as break crop

Studies conducted at Modipuram reveals that growing berseem or cowpea as green fodder during rabi/kharif, respectively at 3 years interval, has significant influence on mitigating the ill effects of continuous rice-wheat cropping system such as reducing sub soil compaction, curbing noxious weed population (*Echinocloa spp* in rice and *Phalaris minor* in wheat) (Table 4) with improved system productivity and soil health.

Table 4 : Effect of legume as break crop on weed intensity (m^{-2}) in rice-wheat system

Treatment	Rice	Wheat
Continuous rice-wheat cropping	23	232
Every 3rd wheat substituted with berseem	13	190
Every 3rd rice substituted with cowpea	09	164

Source: PDCSR Annual Report

Legume as green manures

In pre-Green revolution era, the practice of green manuring was considered indispensable for improving crop productivity. However, with the easy availability of fertilisers and adoption of intensive cropping systems, the practice of green manuring was almost given up. In recent years, with an indication of declining trend in productivity due to indiscriminate use of only chemical fertilisers (NPK), there has been revival of interest in green manuring. Green manuring is an age old practice and may be done with non-grain legumes such as *Sesbania* (dhanicha), *Crotolaria* (Sunnhemp), *Stylosanthes, Centrosema,* and *Desmodium* or with grain legumes such as mungbean, urdbean, cowpea, pigeonpea etc. In addition, loppings of perennial muti-purpose woody legumes, such as, *Gliricidia sepium, Cassia siamea* and *Leucaena leucocephala* (subalool) are widely added to rice fields in South India.

Green manures can add 60-120 kg N ha^{-1} and in many situations can meet the entire N demand of crop to which they are applied. Comparative studies at Pantnagar on nutrient accumulation and decomposition of two *Sesbania* species (*S. cannabina, S. rostrata*), cowpea (*Vigna unguiculata*) and sunnhemp (*Crotolaria juncea*) showed that *S. cannabina* accumulated the highest amount of N 119.4 kg ha^{-1} at 45 days stage. However, for a 60 day crop NPK accumulation was highest for *C. juncea* (160.3 kg N+20.5 kg P + 127.2 kg k ha^{-1}) (Kumar *et al.* 2007). Multi-location trials under AICRP on cropping systems of ICAR have shown that 50% substitution of N is possible by green manuring in rice (*Kharif*) in the rice-wheat system (Hegde, 1998) and rice-rice cropping system. In the same series of trials, application of 50% N though green manure and 50% NPK to *kharif* rice and 100% (recommended dose of fertilizer NPK) to *rabi* rice produced 3.5-11.8% more grain than 100% RDF to rice crops. Partial substitution of fertilizers with *Sesbania* green manure rather produced greater yields of rice and wheat, compared with application of fertilizer alone, at a number of locations (Table 5).

Table 5: Effect of IPNS using green manure (GM) on the productivity of rice-wheat system in the LTEs under AICARP

Treatments		Grain yield (t ha^{-1})		
Rice	Wheat	Rice	Wheat	Total
R. S. Pura (mean of 07 years)				
100% NPK	100% NPK	4.65	3.26	7.92
50% NPK + 50% N (GM)	100% NPK	4.63	3.18	7.81
75% NPK + 25% N (GM)	75% NPK	4.79	3.04	7.83
Ludhiana (mean of 07 years)				
100% NPK	100% NPK	5.97	4.32	9.92
50% NPK + 50% N (GM)	100% NPK	6.05	4.17	10.22
75% NPK + 25% N (GM)	75% NPK	6.33	3.89	10.22
Pantnagar (mean of 07 years)				
100% NPK	100% NPK	4.42	3.51	7.93
50% NPK + 50% N (GM)	100% NPK	4.16	3.50	8.66
75% NPK + 25% N (GM)	75% NPK	4.26	3.18	7.44
Kanpur (mean of 06 years)				
100% NPK	100% NPK	5.16	4.64	9.80
50% NPK + 50% N (GM)	100% NPK	4.98	4.69	9.67
75% NPK + 25% N (GM)	75% NPK	5.26	4.47	9.73
Faizabad (mean of 07 years)				
100% NPK	100% NPK	5.30	3.67	8.97
50% NPK + 50% N (GM)	100% NPK	4.37	3.61	9.73
75% NPK + 25% N (GM)	75% NPK	4.41	3.23	7.64
Varanasi (mean of 06 years)				
100% NPK	100% NPK	4.40	4.11	8.51
50% NPK + 50% N (GM)	100% NPK	4.00	4.21	8.24
75% NPK + 25% N (GM)	75% NPK	4.34	3.96	8.30
Sabour (mean of 05 years)				
100% NPK	100% NPK	3.90	3.19	7.09
50% NPK + 50% N (GM)	100% NPK	3.80	3.21	7.01
75% NPK + 25% N (GM)	75% NPK	3.83	3.16	6.99
Kalyani (mean of 07 years)				
100% NPK	100% NPK	3.14	2.58	5.72
50% NPK + 50% N (GM)	100% NPK	3.39	2.98	6.37
75% NPK + 25% N (GM)	75% NPK	3.34	2.62	5.96

Source: AICARP/AICRP-CS Annual Reports

Despite N saving, yield increase and improvement in soil physical and chemical properties the area under green manure has declined over years in several states. The main reason for decline in area under green manure is the increase in area under irrigation, making farmers to go for remunerative crops. The solution to this problem lies in growing a dual purpose legume (for grain as well as for green manure). Considerable research has been conducted on growing mungbean during summer (mid April -mid June) taking a picking (about 0.5 t ha^{-1} grain) and incorporating the residue in rice-wheat cropping system. Improvement in soil fertility with this practice is also of the same order as with *Sesbania* green manure. This practices gaining popularity in the rice-wheat belt because the mungbean grain fetches a good price and provides the farmer money for investing in inputs needed for the rice crop. Also it adds to the pulse production in the country. Even if 50% of the area under rice-wheat croppings system which is estimate at 10 M ha adopts this practice it adds 2.5 Mt to the pulse kitty of the country.

Alternatively, short-duration grain legumes like green gram can be grown in summer and after picking the pods green biomass may be incorporated to the field. This is as advantageous as *Sesbania* green manure, besides producing a grain yield of 0.8-1.2 t ha^{-1} (Table 6). A green manure crop should be turned one-day prior to transplanting to get maximum advantage. Even under rainfed conditions, where farmers mostly take a *rabi* crop short duration dual purpose legumes can be grown. For example, Sharma *et al.* 1995 showed that incorporation of mungbean and urdbean residues make available 60-94 kg N and 82-115 kg N ha^{-1}, respectively for the succeeding wheat.

Table 6 : Comparative performance of green manuring and legume residue incorporation in rice-wheat system

Treatment	Grain yield (t ha^{-1})		
	Rice	Wheat	Total
Faizabad (Mean of 3 years)			
Green manuring	5.14	4.91	10.05
Green gram residues	5.15	4.79	9.94
Fallow	4.70	4.55	9.25
Chiplima (Mean of 02 years)			
Green manuring	4.19	2.21	6.40
Green gram residues	4.16	2.12	6.29
Palampur (Mean of 02 years)			
Green manuring	4.82	2.76	7.58
Green gram residues	5.00	3.07	8.07
Kharagpur (Mean of 2 years)			
Green manuring	3.65	2.96	6.61
Green gram residue	3.86	2.94	6.80

Source: Hegde (1992)

Green manure may also be sown along with the main crop and incorporated at some stage during the growing period of the main crop. Such a practice may be referred to as live green manure as a corollary to live mulch. For example, in direct seeded rice-wheat cropping system at Malan (Himachal Pradesh) (Mankotia, 2007), green manure crops, namely, cowpea, dhaincha, sunnhemp and soybean were sown along with rice in the first fortnight of July and incorporated in soil 30-35 days after sowing. When live green manured, rice received only 50% of RDF. Rice yields obtained were 78% with *Sesbania*, 81% with sumnhemp, 86% with soybean and 96% with cowpea as compared to that obtained with 100% RDF (90 N, 40 P and 40 K). Thus, cowpea live green manure can replace 50% RDF (recommended dose of fertilizer) for rainfed rice. Live green manure plots gave as good yield of wheat as obtained with 100% RDF.

Legumes in rotation

Different legumes have the capacity to leave behind different amounts of N for the succeeding crop. The perceptible increase in the yield of pearl millet was obtained by Singh and Singh (1977) on the basis of a long term study they reported that cultivation of green gram in rotation with pearl millet supplied with 20 kg N ha^{-1} gave similar yield as with direct application of 40 kg N ha^{-1}. In other words, growing of legume had an effect equivalent to 20 kg N ha^{-1}. However, there are differences on such legume effect with different grain legumes. For instance, Singh *et al.* (1985) observed that rotation of pearl millet with green gram or cluster bean was better than its rotation with moth bean. Reddy *et al.* (1993) reported that the yield and total N uptake of sorghum was maximum after green gram cultivation. Kathju *et al.* (1987) reported that the yield levels of pearl millet under cultivation of green gram, moth bean and cluster bean for past three years were much higher than could be obtained even with 80 kg N ha^{-1}. In rapeseed-mustard inclusion of leguminous crop in the sequence may help to economize N fertilizer requirement to the extent of 20-40 kg N ha^{-1}. At Bawal, green gram - mustard sequence produced the maximum seed yield of mustard and was at par with cluster bean-mustard. Both the crop sequences resulted in significantly superior over other crop sequences. Therefore, in a country like ours where the average consumption of plant nutrients from chemical fertilizers on national basis is very low, the residual fertility buildup due to legumes is obviously a major contribution, which must be fully exploited.

Possibility of legumes in new niches

Legumes can be grown under new niches such as wasteland, reclaimed soils and rice fallow land by efficient crop establishment technique and watershed management approach by replacement of less remunerative crops or intensification. To sustain and improve rice productivity, pulses have to be introduced in rice based cropping systems in different rice growing areas in Peninsular India including the deltaic areas of important river *viz,* Krishna, Godavari, Kaveri, etc. For this, short duration pulses can play vital role. For example, rice fallows occupy about 15m ha. Appropriate exploitation of these areas is possible through introduction of pulses like urd bean, moong bean. In A.P. powdery mildew resistant urd bean variety LBG 17 with matching agro-technology has expanded its acreage many fold is most viable example. Similarily, growing pigeonpea on priferal bunds of rice field in Bundelkhand region is an innovative technological option.Intercropping of pulses with rice is another possible way of crop intensification in the system. Research in the Bastar plateau zone has shown that rice + blackgram in the ratio of 2:1 can give yield of 2.4 t ha^{-1} of rice and around 0.25 t ha^{-1} of black gram. Similarly rice + pigeonpea on raised bed in the ratio of 3:1 can yield 1.0 t ha^{-1} of rice and around 0.6 t ha^{-1} of pigeonpea. Introduction of extra short duration pigeon pea (ICPL 88039) in north India has open the avenue for timely sowing of succeeding wheat crop along with better pulse productivity as compared to exhisting UPAS 120. Development of extra-short duration varieties of pigeon pea chickpea may open a new avenue for their cultivation for green pods during September-December without sacrificing major cereal crops. With all these approach an estimated 2.5 million ha additional area can be brought under different pulses through cropping system manipulation, crop diversification and multiple cropping systems.

Conclusion

Since, last four decades the indiscriminate use of high analysis inorganic fertilizers in intensive cropping systems has led the multinutrient deficiencies in soils, which adversely affected the yields and quality of production. The evidences coited in this chapter, envisaged that inclusion of legume may prove a better option for sustaining system productivity in terms of soil fertility restoration and economic gain in intensively cultivated areas. The frequently documented stagnation or even decline in the productivity of cereal based cropping system can be reversed by diversifying cereal crop with legume as substitute crop, catch crop or green manure crop during lien period. There is further, scope for introduction of legumes in new niches such as wasteland, reclaimed soils and rice fallow land etc. by efficient technological options as a replacement of less remunerative crops or intensification.

References

Bhandari, A.L. Ladha, J.K., Pathak, H., Padre, A.T., Panre, D. and Gupta, R.K. (2000). Yield and soil nutrient changes in long term rice-wheat rotations in India. *Soil Science Society of American Journal,* **66** : 162-170.

Biswas, B.C. Sarkar, M.C., Tanwar, S.P.S., Das, S. and Kaliue, S.P. (2004). Sulphur deficiency in soils and crop response to fertilizer sulpher in India. *Fertiliser News,* **38** (10): 1318, 21-28 & 31-33

Carter, M.R., Gregorich, E.G., Andreson, D.W., Doran, J.W., Janzen, H.H. and Pierce, F.J. (1997). Concept of soil quality and their significance. In: *Soil Quality for Crop Production and Ecosystem Health,* (Gregorich, E.G. and carter, M. (Eds.).Elsevier Science Publishers Amsterdam, Netherlands.

Dauley, H. S., Henry, A. and Bhati, T. K. (1986). Studies on intercropping system. AICRPDA, Jodhpur, pp. 5-7

Dhingra, K. K., Dhillon, M. S., Grewal, D. S. and Sharma, K. (1991). Performance of maize and mungbean intercropping in different planting patterns and row orientation. *Indian Journal of Agronomy,* **36** : 207-212.

Dwivedi, B.S. Shukla, Arvind K., Singh, V.K. and Yadav, R.L. (2003). Improving nitrogen and phosphorus use efficiencies through inclusion of forage cowpea in the rice-wheat system in the Indo-Gangetic Plains of India. *Field Crops Research,* **80** : 167-193.

Dwivedi, B.S., Singh, D., Chionkar, P.K., Sahoo, R.N., Sharma, P.K. and Tiwari, K.N. (2006). Soil fertility Evaluation – a potential tool for balanced use of fertilizers, pp 1-60, IARI, New Delhi / PPIC-India Programme, Gurgaon.

Dwivedi, B.S., Singh, V.K., Shukla, A.K. and Yadav, R.L. (2003). In: NATP Progress Report, PDCSR, Modipuram

Harris, R.F. and Bezdicek, D.F. (1994). In: *Proceedings of International Conference on Soil, Water Environmental Quality – Issue and Strategies,* SSSA special Publication No. 35 pp. 23-26, Madison WI.

Hegde, D.M. (1992). Cropping Systems Research Highlights, Modipuram Meerut-2500110, Uttar Pradesh, India: Project Directorate for Cropping Systems Research, pp. 40.

Hegde, D.M., Dwivedi, B.S. (1992). Nutrient management in rice-wheat cropping system in India. *Fertilser News* **37** (2), 27-41.

Hegde, D.M. (1998). *Indian Journal of Agricultural Sciences,* **68** : 144-148.

Kathju, S., Aggarwal, R. K. and Lahiri, A. N. (1987). Evaluation of diverse effects of phosphate application on legumes of arid area. *Tropical Agriculture,* **64** : 91-96

Kumar, A., Mahapatra, B. S., Misra, A., Patro, H. K. and Singh, S.P. (2007). *Indian Journal of Agricultural Sciences,* 77 : 273-275.

Lal, S. and Singh, M.P. (1990). Pulses in rice-based cropping sequences. *Indian Farming,* **39** (4): 10-12.

Lal, R. and Kimble, J.M. (2000). In : *Proceeding of International Conference on Managing Natural Resources for Sustainable Agricultural Production in the 21st Century,* pp. 116-125, ISSA, New Delhi.

Mankotia, B.S. (2007). *Indian Journal of Agricultural Sciences,* **77** : 512-514.

Misra, D. K. (1971). Agronomic investigation in arid zone. In : *Proceeding of Symposium on Problems of Indian Arid Zones*, pp. 165-169.

Motsara, M.R. (2002). Available nitrogen, phosphorus and potassium status of Indian soils as depicted by soil fertility maps. *Fertilizer News*, **47** (8) : 15-21.

Parihar, S.S. (2000). In: *Proceeding International Conference on Managing Natural Resources for Sustainable Agriculture Production in the 21st century*, pp. 126-137, ISSA, New Delhi.

Powlson, D.S. (1994). Quantification of Nutrient cycles using long-term experiments pp 97-115. In: *Long Term Experiments in Agriculture and Ecological Sciences* (Leigh, R.A and Johnson, A.E (Eds) Walling ford U.K., CAB International.

Ready, G. S., Das, S. K. and Singh, R. P. (1993). Prospects of green leaf manuring as an alternative to fertilizer nitrogen in dryland farming. In: Recent Advances in Dryland Agriculture (ed) L. L. Somani. Scientific Publishers, Jodhpur.

Sharma, P.D. (2008). Nutrient Management – Key to food and health security. *Journal of Indian Society of Soil Science*, **55** : 395-304.

Sharma, P.K., Verma, T.S. and Bhagat, R.M. (1995). *Soil Use and Management*, **11** : 199-203.

Singh, A. K. (2008). Soil Resource Management – key to food and health security. *Journal of Indian Society of Soil Science*, **55** : 348-357.

Singh, M.V. (2001). Evaluation of current micronutrient stocks in different agro-ecological zones of India for sustainable crop production. *Fertiliser News*, **46** (2) : 25-28, 31-38 & 41-42.

Singh, S.D., Bhandari, R.C. and Aggarwal, R.K. (1985). Long term effects of phosphate fertilizers on soil fertility and yield of pearl millet grown in rotation with grain legume. *Indian Journal of Agricultural Sciences*, **55** : 274-278

Singh, V.K., Dwivedi, B.S., Shukla, Arvind K., Chauhan, Y.S. and Yadav, R.L. (2005). Diversification of rice with pigeonpea in a rice-wheat cropping system on a Typic Ustochrept: effect on soil fertility, yield and nutrient use efficiency. *Field Crops Research*, **92** : 85-105.

Singh, V. K., Gangwar, B., (2011). Sulphur management in crops and cropping systems for sustainable production. PDFSR Bulletin No. 1, pp. 70. Project Directorate for Farming Systems Research, Modipuram, Meerut, India.

Singh, V. K. and Mishra, R. P. (2010). Integrated nutrient management in transplanted rice-wheat system. PDCSR Annual Report 2007-2008, pp. 44-50, PDCSR, Modipuram, Meerut, India.

Singh, V.K. and Sharma, B.B. (2001). Productivity of rice as influenced by crop diversification in wheat-rice cropping system on Mollisols of foothills of Himalayas. *Indian Journal of Agricultural Sciences*, **71** (1) : 5-8.

Singh, V.K. and Sharma B.B. (2002). Economic evaluation of rice based cropping systems in foothills of Himalayas. *Indian Journal of Agronomy*, **47** (1) : 12-19.

SSSA (1995). *Soil Science Society of America Agronomy News*, WI. pp.7, June 1995, Madison.

Swaminathan, M.S. 2006. In: *Proceeding of International Conference Soil, Water Environment Quality – Issues and Strategies*, pp 6-8, ISSS, New Delhi.

Swarup, A. and Wanjari, R.H. (2000). Three Decades of AICRP on LTFEs to study change in soil quality, crop production and sustainability, pp. 1-59, ISSS, Bhopal.

Tiwari, K. N., Sharma, S. K., Singh, V. K., Dwivedi, B. S., Shukla, Arvind K., (2006). Site-specific nutrient management for increasing crop productivity in India: Results with rice-wheat and rice-rice system. pp 92. PDCSR Modipuram and PPIC India Programme, Gurgaon.

Yadav, R.L., Singh, V.K., Dwivedi, B.S. and Shukla, Arvind K. (2003). Wheat productivity and N use-efficiency as influenced by inclusion of cowpea as a grain legume in rice wheat system. *Journal of Agricultural Science*, Cambridge, **141** : 213-220.

□□□

System Based Integrated Nutrient Management, 2012
© B. Gangwar & V.K. Singh (eds.), pp. 111-122
New India Publishing Agency, New Delhi (India)
e-mail : info@nipabooks.com; website : www.nipabooks.com

CHAPTER 8

Integrated Nutrient Management in Rice-Wheat Cropping System

YADVINDER SINGH

Rice-wheat is a dominant cropping system in the Indo-Gangetic Plains covering approximately an area of 10.5 million ha. Modern agricultural production practices have emphasized the widespread use of fertilizers as a source of nutrients. The continuous use of high levels of chemical fertilizers over a prolonged period, decline in area under legumes and reduction in the use of organic manures has resulted in soil degradation and environmental pollution. It is estimated that the gap between nutrient removal by crops and addition through fertilizers in India will remain at level of 8-10 million tones per annum. Such negative nutrient balances are depleting the soils of their nutrient reserves resulting in acute deficiencies of nutrients. Integrated plant nutrient supply (IPNS) is an important component of sustainable agricultural intensification. The goal of integrated nutrient management (INM) is to integrate the use of all natural and man-made sources of plant nutrients, so as to increase crop productivity in an efficient and environmentally benign manner, without diminishing the capacity of the soil to be productive for present and future generations. The INM is made up of components which possess great diversity in terms of chemical and physical properties, nutrient release efficiencies, positional availability, crop specificity, and farmers' acceptability. Since organic manures cannot meet the total nutrient needs of modern agriculture, integrated use of nutrients from fertilizers and organic resources seems to be a need of the time. For INM to make desirable progress and find wide acceptance, nutrient supply packages for important agro-ecological environments need to be developed. These should be technically sound, practically feasible, economically attractive and socially acceptable. Integrated plant nutrient systems can ensure long-term sustainability of agricultural growth through improvement in soil health and will significantly reduce the needs for chemical fertilizers. The complementary use of chemical fertilizers and organic fertilizers may increase the efficiency of chemical fertilizers in order to maintain a high level of crop productivity. Because of low primary nutrient content and thus the need for large applications per unit area, farmers and policy makers are often reluctant to adopt and promote the use of organic fertilizers. The different components of INM possess great diversity in terms of chemical and physical properties and nutrient release patterns.

Potential and composition of organic manures

Organic manures such as FYM, composts and crop residues supply low quantities of major plant nutrients. Oil cakes, slaughterhouse wastes, poultry manure, etc. are relatively richer in nutrient contents. Town composts contain higher concentration of nutrients than rural composts. The nutrient concentration in organic manures depends on the type of animal, diet, collection and storage of manure, amount of bedding manorial used and any possible manure treatment. Sharma *et al.* (1978) analyzed 90 randomly selected samples of FYM and found that total N content ranged from 0.39-1.08% (mean value of 0.66%), total P from 0.045-0.198% (mean value of 0.102%) and total K 0.58-1.63% (mean value of 0.99%). According to an estimate, the annual production of dung and urine from bovine population in India is 1228 and 800 m.t., respectively. If the entire wet dung and urine excreted by bovines are conserved for manurial purposes, its potential for supplying major plant nutrients has been worked out as 6.96 m.t. of $N+P_2O_5+K_2O$ (Gaur, 1992). However, availability is often constrained as most farmers use the dung as fuel and most of the nutrients passed into the urine are lost under the traditional methods of recycling. Estimates of animal dung used as domestic fuel in rural areas have increased from 30% in 1970's to 75% in the early 1990's. Transition period of 45-60 days between harvesting of wheat and transplanting of rice could be utilized by planting fast growing legumes as green manure crops. Estimated current potential of crop residues is 400 m.t. and rice and wheat straws account for 70% of crop residues generated in India. Crop residues can contribute about 8,74 m.t. of NPK. Apart from supplying major nutrients, organic resources are rich sources of secondary and micronutrients. Biogas plants serve dual purpose as they provide fuel as well as good quality manure. Therefore, construction of biogas plants by the farmers should be encouraged in the country. An estimate of nutrient potential of organic manures, crop residues, and industrial wastes showed that these sources can provide approximately 6.1 mt of N, 2.3 mt of P and 6 mt of K annually (Juwarkar *et al.*, 1992).

Mineralization-immobilization turnover (MIT)

Manure N generally comprises two major fractions of agronomic importance, ammoniacal N and organic N. The relative contents of these fractions in manure vary depending on manure collection, handling and storage procedures. The concentration of ammoniacal N in manures available in India is very low due to its loss during improper handling and storage, and low quality of feed than that used for animals in the advanced countries. Some estimates suggest that up to two thirds or more of excreted N may be lost before application in the field. Manure N use efficiency for crop production could be increased significantly if the availability of N in these two fractions was better understood. Manure N present in organic form must be mineralized to the inorganic form before plants can take it up. Laboratory studies by Flowers and Arnold (1983) and Castellanos and Pratt (1981) and Yadvinder Singh et al. (1988) suggested that N immobilization occurred during the first 3-4 weeks decomposition period of manure. The (MIT) of soil manure and soil N availability in relation to crop response in the field has not been satisfactorily been quantified. The C/N ratio has been suggested to have a major influence on MIT in soils, although its shortcomings have been recognized. Studies by Kirchman (1985) suggested that the C/N ratio of 15 may be critical for aerobically digested organic manures. For wastes with C/N ratio greater than 23, a lag period occurs as carbon is metabolized and lost, after which mineralization occurs with a rate constant of about 0.0077 day^{-1} (at 35 ^{0}C). The rate of mineralization was affected by method of application, temperature, soil moisture and soil pH. Nitrogen mineralization rate was lower in surface applied manure than for incorporated manure. It has been demonstrated that MIT is time dependent. That is, as decomposition of an organic material in soil proceeds, a net decrease in mineral N will evolve to a net release of mineral N. If such is the case, the question arises as to how net immobilization of mineral N followed by net mineralization will affect crop response in the field. If MIT is slow, a crop may not derive any benefit from the immobilized N during a normal growing season. This probably underlines the importance of time of incorporation of manure before seeding/ planting of crop. Chescheire *et al.* (1986) summarized data of other researchers and reported that the availability of organic N ranged from 0 to 64% depending on manure source, extent of decomposition and water content (e.g. slurry versus dry manure). Van Fassen and Van Dijk (1987) found that 70-100% of the organic N in poultry manure was mineralized in soil during 6 months of incubation, whereas only 38% of added organic N from swine and cow manure was apparently mineralized. In addition, N release following manure application depended on the soil type. For example, N mineralized with cow manure ranged from 13% in a silty to 51% in a clay soil.

Approaches employed for the evaluation of organic manures

The nutritional value of organic manures has traditionally been estimated by two approaches, the decay series or the fertilizer equivalence. The decay series approach has been widely adopted as a means to estimate both initial and residual availability of N in organic wastes. A decay series is essentially a quantitative estimate of the amount of N that will be mineralized from an organic waste over a period of several years, and is usually based on laboratory N mineralization studies chemical indicates have also been evaluated for predicting nutrient supply from organic wastes.

Decay series

When manure is applied to the same field year after year, the availability of N it contains becomes an important factors in determining the application rate. Pratt *et al.* (1973) developed the concept of a decay series for estimating available N from organic manures. The concept recognizes two major factors: (1) The variable amount of inorganic N in different manures that is available to the first crop (2) the gradual mineralization of the organic N over several years. For example, a series of decay constant of 0.35, 0.15, 0.10 and 0.05 for FYM indicates that 35% of N in the manure becomes available in the first year, 15% of the residual N becomes available in the second year, 10% in the third year, and 5% in the fourth year and all succeeding years. Organic manures containing high percentages of N have more rapid decay rates (e.g. decay constants for poultry manure are 0.90-0.10-0.05). There is no entirely satisfactory procedure for measuring the parameters in the decay series directly. The parameters of these series are based on judgements, with a minimum of experimental data, and probably only apply to the fairly specific situations for which they were developed. While the use of decay series is conceptually sound, it is obvious that many factors can affect the success of this approach, including heterogeneity of the wastes, annual variations in climate, and management effects (tillage, irrigation, etc.). However, they are a major development because they provide a logical framework for more experimental work and for summarizing available data. Multiyear field calibration studies are essential to verify a decay series for organic wastes. Further as the decay series estimates only the amount of N that will become available, some techique to adjust waste application rate account for N the potential N losses by volatilization, denitrification, or leaching would be required to optimize N application and thus yield.

Fertilizer N equivalence

Another method of evaluating the manure is to compare it in terms of fertilizer N equivalents. That is, estimate the amount of fertilizer N that 1 t of manure replaces in terms of crop production. The fertilizer equivalence approach determines N availability in organic manures more empirically. Field studies comparing several rates of fertilizer N and organic manures are used to determine the amount of total N in organic manure needed to obtain yields or N uptake by a crop equivalent to that obtained with fertilizer N. Results are expressed in equivalent rates (kg N ha^{-1}) or as percentage of total N. Recent examples of this approach can be found in the studies by Yadvinder-Singh *et al.* (1995) and Bijay-Singh *et al.* (1997). From a three field study with rice-wheat rotation, Yadvinder-Singh *et al.* (1995) reported that fertilizer equivalence for cattle manure based on grain yield of rice ranged from 42-52% of manure total N. Bijay-Singh *et al.* (1997) reported that fertilizer equivalence for poultry manure ranged from 86-146 kg N ha^{-1} depending on the amount of manure N applied. Evidently, the only property of manure in these experiments that is influencing yield is the N furnished by the manure. The fertilizer N equivalents cannot be calculated when the factors limiting yields with and without manure are different.

Crop responses

The agronomic value of organic manures as fertilizer is determined by their ability to increase the yield or quality of crops. There are considerable data in literature that demonstrate the effect of organic materials on crop yields (Gaur *et al.*, 1984). Crop yield responses to addition of organic materials are highly variable and are dependent upon the crop, soil type, climatic conditions, management practices and the quality of the organic manure used. Combining of basic information of N cycling in manure amended soils and field experiments to develop organic manure management practices that can result consistently in comparable yields as obtained with fertilizer N. Information needs to be collected for most organic sources to identify manure application rates required to approximate a fertilizer N addition.

Pre-dominant organic sources

Farmyard manure

Farmyard manure is most commonly used organic manure. Experiments showed that rice yield from combined application of 12 t FYM ha^{-1} and 80 kg Nha^{-1} was equal to an application of 120 kg N ha^{-1} as fertilizer (Maskina et al. 1988a). FYM also gave residual effect equivalent to 30 kg N and 30 kg P_2O_5 ha^{-1} on the yield of succeeding wheat. From large number of trials conducted in different zones of India it was found that fertilizer N equivalent of one tone of FYM averaged 4.4 kg N ha^{-1} as fertilizer in rice (Leelavathi *et al.*, 1986). Efficiency of N added from FYM as compared with urea, ranged from

42 to 53 % in rice (Yadvinder Singh *et al.*, 1995). Significant residual effects of organic manures applied to rice have been reported in the following wheat. Kulkarani *et al.* (1978) showed that combined use of 12 t FYM and 60 kg N ha^{-1} gave rice yield equivalent to 120 kg N ha^{-1}, thereby showing an economy of 60 kg N ha^{-1}. Prasad (1994) reported that application of FYM and composts to rice in combination with 13.1 kg P ha^{-1} gave higher yields than 26.1 kg P ha^{-1} as superphosphate on a calcareous soil. Thus, FYM could substitute 50% of the P requirement of rice and residual effect equivalent of 13.1 kg P ha^{-1} was observed in the following wheat crop. Combined application of organic manures and chemical fertilizers generally produces higher yields than when each is applied alone. Experiments conducted at Varanasi showed that application of 75% N as chemical fertilizer plus 25% as FYM or compost produced higher sustainable rice yields than the application of whole of N as chemical fertilizer (Singh *et al.*, 1994).

Poultry manure and pig manure

The value of poultry manure as a nutrient source for the production of cereal and vegetable crops has long been recognized. Sims (1987) reported that maize yield did not differ significantly between poultry manure and ammonium nitrate sources of N. Maskina *et al.* (1988b) reported greater efficiency of poultry manure than FYM or piggery manure in rice. Rate of N mineralization from poultry manure is much faster than FYM. In a laboratory study, about 45% of total in poultry manure was mineralized in four weeks as compared to 12% from FYM (Yadvinder-Singh *et al.*, 1988). Poultry manure contains high amount of uric acid and urea substances that readily release NH_4^+-N.

Experiments on the agronomic efficacy of poultry manure showed that 4 t poultry manure along with 60 kg N ha^{-1} produced rice yield similar to that obtained with 120 kg N ha^{-1} as urea, thereby resulting in N economy of 60 kg N ha^{-1} in rice (Yadvinder-Singh and Meelu (1995). Studies conducted by Bijay-Singh *et al.* (1997) showed that poultry manure-N was almost as efficient as urea-N in increasing yield and N uptake of rice. Application of poultry manure at 6t ha^{-1} to rice substituted 60 kg N ha^{-1} in rice and residual effect equivalent of 30 kg N + 30 kg P_2O_5 ha^{-1} was observed in the following wheat (Yadvinder-Singh *et al.*, 2009).

Biogas slurry

Biogas slurry is the end product of biogas plants when organic materials are converted into CH_4 and CO_2. Biogas slurry is generally richer in N than FYM containing 1.2-2.0% N, 0.5-0.7%P and 0.5-1.0%K (dry weight basis) along with significant quantities of micronutrients. Biogas slurry may be applied directly with irrigation water or after drying as manure. Anaerobic digestion of organic wastes conserves nutrients needed for crop production. One ton of biogas slurry provides 44 kg nutrients as compared to 19 kg through FYM. Studies showed that application of 20 t ha^{-1} of wet slurry (5% dry matter) each with first and second irrigation in conjunction with 60 kg N ha^{-1} increased wheat yield, which was at par with that obtained with 120 kg N ha^{-1}. Application of 15 t dry slurry ha^{-1} in rice showed N economy of 62.5 kg N ha^{-1} in rice (Gupta *et al.*, 1986).

Enriched compost

The enrichment of organic manures is of practical interest because demerits of bulky organic manures can be overcome by their enrichment. It increases the nutrient content of manures, reduces the bulk to be handled per unit of nutrients, and offers a potential for the utilization of insoluble materials such as low-grade phosphate rocks. Mishra (1992) described a technology for the preparation of P-enriched compost from crop residues, animal feed wastes, grasses, weeds, tree leaves, cattle dung, biogas slurry and Mussoori rock phosphate, known as phosphocompost. It contains between 2.6-3.5% P depending on the degree of decomposition. The total N content of phosphocompost is lower (0.82%) as compared to that of corresponding compost (1.15%) due to the dilution effect. The N content of the phosphocompost could be increased to 2% by the addition of urea-N and iron pyrite. Out of the portion of animal dung mobilized towards agriculture, a sizeable amount of nutrients is lost because of improper storage, handling and incorporation technique. Singh *et al.* (1981) reported that enriched compost (2% N + 1.29% P) when applied to supply 26.2 kg P ha^{-1} proved as effective as superphosphate in field trials on wheat. Its application also saved 30 kg N ha^{-1} in wheat.

Green manures

Leguminous green manure crops can be raised during the transition period of 45-60 days available after the harvest of wheat before transplanting of rice. *Sesbania aculeata*, sunnhemp and cowpea are the most promising green manures (Yadvinder Singh *et al.*, 1991). Biomass and N contributions by sesbania green manure in 50-60 days are about 4-5 t dry matter more than 100 kg N ha (Meelu *et al.*, 1994). Crop yield responses to integrated use of mineral nitrogenous fertilizers and green manures are found to be highly variable and as much as 33-75% of fertilizer N can be saved by green manuring (Bijay Singh *et al.*, 1992). Short duration pulse crops (mung bean and cowpea) are promising dual purpose legumes for grain and green manuring. Generally 6 t Azolla biomass ha^{-1} increased rice yield comparable to 20-30 kg fertilizer N. It is possible to raise Azolla three times for the same rice crop and in this way, the total N input from Azolla can go up to 75-90 kg N ha^{-1} (Meelu *et al.*, 1994). However, Azolla is very sensitive to P deficiency and very high or low temperatures adversely affect its growth.

Agro-industrial wastes

In the wake of technological growth waste accumulation enhanced and disposal has assumed serious dimensions not only in the western world but also in India. The wastes are of different kinds including domestic wastes, city garbages, vegetable wastes, rice mill and sugar mill wastes, sewage effluents, and sludges, distillery wastes. Heavy metal hazards are caused through sewage sludge: Cadmium, Pb, Ni, Cr, Zn, Cu, Al, Mn. Cadmium is a highly toxic element, which is translocated from soils to plants. Besides, India ranks third among the coconut producing countries of the world with production potential of more than 6000 million nuts. The coir pith is a waste material and it accumulates in large amounts in the factories. Coir pith is having pH 5.4 contains 0.25% N and 0.8% K. After composting it can be used as manure.

Pressmud cake

In the recent past, many sugar industries have been established in the country. About 8.9 million tons of pressmud cake (PMC) is produced annually from the sugar industries in India. The carbonation process is used mainly by the old industrial units, which are gradually changed to sulphitation. Carbonation press mud cake (PMC) contains high amounts of $CaCO_3$ and therefore, is not recommended for alkaline soils. The PMC produced from the sugar industries employing sulphitation process contains about 1.8-2.25% N, 0.8-1.2% P, 0.4-0.6% K in addition to several other plant nutrients; 2.5 g Fe kg^{-1}, 1.5 g Mn kg^{-1}, 0.27 g Zn kg^{-1}, and 0.13 g Cu kg^{-1}. Experiments conducted in Punjab showed that application of PMC (5 t ha, dry weight basis) along with 60 kg fertilizer N ha^{-1} produced rice grain yield equivalent to that produced in recommended fertilizer treatment of 120 kg N ha^{-1} (Yadvinder Singh *et al.*, 2008). Application of 80 kg N ha^{-1} and 30 kg P_2O_5 ha^{-1} on PMC plots produced wheat yield equivalent to 120 kg N ha^{-1} and 60 kg P_2O_5 ha^{-1} on non-amended plots. The residual effect of PMC in the following wheat was thus equivalent to 40 kg N and 30 kg P_2O_5 ha^{-1}. Regular application of PMC for 4 years caused significant increase in organic carbon and available P content of the soil. The sharp increase in P availability in PMC-amended plots and maintaining high levels over long periods will be beneficial in meeting crop P requirements.

The sharp increase in P availability in PMC-amended plots and maintaining high levels over long periods will be beneficial in meeting crop P requirements. Long-term application of PMC may even replace the whole fertilizer P needs of the rice-wheat system.

Organic manures and soil fertility

The use of organic manures not only helps to substitute partly for chemical fertilizers but also improves overall soil productivity through beneficial effects on physical, chemical and biological properties of soil. Several workers (Maskina *et al.*, 1988 a, b, Bijay-Singh *et al.*, 1997) have reported increases in organic C content of soil, available nutrient contents with the application of organic manures in rice. Organic manures also play a significant role in correcting micronutrient balance in rice-wheat system. Long-term fertility experiments in India showed a declining trend in productivity with the application of NPK fertilizers alone (Nambiar and Abrol, 1994). Integrated use of organic manures and chemical fertilizers has been found to be promising not only in maintaining higher productivity but also for providing stability in crop production. The effects have been more pronounced in case of acidic than normal soils. Since FYM was added on top of the recommended NPK application through chemical fertilizers, its impact on crop yields can be attributed largely to beneficial effect on soil properties, soil moisture, better nutrient availability, more favourable conditions for root growth and supply of micronutrients. Organic manures can influence the several microbiological processes in soil such as urea hydrolysis, nitrification, denitrification, etc. Organic manures also promote the growth and multiplication of useful microorganisms in the soil due to the fresh supply.

Conclusions and agenda for future research

The integrated use of green manures, organic manures, and inorganic fertilizers in rice-wheat system has a great potential to off-set the heavy requirements of chemical fertilizers, to achieve maximum yields and to sustain the crop productivity on long term basis. Some estimates of fertilizer equivalents of organic materials are available but they are based on limited data and for few soils and agro-ecological regions. There are a number of gaps in our knowledge for developing quantitative estimates on different aspects of integrated nutrient management for specific environments. Priorities for future research are as under:

1. Efficient management of organic materials as nutrient source requires a knowledge of their chemical characteristics, and several aspects of their transformations in soil.
2. A large fraction of manure N is lost from the soil-plant system. Proper storage, composting, application techniques of manures is needed to avoid nutrient losses and conserve maximum nutrients. Composting helps in reducing N losses via leaching or denitrification.
3. Research on developing cultures of micro-organisms and techniques which hasten the process of composting in order to produce good quality

of compost will be useful. Role of earthworms in digesting organic wastes needs to be evaluated.

4. Estimation of the percentage contribution of organic sources to soil nutrient budget that will be available during the current growing season and residual nutrients from their applications.

5. Breeding and selection of superior N_2- fixing legume species and cultivars for green manuring, short duration pulses and fodder legumes with fast growth to provide highest possible biomass and N input in short duration.

6. Calculation of fertilizer equivalents of different organic sources for different soils and agro-ecological regions. Development of nutrient supply packages with optimum application rates of organic manures and inorganic fertilizers.

7. Role of organic materials in improving fertilizer use efficiency and utilization of low grade rock phosphate needs to be investigated.

8. Quantification of nutrient losses from organic materials and devising methods to increase the efficiency of their use.

9. Efficient management of crop residues in rice-wheat system should be investigated. There is a need to study the effect of co-incorporation of green manure and crop residues on crop and soil productivity.

10. Evaluation of long term benefits of organic materials on physical, chemical and biological properties of soils and sustainability of the rice-wheat system.

11. Economic evaluation of the each integrated nutrient management technology and identification of constraints in the adoption of new technology.

References

Bijay-Singh, Yadvinder-Singh, Maskina, M.S. and Meelu, O.P. (1997). The value of poultry manure for wetland rice grown in rotation with wheat. *Nutrient Cycling Agroecosystem*, **47** : 243-250.

Bijay-Singh, Yadvinder-Singh and Khind, C.S. (1992). Fertilizer management in soils amended with green manures. *Fertilizer News*, **37** (8): 41-45.

Castellanos, J.Z. and Pratt, P.F. (1981). Mineralization of manure N- correlation with laboratory indexes. *Soil Science Society of America Journal*, **45**: 354-357.

Chae, Y.M. and Tabatabai, M.A. (1986). Mineralization of nitrogen in soils amended with organic wastes. *Journal of Environment and Quality*, **15**: 193-198.

Flowers, T.H. and Arnold, P.W. (1983). Immobilization and mineralization of nitrogen in soils incubated with pig slurry or ammonium sulphate. *Soil Biology and Biochemistry*, **15** : 329-335.

Kulkarni, K.R., Hukkeri, S.B. and Sharma, O.P. (1978). Fertilizer response experiments on cultivators field in India. In: *India/ FAO/ Norway Seminar on the development of Complementary Use of Mineral Fertilizers and Organic Materials*. pp. 67-72. Ministry of Agriculture and Food, Krishi Bhavan, New Delhi.

Kumar V. and Mishra B. (1991). Effect of two types of press mud cake on growth of rice and maize, and soil properties. *Journal of Indian Society of Soil Science*, **39** : 109-113.

Leelavathi, C.R., *et al.* (1986). Revised yardsticks of additional production of rice due to improvement measures. IASRI, New Delhi. 1108 pp.

Maskina, M.S., Bijay Singh, Yadvinder Singh and Meelu, O.P.(1988a). Fertilizer requirement of rice-wheat and maize-wheat rotations on coarse textured soils amended with farmyard manure. *Fertilizer Research*, **17** : 153-64

Maskina, M.S., Bijay Singh, Yadvinder Singh and Meelu, O.P. (1988b). Response of wet land rice to fertilizer N on a soil amended with *Cattle, Poultry and Pig Manures. Biol. Wastes*, **26** : 1-8.

Meelu, O.P., Yadvinder-Singh and Bijay-Singh (1994). Green Manuring for Soil Productivity Improvement. *FAO World Soil Resources Reports* No. **76**. : 123 pp.

Mishra, M.M. (1992). Enrichment of organic manures with fertilizers. In.: *Non- traditional Sectors for Fertilizer Use*. Ed. H.L.S. Tandon. pp. 48-60. Fertilizer Development and consultancy organization, New Delhi.

Sims, J.T. (1987). Agronomic evaluation of poultry manure as a nitrogen source for conventional and no-tillage corn. *Agronomy Journal*, **79** : 563-570.

Sluijsmans, C.M.T. and Kolenbrander, G.J. (1977). The significance of animal manure as a source of nitrogen in soil. *Proc. Int. Seminar on Soil environment and fertility management in intensive agriculture*. Kyoto, Japan. pp. 403-411. The Society of the Science of Soil and manure, Japan.

Van Fassen,H.G. and Van Dijk, H. (1987). Manure as a source of nitrogen and phosphorus in soils. In: *Animal Manure on Grasssland and Fodder Crops. Fertilizer or Waste*?

Yadvinder-Singh, Khind, C.S. and Bijay-Singh (1991). Efficient management of leguminous green manures in wetland rice. *Advances in Agronomy*, **45** : 135-189.

Yadvinder-Singh, Bijay-Singh, R.K. Gupta, J.K. Ladha, J.S. Bains and Jagmohan-Singh (2008). Evaluation of press mud cake as a source of nitrogen and phosphorus for rice-wheat cropping system in the Indo-Gangetic plains of India. *Biology and Fertility of Soils*, **44** : 755-762.

Yadvinder-Singh, Bijay-Singh, Maskina, M. S. and Meelu, O.P. (1988). Effect of organic manures, crop residues and green manure (*Sesbania aculeata*) on nitrogen and phosphorus transformations in a sandy loam at field capacity and waterlogged conditions. *Biology and Fertility of Soils*, **6** : 183-187.

Yadvinder-Singh, Bijay-Singh, Maskina, M.S. and Meelu O.P. (1995). Response of wetland rice to nitrogen from cattle manure and urea in a rice-wheat rotation. *Tropical Agriculture* (Trinidad) **72** : 91-96.

Yadvinder-Singh and Meelu, O.P. (1995). Integrated nutrient management in rice. *Progressive Farming* (PAU). May Issue: 12-13 & 27.

Yadvinder-Singh, R.K. Gupta, H.S. Thind, Bijay-Singh, Varinderpal-Singh, Gurpreet-Singh, Jagmohan-Singh, and J.K. Ladha (2009). Poultry litter as a nitrogen and phosphorus source for a rice-wheat system. *Biology and Fertility of Soils,* **45** : 171-175.

□□□

System Based Integrated Nutrient Management, 2012
© *B. Gangwar & V.K. Singh (eds.), pp. 123-154*
New India Publishing Agency, New Delhi (India)
e-mail : info@nipabooks.com; website : www.nipabooks.com

CHAPTER 9

Integrated Nutrient Management in Oilseeds – Based Cropping Systems

D.M. HEGDE

Oilseeds serve as rich source of food, feed, energy, employment and commerce. The productivity of oilseeds in India (1006 kg ha^{-1} during 2008-09) is very low when compared with the world productivity (1878 kg ha^{-1} during 2008). The oilseeds production in the country often suffers from a high degree of variation in annual production owing to their predominant cultivation under low and uncertain rainfall situations which is further handicapped by input starved conditions with poor crop management. Nutrient and water are the key inputs for crop production. Improving efficiency and factor productivity under complexities of diminishing quality resources and increasing ecoawareness are critical for sustainable oilseeds production.

The vegetable oils consumption in the country is steadily rising and has sharply increased in the last couple of years touching around 14.1 kg $head^{-1}$ $year^{-1}$ during 2009-10. As per DAC-Rabo Bank, India's per capita consumption is likely to reach at least 16.38 kg $year^{-1}$ by 2020 and for a projected population of 1331 million, vegetable oil requirement will be 21.8 million tonnes. This is equivalent to about 66 million tonnes of oilseeds assuming that there is no change in the proportion of different oilseeds produced in the country in the next ten years. Assuming about 20% of vegetable oils contribution from sources like ricebran, cotton seed, coconut, oilpalm, etc., still the country needs to produce about 55 million tonnes of oilseeds by 2020. Considering that the oilseeds production during 2008-09 was just 27.56 million tonnes, the country needs to double oilseeds production in next 10 years requiring a growth rate in excess of 6%.

Nutrient needs of oilseed crops

Oilseed crops are energy rich crops and obviously the requirement of major nutrients including secondary and micronutrients is very high (Table 1 & 2). Considerable variations in nutrient removal occur from situation to situation depending on crop productivity, soil fertility, amount of crop residues left over in the field and their composition. Some of the widely adopted cropping systems involving oilseeds may remove as much as 400 to 900 kg nutrients ($N+P_2O_5+K_2O$ ha^{-1}) under high productivity conditions (Table 3). The estimated nutrient removal by nine oilseed crops during 2008-09 comes to about 3.69 million tonnes (Table 4). At the production levels of 55 and 66 million tonnes of oilseeds, the nutrient removal will be of the order of 6.680 and 8.017 million tonnes, respectively (Table 5). Unless the soils are replenished with all the nutrients taken up by the crops, there will be persistent nutrient exhaustion posing a great threat to sustainable oilseeds production. Since chemical fertilisers alone will not be able to meet the total nutrient needs, integrated use of all the potential sources of nutrients seems to be the only option to maintain soil fertility and crop productivity.

Table 1 : Average major nutrient removal by oilseed crops $tonne^{-1}$ yield under field conditions

Crop	Kg uptake t^{-1} produce					
	N	P_2O_5	K_2O	S	Ca	Mg
Groundnut	58.1	19.6	30.1	7.9	20.5	13.3
Mustard	32.8	16.4	41.8	17.3	42.0	8.7
Raya	64.5	20.6	53.4	16.0	56.5	9.5
Taramira	70.0	26.0	61.1	20.7	19.3	9.3
Soybean	70.7	30.9	57.7	6.7	14.0	7.6
Safflower	38.8	8.4	22.0	12.6	-	-
Sesame	51.7	22.9	64.0	11.7	37.5	15.8
Sunflower	63.3	19.1	126.2	11.7	68.3	26.7
Linseed	60.0	18.6	54.0	5.6	31.2	13.1
Castor	40.0	9.0	16.0	-	-	-

Source : Tandon and Sekhon, 1988; Pasricha and Tandon, 1993; Hegde, 1998.

Table 2 : Average micronutrient removal by oilseed crops tonne^{-1} yield under field conditions

Crop	g uptake t^{-1} produce			
	Zn	Fe	Mn	Cu
Groundnut	109	2284	93	36
Mustard	100	1123	95	17
Raya	58	635	169	21
Taramira	16	708	153	29
Soybean	77	346	83	30
Sesame	168	793	115	117
Sunflower	47	1075	-	-
Linseed	46	664	177	30

Table 3 : Nutrient uptake in some cropping systems involving oilseeds

Cropping system (t ha^{-1})	Nutrient uptake (kg ha^{-1})			
	N	P_2O_5	K_2O	Total
Sorghum (3.6) - Groundnut (2.2)	216	74	184	474
Maize (3.0) - Mustard (2.0)	127	70	177	374
Cotton (2.0) - Sunflower (1.8)	203	90	376	669
Soybean (1.5) - Potato (26.0) - Wheat (4.0)	305	177	345	827
Sorghum (4.0) - Sunflower (1.5) - Groundnut (2.0)	301	121	385	807
Rice (4.8) - Rice (5.7) - Groundnut (2.0)	316	153	369	838
Rice (4.7) - Rice (5.6) - Sesame (0.7)	243	131	354	728
Sorghum (F) (18.9) - Potato (29.0) - Sunflower (1.8)	296	121	481	898
Maize (2.8) - Potato (27.0) - Sunflower (1.9)	296	110	479	885
Soybean (1.9) - Wheat (4.8)	252	100	267	619
Sunflower (1.6) - Groundnut (3.4)	294	97	304	700

Integrated nutrient supply and management in oilseeds (INSAM)

INSAM is an age old practice but its importance was not very much realised in pre-green revolution era due to low nutrient demand of traditional agriculture. This approach aims at efficient and judicious use of all the major sources of plant nutrients in an integrated manner, so as to get maximum economic yield without any deleterious effect on physico-chemical and biological properties of the soil. Thus, the basic concept underlying the principles of INSAM is the maintenance and possible improvement in soil fertility for sustained crop productivity on a long-term basis.

Table 4 : Estimated nutrient removal by oilseed crops during 2008-09

Crop	Production (000 t)	Nutrient uptake (000' t)			
		N	P_2O_5	K_2O	Total
Groundnut	7168	417	140	215	72
Rapeseed-mustard	7201	236	118	417	771
Soybean	9905	701	311	571	1583
Sunflower	1158	74	23	172	269
Sesame	640	33	14	42	89
Safflower	189	8	2	4	14
Niger	117	9	2	7	18
Castor	1171	51	10	19	80
Linseed	169	11	4	8	23
Total	27719	1540	624	1455	3619

Table 5 : Estimated nutrient removal by oilseed crops at targeted production by 2020

Nutrient (000 t)	Target 55 mt	Target 66 mt
N	3011	3613
P_2O_5	1175	1411
K_2O	2494	2993
Total	6680	8017

INSAM as a concept and farm management strategy embraces and transcends from single season crop fertilization efforts to planning and management of plant nutrients in crop rotations and farming systems on a long-term basis for enhanced productivity, profitability and sustainability. It aims to:

1. Enhance crop and soil productivity through a balanced use of mineral fertilizers combined with organic and biological sources of plant nutrients to ensure sustainability of the production systems.
2. Improve the capital stock of plant nutrients in the soil, and
3. Improve the efficiency of plant nutrient use, limit losses of nutrients to the environment and promote environmental security.

The necessity of promoting and adopting the wider concept of INSAM arises from past mineral fertilization based practices and lessons and thus results in paradigm shifts in several aspects of farming (Finck, 1995): from individual crop nutrient requirements to optimum use of nutrient sources; from static nutrient balances to nutrient flows (fluxes) and nutrient cycles; from least concern for environmental impact to due attention to the unwanted side effects of fertilization (on soils, weed growth, crop diseases, etc., pollution of water and air); from the first years' nutrient effects to long-term effects (residual nutrient effects, fate of unused nutrients, storage, carry over); from the narrow concern for yield effects to resistance of crops against stress conditions (dry, cold, salty, alkaline, toxicity, pollution); from the assumption of ideal growth conditions to an awareness of not or hardly controllable growth limiting factors and production risks; from exploitation of soil fertility to its improvement or maintenance; from the neglect of protective restrictions to awareness against dangerous or even toxic elements; from productivity to productivity and sustainability; from quantitative use to both quantitative and qualitative aspects of economy and efficiency in nutrient use; from fertilizer use promotion to knowledge intensive nutrient management; and from emphasis on direct effect of fertilizers to emphasis on synergy and interactive effects of crop-water-nutrient.

INSAM has assumed great importance in recent years in India because with high nutrient turnover in soil-plant system under intensive farming, neither the chemical fertilizers nor organic/biological sources alone can achieve production sustainability. Long-term studies carried out in India mostly in cereal based cropping systems have indicated that even with so called balanced use of NPK fertilisers, high yields could not be maintained over years because of emergence of secondary and micro-nutrient deficiencies and deterioration in soil physical environment (Nambiar and Abrol, 1989; Hegde and Dwivedi, 1993). Nevertheless, integrated use of fertilizers and organic manures/ biological sources helped in maintaining yield stability in most agroecoregions through correction of marginal deficiencies of secondary and micronutrients, enhancing efficiency of applied nutrients and providing favourable soil physical conditions. The interactive advantages of combining organic and inorganic sources of nutrients in INSAM have proved superior to the use of its each component separately (Roy, 1992; Hegde, 1998).

Key concepts of INSAM

INSAM enhances soil productivity through a balanced use of soil nutrients, chemical fertilisers, combined with organic sources of plant nutrients, including bioinoculents and nutrient transfer through agro-forestry systems and has adaptation to farming systems in both irrigated and rainfed agriculture. INSAM incorporates the underlying relationship between use of plant nutrients economic feasibility and maintenance of environmental quality. Operationally, it focuses first on the seasonal or annual cropping system rather than on an individual crop, secondly on the management of nutrients in the whole farming system and thirdly on the concept of village or community areas and watersheds rather than individual fields through the nutrient budgeting approach, soil fertility maps and analysis of practices which may be contributing to nutrient losses or inefficient plant nutrient management and their remedy.

The following key concepts are fundamental to INSAM and should be built into the strategy and techniques recommended for promoting INSAM as one of the tenets of sustainable oilseeds production:

1. Loss of soil productivity is of greater concern than loss of soil fertility.
2. Adoption of soil conservation practices and improved organic matter management practices are crucial for maintaining soil productivity.
3. Inclusion of legume species in the cropping system and rotations as grain, forage and green manure crop.
4. Nutrient management cannot be dealt with in isolation but should be managed as an integral part of a productive farming system.
5. The synergy between best water management practices (rainwater under rainfed dryland conditions and irrigation water under irrigated conditions) and best nutrient management practices must be optimized.
6. The full benefits from the supply of plant nutrients can be realised by farmers only after they have made improvements in the biological, physical and hydrological properties of the soils and removed soil related constraints.
7. It is necessary to identify the socio-economic constraints in the adoption of INSAM at community level and devise appropriate policy interventions and improvements.
8. Promotion of INSAM must be bottom-up rather than top-down in orientation, planning and implementation with the full involvement and participation of the farmers and local communities.

The major components of INSAM are fertilisers, organic manures (FYM/compost), legumes, crop residues/recyclable wastes, and biofertilisers. These sources possess great diversity in terms of chemical and physical properties, nutrient release efficiencies, positional availability, crop specificity and farmer acceptability. Therefore, the combination of different components to ensure optimum supply of nutrients to a given cropping system involving oilseeds depends on several ecological and socio-economic factors.

Nutrient deficiencies in oilseed growing areas

About 72% of the area under oilseeds is rainfed. Rainfed lands are prone to soil degradation through various means. Soil organic matter besides playing a key role as a reservoir of nutrients also impacts tilth, root growth and moisture conservation. Thus, maintenance of optimum organic matter level is imperative to avoid multiple nutrient deficiencies. The general extent of nutrient deficiencies in dryland areas which are directly relevant to oilseeds production have been summarized by Katyal and Reddy (1997) and others (Table 6).

1. Nitrogen is universally deficient in semi-arid tropics. Phosphorus deficiency is widespread and constrains the productivity of rainfed crops considerably (Swindale, 1982). Because of low productivity of rainfed crops, delineation of P-deficient areas on the basis of soil test values alone has been erratic. This situation is also associated with other nutrients. Potassium deficiency may be a problem in coarse textured soils in some red and lateritic soils and in soils where K applications are ignored continuously as in case of castor production in Telangana region of Andhra Pradesh for long time.
2. Sulphur deficiencies have been reported to occur in more than 250 districts (Motsara, 2002; Tandon and Messick, 2002). On an average, 17% of the soils in Gujarat, many parts of Karnataka and large areas in the country as a whole are reported to be S deficient. Sulphur deficiency can be a severe constraint because oilseeds have a particularly high requirement of S and in general 10-12 kg S uptake for each tonne of oilseed harvested.

Table 6 : Nutrient status of Indian soils and their extent of deficiencies

Nutrient	Soil fertility status
N*	Low in 63% samples, medium in 26%, high in 11% samples
P*	Low in 42% samples, medium in 38%, high in 20% samples
K	Low in 13% samples, medium in 37%, high in 50% samples
S	Deficiencies of variable degree found in over 250 districts
Mg	Kerala, other southern and north-eastern states, very acid soils are deficient
Zn	48% of 250000 soil samples analysed found to be deficient
Fe	12% samples tested deficient. Important on upland calcareous soils
B	33% soil samples tested deficient. Important in Bihar, Karnataka, Jharkhand, Madhya Pradesh, Uttar Pradesh, Chattishgarh and West Bengal.

* Based on 3.65 million soil samples analysed during 1977-1999 (Motsara, 2002)

3. Several workers have reported widespread Zn deficiency in Indian soils (Katyal and Rattan, 1993). Micronutrient status of the soils is linked with soil organic matter status. Soil test values do not clearly indicate the influence of organic matter level on reduction of Cu, Mn and Fe availability. However, some dicots like groundnut can exhibit Fe deficiency (chlorosis), which is often unrelated to iron availability levels. Deficiency of Zn in sandy loam soils, Fe in calcareous black soils and B in heavy rainfall light soils has been reported.

Chemical fertilisers in INSAM

Fertiliser is the important component of INSAM in oilseeds. Inadequate and/or imbalanced use of fertilisers has been identified as one of the critical constraints holding oilseeds production. Yield increases ranging from 26 to 300% with fertiliser application alone have been recorded in rainfed areas (Subba Rao, 1994). With fertilisers on an average contributing upto 50% rise in production, their importance in any strategy to meet the challenges on rising demand for oilseeds can hardly be emphasized. Our dependence on fertilisers as a source of nutrients even in oilseeds is increasing because of the need to supply large amounts of nutrients. While the use of fertilisers is the quickest and surest way of boosting oilseeds production, the following issues related to present fertiliser scenario in the country needs special attention.

Inadequate domestic availability: The total plant nutrient uptake in 2008-09 was estimated to be around 34-35 million tonnes as against the fertiliser consumption of around 24.91 million tonnes, leaving a gap of around 10 million tonnes. Although a part of this gap is bridged from the non-chemical sources like organic and biofertilisers, still a substantial gap is expected to be bridged by depletion of soil nutrient reserves. The fertiliser needs of oilseeds at a targeted production of 55 and 65 million tonnes works out to nearly 2.795 and 3.356 million tonnes, respectively.

Low fertiliser use efficiency: Hardly 30-40% of nutrients applied through fertilisers are utilized by the crops and the remaining are lost through various pathways such as leaching, volatalization, surface runoff, soil erosion and fixation. Less than 15% of nutrients absorbed by oilseed crops is contributed by fertilisers (Table 7) while the rest is contributed from soil reserves, organic manures, biological sources, residues and wastes. The declining efficiency of fertilisers is of great concern. In a country like India, where nearly 75% of the farmers operate small and marginal holdings and are resource poor, fertiliser nutrient wastage arising from inefficient use will prove prohibitively expensive and thus can be ill-afforded. Further, any inefficient use of fertiliser nutrients can produce adverse effects on the environment casting additional doubt on the long-term sustainability of oilseeds production. In-efficient use of fertiliser

Table 7 : Estimated fertiliser use and its contribution to nutrient removal in oilseed crops during 2008-09

Nutrient	Uptake (000't)	Fertiliser use (000't)	Fertiliser use efficiency (%)	Contribution to uptake	
				Fertiliser (%)	Others (%)
N	1561	588	50	18.5	81.5
P_2O_5	635	472	30	22.3	77.7
K_2O	1442	167	50	5.8	94.2
Total	3668	1227		14.5	85.5

nutrients is likely to be triggered also by policies favouring lower prices of certain nutrients and leaving others non-attended. There are several well-proven scientific practices through which the efficiency of applied fertilisers can be increased which need to be popularized among oilseed farmers.

Use of high analysis fertilisers: Traditional low analysis fertilisers like ammonium sulphate, single super phosphate, calcium ammonium nitrate, potassium sulphate which used to supply NPK in past, also added inadvertently considerable amounts of secondary and micronutrients. At present, emphasis is laid on use of chemically pure, high analysis fertilisers such as urea, diammonium phosphate and muriate of potash which are practically devoid of nutrients other than NPK. Continuous use of these fertilisers coupled with heavy off take of secondary and micronutrients by exhaustive production systems has led to emergence of multinutrient deficiencies. Now, deficiency of secondary nutrient like S and micronutrient like Zn are reported from large areas growing oilseeds besides deficiencies of NPK (Aulakh and Pasricha,1997; Hegde, 2000).

Response of oilseeds to fertilisers: For leguminous oilseeds like groundnut and soybean, the main nutrients from fertiliser management point of view are P, S, Ca and Zn as these crops can meet a large part of their nitrogen needs through biological N fixation. For rapeseed-mustard, application of N, P, S and in increasing cases Zn and to some extent B play a key role in stepping up the yields. For sunflower and safflower, the major requirements are N, P and S in certain cases. In coarse textured soils at high yield levels and where leaching is of major concern, K application becomes critical. Iron assumes importance in alkaline calcareous soils and Mo in very acid soils, particularly for groundnut and soybean. Boron is important for sunflower also.

All oilseed crops have shown response to fertiliser nutrients. Significant increases in yields due to application of one or more nutrients in different agro-ecoregions have been reported by several workers (Tomer *et al.*, 1980; Daulay and Singh, 1982; Ankineedu *et al.*, 1983; Pasricha and Aulakh, 1986; Pasricha *et al.*, 1987; Agasimani and Hosmani, 1989; Saini *et al.*, 1989; Takkar and Nayyar, 1989; Tandon, 1990; Pasricha and Tandon, 1993; Gangasaran and Rana, 1994; Subba Rao *et al.*, 1995; Jain *et al.*, 1998; Malligawad *et al.*, 2000; Hegde and Sudhakarababu, 2001). There has also been very good response to sulphur and micronutrients (Table 8 & 9). The data from frontline demonstrations conducted during 1988-89 to 1997-98 in various crop growing situations on farmers' fields indicated a mean response of 9 to 48% under rainfed conditions and 14 to 45% under irrigated conditions in different oilseed crops (Table 10). Various reports on response of oilseeds to fertiliser clearly indicate that in several parts of the country, there are nutrient deficiencies and/or response (hidden hungers) of N, P, K, S and the micronutrients like Zn, Mn, Fe and B (in a limited way in a few pockets) in one or the other oilseed crops (Singh, 1999; 2001). Thus, sustainability considerations demand that these nutritional disorders should be taken care of in the management of oilseed crops.

Table 8 : Increase in oilseed yield due to S application at farmers' fields

Crop	State	No. of field trials	Yield in control (No S) (kg ha^{-1})	Response (kg ha^{-1}) at		Response at 40 kg S ha^{-1}	
				20 kg S ha^{-1}	40 kg S ha^{-1}	%	VCR
Mustard	Bihar	7	1400	283	414	29.6	15:1
	Gujarat	3	1380	221	-	-	-
	M.P.	3	1616	316	553	34.2	16:1
	Total	13	1445	273	450	31.1	15:1
Raya	Haryana	7	1607	352	591	36.8	30:1
	Punjab	10	942	331	472	50.1	24:1
	Total	17	1216	340	521	42.8	26:1
Gobhi sarson	Punjab	5	888	448	631	71.1	32:1
Linseed	M.P.	3	1014	302	316	31.2	27:1
Soybean	M.P.	4	1972	492	565	28.9	18:1
Groundnut	Gujarat	5	1760	154	204	11.6	7:1
	A.P.	8	1392	187	401	28.8	14:1
	Total	13	1533	174	335	21.9	12:1
Sunflower	Punjab	1	1800	500	630	35.0	23:1
	Total	56	1366	314	467	34.2	20:1

VCR = Value cost ratio

Table 9 : Response of oilseeds to micronutrients in India

Micronutrient	Crop	No. of experiments	Response (kg ha^{-1})	
			Range	Average
Boron	Groundnut	11	50-40	210
	Linseed	2	110-140	120
	Sunflower	1		520
	Mustard	2	210-310	260
	Sesame	1		670
Iron	Groundnut	10	5-700	390
	Sunflower	3	460-80	550
	Soybean	3	210-1000	340
Manganese	Groundnut	1		110
	Sunflower	1		550
	Sesame	1		430
	Soybean	2	30-1030	310
Zinc	Groundnut	83	210-470	320
	Soybean	12	160-300	360
	Mustard	11	140-260	270
	Linseed	5	150-200	160
	Sunflower	8	150-200	240
	Sesame	6	80-150	110

Source: Singh (2001)

Table 10 : Productivity improvement through fertilisation in different oilseed crops

Crop	Productivity improvement (%)			
	Rainfed		Irrigated	
	Mean	Range	Mean	Range
Groundnut	9	8-10	14	11-15
Rapeseed-mustard	-	-	37	11-84
Sunflower	34	32-48	27	24-74
Sesame	42	28-65	-	-
Safflower	43	19-138	40	25-42
Castor	41	23-59	18	7-69
Linseed	38	24-54	45	9-71
Niger	48	23-80	-	-

Fertiliser management on a system basis: Cropping system rather than an individual crop and the farming system rather than an individual field are the focus of attention in INSAM. The low level of utilization of nutrients supplied through fertilisers and manures calls for choosing appropriate combination of sequence of crops to effectively utilize the nutrients for long-term sustainability (Sharma *et al.*, 1999). Specific attention needs to be given to harness the residual effect of fertilisers containing P (Table 11) and S (Table 12 & 13) so that, application of these nutrients can be phased to get maximum benefit. In wheat-groundnut, maize-groundnut and *raya*-groundnut cropping systems, groundnut crop was observed to thrive on residual effect when wheat, maize and *raya* received recommended fertilisers (Subba Rao, 1994). It is also possible to get a changing pattern of response to applied fertiliser nutrients even with balanced fertilization due to build up/ depletion of several nutrients which have a direct bearing on crop response and productivity of oilseeds. Therefore, there is a strong need to continuously monitor the soil for possible changes in nutrient status so that those nutrients which are in short supply can be supplemented to increase overall nutrient use efficiency.

Table 11 : Residual effect of P applied to first crop of groundnut on pod yield and oil content of third crop of groundnut in groundnut-maize-groundnut sequence

P_2O_5 (kg/ha^{-1})	Pod yield (kg ha^{-1})	Oil content (%)
0	15.6	42.8
20	17.7	44.3
40	18.9	45.9
60	21.0	45.8
CD (P = 0.05)	1.0	0.9

Table 12 : Direct and residual effect of S in rice-sunflower sequence (Mean data of 2000-01 and 2001-02)

Treatment	Rice yield (kg ha^{-1})	S use efficiency (kg kg S^{-1})	Sunflower seed yield (kg ha^{-1})	S use efficiency (kg kg S^{-1})	Sunflower oil yield (kg ha^{-1})	Value : Cost ratio		
						Rice	Sunflower	Total
NPK	6043	-	1421		568			
$NPKS_{15}$	6907	57.6	2124	46.9	862	58.7	102.8	161.5
$NPKS_{30}$	7338	43.2	2245	27.5	910	42.4	60.7	103.1
$NPKS_{45}$	7705	36.9	2255	18.5	958	36.6	41.0	77.6
CD(P=0.05)	790		480		218	-	-	-

Table 13 : Direct and residual effect of S in sunflower-groundnut sequence (Mean data of 2000-01 and 2001-02)

Treatment	Sunflower yield (kg ha^{-1})	S use efficiency (kg kg S^{-1})	Sunflower oil yield (kg ha^{-1})	Ground-nut yield (kg ha^{-1})	S use efficiency (kg kg S^{-1})	Ground-nut oil yield (kg ha^{-1})	Value : Cost ratio		
							Sunflower	Groundnut	Total
NPK	976		325	2831		1076			
$NPKS_{15}$	1085	7.3	375	3284	30.2	1305	16.6	65.6	82.2
$NPKS_{30}$	1204	7.6	416	3354	17.4	1307	16.7	37.8	54.5
$NPKS_{45}$	1402	9.5	501	3572	16.5	1407	20.6	35.7	56.3
CD(P=0.05)	145	-	-	452	-	-	-	-	-

Organic manures

Organic manures have been time tested materials for improving the fertility and productivity of the soils. It is only during the last 25 years that these have been incorporated into INSAM for intensive cropping sequences in contrast to the subsistence level of production of the past. Use of organic manures has been continuously declining in Indian agriculture due to a variety of reasons. We have ignored the exploitation of the potential of organic manures and their synergistic effect with chemical fertilisers for increasing production, sustainability of agriculture and improving soil health and environmental security. The potential nutrient availability from major organic sources is estimated at 10.5 to 16.2 million tonnes, out of which only 3.9 to 5.7 million tonnes is available for agriculture use (NAAS, 1996). Vast potential lies in the country for production and use of FYM and composts. The potential of rural and urban compost in India is estimated to be 600 and 16 million tonnes respectively, out of which slightly less than 50% is actually produced. Thus, the average consumption has been about 2 tonnes ha^{-1} year^{-1}. Even this much is not available for oilseeds as organic manures are mostly diverted to high value crops like fruits, vegetables, potato, sugarcane and plantation crops.

Despite limited data, the role of organic manures in oilseeds production is being increasingly recognized. In groundnut, application of FYM @ 7.5 t ha^{-1} increased the pod yield by 60% over control and 27% over 25:50:25 NPK, kg/ha (Agasimani and Hosmani, 1989). Phosphorus enriched manure had significant influence on the yield of groundnut (Das *et al.*, 1992). Poultry manure appeared to increase the pod yields of groundnut much more than that done by FYM and pig manure when supplied along with SSP or SSP + RP (Das *et al.*, 1992). Sagare *et al.* (1992) reported that highest pod yield of

1.65 t ha^{-1} and oil yield of 0.81 t ha^{-1} was obtained from application of 25 kg N + 50 kg P_2O_5 ha^{-1} as enriched FYM. With enriched FYM, NPK uptake also increased. Kathmale *et al.* (2000) reported the possibility of substituting 50% recommended dose of fertilisers to groundnut by application of 5 t FYM ha^{-1}. Beneficial effect of organic manures like FYM and poultry manure in groundnut was also reported by several workers in recent years (Mehta *et al.*, 1995; Balasubramanian, 1997; Talashikar *et al.*, 1997; Malavia *et al.*, 1998; Pattar *et al.*, 1999; Mishra, 2000; Kumaran *et al.*, 2001). Sukhija *et al.* (1985) found beneficial effect of application of 50 kg N + FYM 5 t ha^{-1} on seed yield and quality of mustard. Significant response to FYM application in mustard was also reported by Sardana and Sidhu (1994); Dhurandher *et al.* (1999); Mahajan *et al.* (1999) and Patel and Shelke (2000). In sunflower, application of FYM 10 t ha^{-1} increased the seed yield by 28.7 and 28.4% at Bangalore and Coimbatore, respectively. Application of compost (Khatik and Dikshit, 2001), vermicompost (Devidayal and Agarwal, 1998) besides FYM (Singh and Singh, 1997; Shakthawat and Bansal, 1999; Shanwad and Agasimani,2001) had significant effect on productivity of sunflower. Application of FYM could also increase the productivity of soybean besides increasing the oil and protein content (Ramamurthy and Shivashankar, 1995; 1996; Mishra *et al.*, 1999; Paneerselvan *et al.*, 2000; Babulkar *et al.*, 2000). At Tikamgarh, Tiwari *et al.* (1995) obtained 28.7% higher yield of sesame due to application of 2.5 t FYM ha^{-1} along with recommended NPK. Singh *et al.* (1997) from Bilaspur observed significant increase in yield of sesame with poultry manure. Significant responses to use of organic manures in sesame was also reported by many other workers (Narkhede *et al.*, 1997; 2001; Duhoon *et al.*, 2001a, b). In castor, 25 to 75% of N needs could be substituted both under rainfed and irrigated conditions through use of different organic sources (DOR, 1992; Baby and Bapireddy, 1998). Even in crops like niger, significant response to FYM application was reported (DOR, 1997).

The results of All India Coordinated Research Project on Long Term Fertiliser Experiments at Jabalpur and Ranchi clearly indicated the benefits of the application of FYM (supplying 38 N, 8 P_2O_5, 30 K_2O, kg ha^{-1}) to soybean super imposed over either 100 or 150% recommended NPK over a 13 to 16/ 17 year period (Anonymous, 1992). The soybean yields with 100% recommended NPK + FYM were greater than those obtained with even 150% NPK in all the years of study.

Organic manures besides supplying nutrients to the current crop, very often leave substantial residual effect on the succeeding crops in the system and this residual effect lasts often for several seasons.

Organic manures are not just sources of nutrients, they have profound, even some times dominant effect on soil physical properties resulting in better structure, greater water retention, more favourable environment for root and pod growth and better utilization of water. The beneficial effects of such properties on crop yields has rarely been given due economic importance.

The composition of organic manures vary greatly. Current recommendations based on casual evaluation of FYM/compost in terms of t ha^{-1} with very little emphasis on the quality aspect are many times misleading. There are proven scientific ways of improving the quality of organic manures which need to be popularised amongst the farmers. The quality can be substantially enhanced by enrichment of the manure during decomposition with cheaper indigenous materials. Rock phosphate enriched compost, popularly known as phospho-compost contains more than 1% P_2O_5 as against 0.3-0.6% in usually prepared rural compost.

Besides FYM and compost, there are several other organic sources with good nutrient potential. However, there is hardly any information on their integration in oilseed production. There is a need to integrate different organic manures available in specific agro ecoregions with chemical fertilisers for cropping systems involving oilseeds.

Crop residues and recyclable wastes

The country has large potential of crop residues and other farm/industrial wastes with good manurial value (Table 14). Recent estimate of crop residues varies from 313 to 356 million tonnes with nutrient potential of 6.3 to 8.7 million tonnes (NPK), (Sarkar *et al.*, 1999). Although most of the crop residues with feed value are needed to support huge livestock population, some residues available in specific locations/situations can be made use of as a source of plant nutrients because either they are not preferred as animal feed or their transport to needy areas is not economically viable. About one-third of the residues produced are available for direct recycling on the land amounting to about 136 million tonnes which can add 3.54 million tonnes of N, P_2O_5, K_2O. (Hegde and Sudhakarababu, 2001). Crop wastes and residues are renewable and readily available but they are limited and scattered. In addition to improving the soil properties, conserving soil moisture and suppressing weeds, application of crop residues may partially substitute the chemical fertilisers. Of the nutrients taken up by cereal crops, on an average, 25% N and P, 50% S and 75% of K is retained in crop residues making them valuable nutrient sources. Gaur and Mukherjee (1979) observed 95.5% increase in pod yields of groundnut over control with the application of 5 t ha^{-1} of wheat straw, which also improved the seed oil content. On acidic soils, the incorporation of biodigested slurry @ 2 t ha^{-1} and Rhizobium treatment had augmented

Table 14 : Nutrient potential of crop residues, agricultural and industrial wastes

Residues	Quantity (million tonnes)	Nutrients year^{-1} (000't)			
		N	P_2O_5	K_2O	Total
Crop residues	273.3	1283.1	1965.6	3903.9	7152.6
Potato haulms	12.0	294.0	12.0	225.0	531.0
Sugarcane trash	30.0	56.0	15.0	99.0	170.0
Forest litter	18.7	99.7	37.4	99.7	236.8
Nonedible oil cakes	0.4	12.0	5.0	12.0	29.0
Sewage sludge	0.5	5.1	2.9	2.8	10.8
Press mud	3.2	33.3	79.4	55.4	168.1
Water hyacinth	73.0	72.0	14.0	84.0	170.0
Domestic waste water	6351.0*	317.6	139.7	190.5	647.8
Industrial waste water	66.2*	2.9	0.9	1.3	5.1

* Million cubic meters/annum Source: Juwarkar *et al.*, 1992; Hegde and Dwivedi,1993.

symbiotic efficiency and resulted in 39.2% higher pod yield of groundnut over control (Ponniah *et al.*, 1991). Application of coirpith inoculated with *Pleuroteus* sp + NPK increased the pod yield whereas uninoculated coirpith was not effective (Nagarajan *et al.*, 1986). In sunflower, incorporation of crop residues after treatment with *Trichoderma viridae* or biogas slurry could reduce the fertiliser needs of succeeding chickpea by 25% with substantial increase in productivity (DOR, 1998). Incorporation of pigeonpea stalks has resulted in 44% increase in yield of sunflower over control and further there was improvement in water holding capacity of the soil (Reddy and Sudhakarababu, 1996b). Beneficial effect of use of mentha waste on mustard (Chand *et al.*, 2001), mustard residues on rice (Vinod Kumar *et al.*, 2000), sorghum + cowpea residues on sunflower (Bharambe *et al.*, 1999) have been reported.

Crop residues, besides supplying nutrients to the current crop, leave substantial residual effect on succeeding crops in the system (Table 15). Organic N is slowly mineralised and about 30% N is generally available to the first crop. About 60-70% P and 75% of K is likely to become available to the first crop and the rest to the subsequent crops (Gaur, 1992). Besides supplying nutrients, crop residues and wastes favourably influence physical, chemical and biological properties of the soil. The amount of stubbles left in the field after the harvest of some of the cereals and pulses may range from 0.2 to 1.5 t ha^{-1}. Before planting next crop, in many cases, these stubbles are collected and burnt as they are difficult to decompose, leading to a significant loss of

Table 15 : Residual effect of rice straw applied to first crop of groundnut on pod yield and oil content of third crop of groundnut in groundnut-maize-groundnut sequence

Rice straw (t ha^{-1})	Pod yield (q ha^{-1})	Oil content (%)
0	16.2	43.4
2	17.8	44.3
4	18.6	45.2
6	20.6	46.0
CD (P=0.05)	1-0	0.9

Source: Reddy *et al.* (1990)

nutrients. There is need for appropriate management practices to make use of these stubbles in INSAM of cropping systems involving oilseeds. The inconsistent crop response to incorporation of crop residues under different situations may be expected due to varying pace of immobilization-mineralization cycle, particularly that of N. There are many recyclable wastes of agricultural and industrial origin such as sugarcane trash, potato haulms, non-edible cakes, tobacco and tea wastes, cotton wastes, press mud, forest litter, water hyacinth, wastes from food processing industries etc. However, there is practically no information on their integration with fertilisers in the oilseeds production.

Legumes in INSAM

Legumes have been known for their soil recuperation power since time immemorial. Legumes can form an important component of INSAM in oilseeds when grown as a sequence crop, intercropped with non-leguminous oilseeds or when introduced as a green manuring crop in a cropping system.

Legumes grown in sequence: Nodulating legumes benefit the succeeding non legume crops in terms of increased yields in comparison with the yield of non legumes or non-nodulating lines of legumes (Wani and Lee, 1992). Such increased yields are attributed to enrichment of soil N due to N fixation and N conserving effect of legumes. However, all such benefits cannot be explained by N effect alone, and several factors *viz.*, improved soil physical, chemical and biological properties, reduced weed and diseases severity and insect infestation etc. are also responsible. If a cereal or a non legume crop succeeds groundnut, 20-25 kg N ha^{-1} can be reduced and no phosphorous application is required if groundnut has already been supplied with P. In case of legumes being succeeding crops, only phosphorous @ 15-20 kg ha^{-1} is suggested (Reddy and Sudhakarababu, 1996b). The nitrogen requirement of mustard in different double cropping systems at Hisar revealed that the highest yield was obtained with 120 kg N ha^{-1}. However, the response to N application in mustard declined after 40 kg N ha^{-1} if the preceding crops were groundnut and pigeonpea, while after pearl millet, there was good response upto 80 kg N ha^{-1} (Kumar and Singh, 1988). A 50% reduction (15-25 kg N ha^{-1}) in N application to safflower was suggested if the preceding crop is a grain legume receiving its full complement of fertilisers (Hegde, 2002).

The long term fertiliser trial on sunflower based sequences at Hyderabad revealed that groundnut and castor crops are efficient in utilising the residual fertility from preceding sunflower. After third cycle of different sunflower based crop sequences, sunflower yields at 50% recommended fertilisers, when rotated with groundnut, were comparable with that at 150% recommended fertilisers in sunflower-sunflower sequence (Reddy and Sudhakarababu, 1996a).

Groundnut and soybean crops being legumes, account for nearly 50% of the area under oilseeds in the country. These crops in sequence with other non legumes, can help in saving fertiliser to the extent of 30 to 40 kg N ha^{-1}. In other non-leguminious oilseeds, inclusion of leguminous crop in the sequence may help to economise N fertiliser requirement to the extent of 20-40 kg N ha^{-1}. In a country where the average consumption of plant nutrients is very low and still lower in oilseeds, the residual fertility buildup due to legumes is obviously a major contribution which must be fully exploited for increasing oilseeds production.

Legumes as intercrops: A leguminous crop like cowpea can be grown as an intercrop for a six week period and incorporated *in situ* as green manure. Results of such a study from Hyderabad for castor were reported by Vijayalakshmi (1984). The data indicated that over years, 10 kg N + green manuring (5 t ha^{-1}) was as good as 40 kg N ha^{-1} in red sandy loam soils. The contribution of intercropped cowpea green manured *in situ* could be upto 30 kg N ha^{-1}. Groundnut + pigeonpea intercropping produces higher yields due to less competition for soil P because pigeonpea as a sole crop or in intercropping system with mung bean and maize prefers A1-P while with groundnut, prefers Ca-P (Subba Rao, 1994). There are also beneficial effect of intercropping leguminous oilseeds like groundnut and soybean in cereal crops (Hegde and Pandey, 1992).

Nitrogen economy through intercropping legumes is yet to be assessed in different cropping systems. Although many legumes have been identified for intercropping in oilseed crops, no information is available on the N economy achieved through these systems.

Legumes as green manures: Before introduction of chemical fertilisers, green manuring with leguminous crops was invariably practiced with most crops.

However, with the easy availability of chemical fertilisers and crop intensification, the practice of green manuring was almost given up. Of late, there has been revival of interest in green manuring with an indication of unsustainable yield levels due to continuous use of only chemical fertilisers. At present, green manuring is practised on about 6-7 million hectares. Sunhemp and sesbania are the most common green manure crops in India, while cluster bean, Egyptian clover or berseem and cowpea are also used occasionally. Large amount of green biomass of narrow C:N ratio when incorporated into the soil during green manuring definitely contributes sizeable amounts of plant nutrients, particularly N and brings improvement in soil physical conditions.

Green manuring in *toria* have shown very promising results (Pasricha *et al.*, 1988). Addition of green manure (5 t ha^{-1}) resulted in *toria* yield (10.48 q ha^{-1}) equal to that obtained with 60 kg N ha^{-1} (10.49 q ha^{-1}). Green manuring with cowpea along with inorganic N led to an additional yield of *toria* ranging from 5.6 to 8.3 q ha^{-1} (Table 16) in *toria*-wheat sequence. Beneficial effect of green manuring in *raya* was also reported (Pasricha *et al.*, 1987). At Ludhiana, green manuring gave similar yield of mustard as that of N alone @ 100 kg ha^{-1}. In maize-mustard system, average direct effect of green manuring on maize was a yield increase of 763 kg ha^{-1}, equivalent to 42 kg N ha^{-1} through chemical fertilisers and a further increase of 145 kg ha^{-1} in mustard yield through its residual effect (Pasricha *et al.*, 1987). In rice-mustard sequence, application of 3 t/ha green manure to rice increased the yield by 12.66q ha^{-1} and increased the yield of mustard by 2.35 q ha^{-1} due to residual effect (Dhurandher *et al.*, 1999). Incorporation of sunhemp or *Leucaena* loppings resulted in yield of safflower similar to that obtained from 37.5:25 N, P_2O_5, kg ha^{-1} and increased the oil and protein content in seeds. Beneficial effect of green leaf manure on groundnut and castor yields was reported by Bheemaiah *et al.* (1998a, b).

While green manuring practice has both direct and residual benefit on oilseeds, the point for consideration is whether the farmers would be willing to forego a crop merely for growing a green manure crop. Many a times, farmers are not willing to spare their meagre land resources and inputs for raising a green manure crop as it neither provides food nor immediate income. Wherever intensive cultivation and growing more than one or two crops is aimed at, green manuring will have no scope in cropping systems. Instead, the oilseeds are better intercropped or be intercropped with other legumes to achieve the same objective of green manuring at least partially. Fresh loppings of perennial leguminous trees such as *Leucaena* grown on hedge rows and field bunds may be used for incorporation into the soil as a source of N for non leguminous oilseeds.

Table 16 : Effect of green manuring on seed yield of *toria* (Mean data of 3 years)

N applied (kg ha^{-1})	Seed yield of *toria* (q ha^{-1})		
	Without green manuring	With green manuring	Increase (q ha^{-1})
0	2.17	10.48	8.31
20	6.10	13.64	7.54
40	8.53	15.62	7.09
60	10.49	16.10	5.61

Source: Pasricha *et al.* (1988)

Biofertilisers in INSAM

Biofertilisers have an important role to play in improving the nutrient supply and nutrient availability in crop production. They help in increasing the biologically fixed atmospheric N and enhancing native P availability to crops. Among the biofertilisers for increasing N supply to oilseed crops, N fixing bacteria *viz., Rhizobium, Azotobacter* and *Azospirillum* are important. The availability of P is improved by P solubilising microorganisms and mycorrhizae.

Rhizobium: There is scope for exploiting the ability of leguminous oilseeds-groundnut (112 to 152 kg N ha^{-1}) and soybean (49 to 130 kg N ha^{-1}) to fix atmospheric N to meet in part or complete N requirement (Wani and Lee, 1992). But these crops too need starter doses of N (15-20 kg ha^{-1}) through fertilisers to be effective as N fixer. Inoculation with efficient *Rhizobium* strain specific to each crop is very essential for the N gains and better crop yields. In groundnut, new *Rhizobium* strains IGR-6 and IGR-40 were found to be tolerant to Thiram and hence seed treatment with fungicide and inoculation of *Rhizobium* can go together (Reddy and Sudhakarababu, 1996b). *Rhizobium* inoculations have proved beneficial where these crops are cultivated for the first time and every time in rice-fallow situations. However, for the *Rhizobium* to be effective, the soils should have no deficiency of P, Cu and Mo.

Response of groundnut to *Rhizobium* inoculation was 272 kg ha^{-1} averaged over 24 trials (ICAR, 1987). Results from Junagadh (Gujarat) indicated 15 to 30% increase in summer groundnut yields with *Rhizobium* inoculation (NRCG, 1988). AT Akola (Maharashtra), seed inoculation with *Rhizobium* + 4 g Mo kg^{-1} seed increased groundnut yield significantly (Wankhade *et al.*, 1992). In *kharif* groundnut, 13 to 53% increase in yields were reported with rhizobial inoculations (Anonymous, 1993). Significant responses to *Rhizobium* use in groundnut were also reported by many other workers (Patel and Thakur, 1997, 1998; Detroja *et al.*, 1997). There are also reports of no significant response to *Rhizobium* inoculation in groundnut (Umamaheshwari *et al.*, 2001). In soybean, *Rhizobium* inoculation along with 30 kg N ha^{-1} produced the highest yield and nitrogen dose less or greater than 60 kg ha^{-1} caused significant reduction in yield (Tiwari *et al.*, 1993). Significant response to rhizobial inoculation in soybean was reported by many workers (Bisht and Chandel, 1991; Dahatonde and Shava, 1992; Vara *et al.*, 1994; Dubey, 1997 and 1998; Pannase *et al.*, 2001). Pelleting of *Rhizobium* inoculated seeds with $CaCO_3$ significantly increased the yield of soybean (Prabhakaran and Sivasubramanian, 1991).

Response to *Rhizobium* inoculation under field conditions have been quite variable and inconsistent depending on soil conditions, quality of inoculum and effectiveness of native population. Use of *Rhizobium* cultures in groundnut and soybean has not spread to the extent desired as there is no quality control in commercially produced inoculants and farmers are not fully convinced of its usefulness.

Azotobacter and *Azospirillum*: Among other important groups of bacteria, *Azotobacter* is free-living N fixer while *Azospirillum* fixes N in loose association with plant roots. Not many reports are available on the usefulness of these microorganisms in augmenting N supplies to oilseeds. In mustard, seed inoculation with *Azotobacter* and *Azospirillum* increased the seed yield equivalent to that obtained with 25 to 30 kg ha^{-1} of fertiliser N and *Azospirillum* was slightly more efficient than *Azotobacter* (Chauhan *et al.*, 1995a, b). In groundnut, significant increase in pod yield was reported due to seed inoculation with *Azospirillum* (Balasubramanian and Palaniappan, 1994) and combined inoculation of *Rhizobium* and *Azospirillum* had no additional advantage. *Azospirillum* seed treatment in sesame and *Azotobacter* in *toria*, niger and sunflower might reduce the N requirement of these crops by almost 50% (Sawarkar, 1997; Duhoon *et al.*, 2001b, DOR, 1997; 2002). In safflower, it was possible to substitute 50% N needs amounting to 20 kg N ha^{-1} by seed treatment with *Azotobacter* or *Azospirillum* (DOR, 2001) under rainfed conditions (Table 17).

Inconsistency in crop response to *Azotobacter* and *Azospirillum* has been more common than *Rhizobium* and the response is dependent on crops and their cultivars, location, season, crop management practices, bacterial strains, soil fertility and native microbial population.

Table 17 : Response of safflower to biofertilisers

Treatment	Seed yield (kg ha^{-1})		
	Solapur	Annigeri	Tandur
Control (No N)	798	1236	551
50% N	929	1606	630
100% N	1170	1765	790
Azotobacter (Seed treatment)	910	1533	597
Azospirillum (ST)	898	1685	650
Azotobacter + Azospirillum (ST)	996	1535	638
50% N + Azotobacter (ST)	1045	1744	654
50% N + Azospirillum (ST)	1065	1714	815
50% N + Azotobacter + Azospirillum (ST)	1075	1785	877
CD (P = 0.05)	78	77	85

Phosphate solubilizers: Most of the soil phosphorous is in unavailable form while the P requirement of oilseed crops is high. In recent years, several strains of P solubilising bacteria and fungi have been isolated. In groundnut, significant increases in pod yield due to seed inoculation with P solubilizers like *Pseudomonas striata* and *Paecilomyces fussispores* have been reported (Mehta *et al.*, 1995). Combined application of PSB inoculation with FYM was better than PSB alone in irrigated groundnut (Balasubramanian and Palaniappan, 1994). Effect of PSB in groundnut was further enhanced when used with FYM (Mehta *et al.*, 1995; Detroja *et al.*, 1997). In soybean, combined use of *Rhizobium* along with PSB had greater effect on productivity than their individual use (Dubey, 1997, 1999; 2001). There are also reports of response to PSB in mustard (Baldevram and Pareek, 1999; 2000), sunflower (Gadag and Alagawadi, 1998; DOR, 2002), sesame (DOR, 1999) and safflower (DOR, 2001).

Vescicular arbuscular mycorrhizae (VAM), a kind of symbiosis between plant roots and certain fungi, substantially enhances P availability and its uptake by the crops and found beneficial particularly on P deficient soils. VAM inoculations increased the yield of groundnut by 14.5% in *kharif* and 27.8% in *rabi* season in sandy loam soils at Tirupati (Anonymous, 1993). In another study, 26.4% increase in soybean and 20.9% increase in groundnut yields were reported with inoculation of VAM, *G. fasciculatum* (Wani and Lee, 1992). However, this increase had decreased to 18.5 and 10.1% respectively when inoculated in combination with *Rhizobium*. Because of problems in isolation and multiplication of pure strains of VAM fungi, it has not been possible to use this potential biofertiliser on a large scale in oilseeds.

Wider acceptability of biofertilisers among the oilseeds farmers is constrained because of inconsistent responses and problems associated with the availability, transport, storage and handling. The production and distribution of different biofertilisers in the country can hardly meet the requirement of less than 10% of the cropped area. The success rate in case of any biofertilisers in terms of significant increase in yield ranges from 30 to 65%. In order to improve the consistency in response, greater research thrust is needed in isolation and multiplication of strains which can compete with indigenous ones and adapt to a wide range of soil and climatic conditions. There is also need to bring biofertilisers under quality control to ensure that inoculums farmers use conform to certain minimum standards.

Conclusions and future research priorities

Integrated nutrient management holds great promise in meeting the growing nutrient demands of oilseed crops in intensive agriculture. It can also help in maintaining production sustainability without deterioration in the quality of plant's environment. Despite many constraints like use of cattle dung as fuel, increasing competitive value of crop residues as animal feed, extra cost and time required for raising green manure crop, poor and inconsistent responses to biofertilisers etc., efforts must be made to develop agroecoregion specific practical recommendations for different oilseeds and oilseeds based cropping systems utilising the already generated data on different components of INSAM. An example of possible INSAM package for meeting the nutrient needs of irrigated sunflower during *Rabi* season in peninsular India is given in Table 18.

Table 18 : Some alternate strategies for meeting nutrient needs through INSAM for *Rabi* sunflower in peninsular India

Option	Nutrients (kg ha^{-1})					Fertilisers	Punjab legume	FYM (4 t ha^{-1})
	N	:	P_2O_5	:	K_2O			
1	60	:	90	:	30	60 : 90 : 30	-	-
2	60	:	90	:	30	40 : 40 : 30	20 : 0 : 0	-
3	60	:	90	:	30	40 : 78 : 10	-	20 : 12 : 20
4	60	:	90	:	30	20 : 78 : 10	20 : 0 : 0	20 : 12 : 20

Nutrient application rates with most oilseed farmers are well below the optimum and various organic sources and biofertilisers can help to take these towards the optimum level and to that extent supplement fertilisers (Tandon, 1992). These sources will be able to substitute part of the fertiliser particularly N, where farmers have reached the optimum level of nutrient application. For most farming conditions, packages of INSAM should be assembled within the overall limits of recommended nutrient application rates and not in a haphazard manner. Blue prints for INSAM need to be developed in a realistic manner and little purpose will be served by loading all available components into a package. The assembling of package must be more in the form of alternative options. The package must integrate the comparative advantages of diverse sources of nutrients for more efficient and cost effective nutrient management.

While possibilities for INSAM are real and attractive, most nutrient packages for high yields of oilseeds required to meet the needs of expanding population from non-expanding area, will continue to be fertiliser driven. This assessment in no way underestimates the importance of other nutrient sources, most of which are gainfully integrable. Many research gaps need to be filled and the following are some of the important future research priorities. There is a need to prepare an inventory of promising organic and biological sources of nutrients available under different agroecoregions involving diverse farming situations. Efforts must be made to assemble agroecoregion specific INSAM packages for various oilseed crops utilising the available information. To fill the information gaps, wherever needed, planning new experiments on well defined bench mark sites is required. There is a need to develop methodologies for quantifying the non-nutrient benefits of using organic sources of nutrients which are important for production sustainability. Different industrial wastes with potential manurial value need to be characterised. More field data on *in situ* decomposition of crop residues including stubbles in relation to soil properties and crop productivity are needed on short, medium and long term basis. Research on INSAM in relation to soil moisture management deserves greater attention and priority for sustainable oilseeds productivity.

Research on residual effect of various organic sources under different environments and management practices needs to be initiated. Biofertiliser technology needs to be refined to make it more acceptable to farmers. Microbial strains which can compete with indigenous ones and work efficiently over a wide range of soil-climatic conditions need to be isolated and multiplied. Large scale on-farm testing of different components of INSAM locally available is necessary to convince the farmers of the effectiveness of INSAM in oilseeds. Mass awareness on conservation of organic source and their recycling needs to be improved and promotional literature should be prepared in local language and made available to the farmers and extension personnel in addition to some adaptive research, demonstrations and specific training programmes.

References

Agasimani, C.A. and Hosmani, M.M. (1989). Response of groundnut crop to farm yard manure, nitrogen and phosphorus in rice fallows in coastal sandy soils. *Journal of Oilseeds Research*, **6** (2): 360-363.

Ankineedu, G., Rao, J.V. and Reddy, B.N. (1983). Advances in fertiliser management for rainfed oilseeds. *Fertiliser News*, **28** (9): 76-90 & 105.

Anonymous, (1992). Annual Report 1987-88/1988-89, AICRP on Long Term Fertiliser Experiment, ICAR, New Delhi.

Anonymous, (1993). Annual Report, Krishna-Godavari Zone, APAU, Hyderabad.

Aulakh, M.S. and Pasricha, N.S. (1997). Role of balanced fertilization in oilseed based cropping systems. *Fertiliser News*, **42** (4): 101-111.

Babulkar, P.S., Wandile, R.M., Badole, W.P. and Balpande, S.S. (2000). Residual effect of long term application of FYM and fertilisers on soil properties (vertisols) and yield of soybean. *Journal of the Indian Society of Soil Science*, **48** (1): 89-92.

Baby, A. and Bapireddy, T. (1998). Integrated nutrient management in rainfed castor. *Journal of Oilseeds Research*, **15**(1): 115-117.

Balasubramanyan, P. (1997). Integrated nutrient management in irrigated groundnut (*A. hypogaea*). *Indian Journal of Agronomy*, **42**(4): 683-687.

Balasubramanian, P. and Palaniappan, S.P. (1994). Effect of combined application of bacterial inoculation with farm yard manure in irrigated groundnut (*Arachis hypogaea*). *Indian Journal of Agronomy*, **39**(1):131-132.

Baldevram and Pareek, R.G. (1999). Effect of phosphorus, sulphur and PSB on growth and yield of mustard. *Agricultural Science Digest*, **19**:203-206.

Baldevram and Pareek, R.G. (2000). Effect of phosphorus, sulphur and phosphate solubilising bacteria on yield, oil content and nutrient uptake by mustard (*Brassica juncea*(L.) Czern and Coss. *Agricultural Science Digest*, **20**(4): 241-243.

Bharambe, P.R., Patil, M.A., Oza, S.R. and Shelke, D.K. (1999). Effect of residue incorporation and irrigation on sunflower yield and soil productivity under sorghum-sunflower cropping system. *Journal of Indian Society of Soil Science*, **47**(1): 169-171.

Bheemaiah, G., Madhusudhan, T., Subramanyam, M.V.R. and Ismail, S. (1998a). Integrated nutrient management in rainfed castor alley cropped with subabul. *Journal of Oilseeds Research*, **15**(1): 130-134.

Bheemaiah, G., Subramanyam, M.V.R., Ismail, S., Sridevi, S. and Radhika, K. (1998b). Effect of green leaf manures and fertilisers application on growth and yield of summer groundnut under different cropping systems. *Indian Journal of Dryland Agricultural Research and Development*, **13**(1): 50-52.

Bisht, J.K. and Chandel, A.S. (1991). Effect of integrated nutrient management on leaf area index, photosynthetic rate and agronomic and physiological efficiencies of soybean (*Glycine max*). *Indian Journal of Agronomy*, **36**: 129-132.

Chand, S., Chattopadhyay, A., Anwar, M. and Patra, D.D. (2001). Mentha distillation waste - A potential source of sustaining soil fertility and crop productivity in mentha-mustard cropping system. *Fertiliser News*, **46**(3); 71-75.

Chauhan, D.R., Paroda, S., Kataria, O.P. and Singh, K.P. (1995a). Response of Indian mustard (*Brassica juncea*) to biofertilisers and nitrogen. *Indian Journal of Agronomy*, **40**(1): 86-90.

Chauhan, D.R., Paroda, S., and Singh, D.P. (1995b). Effect of biofertilisers, gypsum and nitrogen on growth and yield of raya (*Brassica juncea*). *Indian Journal of Agronomy*, **40**(4): 639-642.

Dahatonde, B.B. and Shava, S.V. (1992). Response of soybean *(Glycine* max) to nitrogen and *Rhizobium* inoculation. *Indian Journal of Agronomy*, **37**(2): 370-371.

Das, M., Singh, B.P., Ram, M. and Prasad, R.N. (1992). Mineral nutrition of maize and groundnut as influenced by P enriched manures on acid alfisol. *Journal of Indian Society of Soil Science*, **40**:580-583.

Dauley, H.S. and Singh, K.C. (1982). Effects of N, P rates and plant densities on the yield of rainfed sesame. *Indian Journal of Agricultural Sciences*, **52**(3): 166-169.

Detroja, K.S., Malavia, D.D., Kaneria, B.B., Khanapara, V.B. and Patel, R.K. (1997). Response of summer groundnut (*A. hypogaea*) to phosphorus, biofertiliser and seed rate. *Indian Journal of Agronomy*, **42**(1):165-168.

Devidayal and Agarwal, S.K. (1998). Response of sunflower (*H.annuus*) to organic manures and fertilisers. *Indian Journal of Agronomy*, **43**(3): 469-473.

Dhurandher, R.L., Pandey, N. and Tripathi, R.S. (1999). Integrated nutrient management in rice-mustard crop sequence under irrigated conditions. *Journal of Oilseeds Research*, **16**(1): 130-131.

DOR, (1992). AICORPO (Castor), Annual Report, Directorate of Oilseeds Research, Rajendranagar, Hyderabad.

DOR, (1997). AICORPO (Sesame and Niger), Annual Report, Directorate of Oilseeds Research, Rajendranagar, Hyderabad.

DOR, (1998). AICORPO (Sunflower), Annual Report, Directorate of Oilseeds Research, Rajendranagar, Hyderabad.

DOR, (1999). AICORPO (Sesame and Niger), Annual Report, Directorate of Oilseeds Research, Rajendranagar, Hyderabad.

DOR, (2001). AICRP (Safflower), Annual Report, Directorate of Oilseeds Research, Rajendranagar, Hyderabad.

DOR, (2002). AICRP (Safflower), Annual Report, Directorate of Oilseeds Research, Rajendranagar, Hyderabad.

Dubey, S.K. (1998). Response of soybean (*Glycine max*) to biofertilisers with and without nitrogen, phosphorus and potassium on swell-shrink soil. *Indian Journal of Agronomy*, **43** (3): 546-544.

Dubey, S.K. (1999). Soil available nutrient balance as influenced by phosphate solubilising agents and phosphate application in soybean on vertisols in dryland. *Annals of Agricultural Research*, **20**(3): 374-376.

Dubey, S.K. (2001). Balance sheet of N and P as influenced by phosphate solubilising bacteria and phosphate fertilization in rainfed soybean grown on vertisols. *Annals of Agricultural Research*, **22**(1): 162-164.

Duhoon, S.S., Jain, H.C., Deshmukh, M.R. and Goswami, U. (2001a). Integrated nutrient management in *kharif* sesame (*Sesamum indicum* L.). *Journal of Oilseeds Research*, **18**(1): 81-84.

Duhoon, S.S., Jain, H.C., Deshmukh, M.R. and Goswami, U. (2001b). Effect of organic and inorganic fertilisers along with biofertilisers on *kharif* sesame (*S. indicum* L.) under different soils and locations in India. *Journal of Oilseeds Research*, **18**(2): 178-180.

Finck, A. (1995). Integrated plant nutrition system. *FAO Fertiliser and Plant nutrition Bulletin*, **12** : 67-82.

Gadag, R. and Alagawadi, A.R. (1998). Response of sunflower to the P solubilising biofertilisers with different sources and levels of phosphorus. *Karnataka Journal of Agricultural Sciences*, **11**(1): 50-55.

Gangasaran and Rana, D.S. (1994). Integrated nutrient management in oilseed production. pp. 124-141. In Proc. of Symposium on 'Integrated Input Management for Efficient Crop Production', TNAU, Coimbatore, 22-25 Feb.1994. Indian Society of Agronomy, New Delhi.

Gaur, A.C. (1992). Bulky organic manures and crop residues. *In* H.L.S. Tandon (Ed). *Fertilisers, Organic Manures, Recyclable Wastes and Biofertilisers*. Fertiliser Development and Consultation Organisation, New Delhi pp.36-51.

Gaur, A.C. and Mukherjee, P. (1979). Note on the effect of straw on the yield of groundnut and the following wheat. *Indian Journal of Agricultural Sciences*, **49**:984-987.

Hegde, D.M. (1992). Cropping Systems Research Highlights: Coordinator's Report 20th Workshop of Project Directorate for Cropping Systems Research, 1-4 June,1992. TNAU, Coimbatore, pp.39.

Hegde, D.M. (1998). Integrated nutrient management for production sustainability of oilseeds - A review. *Journal of Oilseeds Research*, **15** (1): 1-17.

Hegde, D.M. (2000). Nutrient management in oilseed crops. *Fertiliser News*, **45** (4): 31-42.

Hegde, D.M. (2002). Safflower - Package of practices for increasing production. Directorate of Oilseeds Research, Rajendranagar, Hyderabad. P.26.

Hegde, D.M. and Dwivedi, B.S. (1993). Integrated nutrient supply and management as a strategy to meet nutrient demand. *Fertiliser News*, **38**(12): 49-59.

Hegde, D.M. and Pandey, R.K. (1992). Profitable intercropping systems for different regions. *Indian Farming*, **41**(12): 16-23.

Hegde, D.M. and Sudhakarababu, S.N. (2001). Nutrient management strategies in agriculture: A future outlook. *Fertiliser News*, **46**(12): 61-66 & 71-72.

ICAR, (1987). Biological N fixation in groundnut. *Technologies for Better Crops*. No.37, ICAR, New Delhi.

Jain, K.K., Sharma, R.S. and Sharma, R.K. (1998). Productivity and economics of *rabi* oilseeds under varying inputs. *Journal of Oilseeds Research*, **15**(1): 93-97.

Juwarkar, A.S., Shende, A., Thawale, P.R., Satyanarayana, S., Deshbratar, P.B., Bal, A.S. and Juwarkar, A. (1992). Biological and industrial wastes as sources of plant nutrients. *In* H.L.S. Tandon (Ed). *Fertilisers, Organic Manures, Recyclable Wastes and Biofertilisers*, Fertiliser Development and Consultation Organisation, New Delhi pp.72-90.

Kathmale, D.K., Khadtare, S.V., Kamble, M.S. and Patil, R.C. (2000). Integrated nutrient management in groundnut(*Arachis hypogaea*)-wheat (*Triticum aestivum*) cropping system in vertisols of Western Maharashtra plains zone. *Indian Journal of Agronomy*, **45**(2): 248-252.

Katyal, J.C. and Rattan, R.K. (1993). Distribution of zinc in Indian Soils. *Fertiliser News*, **38**(3): 15-26.

Katyal, J.C. and Reddy, K.C.K. (1997). Plant nutrient supply needs: Rainfed crops. In J.S. Kanwar and J.C. Katyal (Eds). Plant Nutrient Needs, Supply, Efficiency and Policy Issues: 2000-2025. National Academy of Agricultural Sciences, New Delhi pp. 91-113.

Khatik, S.K. and Dikshit, P.R. (2001). Integrated use of organic manures and inorganic fertilisers on yield, quality, economics and nutrition of sunflower grown in Haplustert clay soil. *Agricultural Science Digest*, **21**(2): 87-90.

Kumar, A. and Singh, Y. (1988). Response of oilseeds to P in Northern India. Paper presented at FAI (NR) / PPCL seminar on 'Importance of P fertiliser with special reference to oilseeds and pulses in Northern India'. Lucknow, 21-22 June,1988.

Kumaran, S., Solaimalai, A., Arulmurgan, K. and Ravisankar, N. (2001). Response of groundnut crop to organic and inorganic fertilisers under irrigated conditions. *Journal of Oilseeds Research*, **18**(1): 123-125.

Mahajan, G., Negi, S.C. and Sardana,V. (1999). Nutrient uptake by wheat + swede rape intercropping system as influenced by sowing methods, FYM and NPK levels. *Annals of Agricultural Research*, **20**(3): 377-379.

Malavia, D.D., Dudhalia, M.G., Vyas, M.N. and Mathukia, R.K. (1998). Direct and residual effect of farm yard manure, irrigation and fertiliser on rainy season groundnut (*Arachis hypogaea*) - summer pearl millet (*Pennisetum glaucum*) cropping sequence. *Indian Journal of Agricultural Sciences*, **68**(2): 117-118.

Malligawad, L.H., Patil, R.K., Kannur, V. and Giriraj, K. (2000). Effect of fertility management practices in groundnut. *Karnataka Journal of Agricultural Sciences*, **13**(2): 299-305.

Mehta, A.C., Malavia, D.D., Kaneria, B.B. and Khanpara, V.B. (1995). Effect of phosphatic biofertilisers in conjunction with organic and inorganic fertilisers on growth and yield of groundnut (*Arachis hypogaea*). *Indian Journal of Agronomy*, **40**(4): 709-710.

Mishra, C.M. (2000). Effect of farm yard manure and chemical fertilisers on the yield and economics of groundnut (*Arachis hypogaea*) under rainfed condition. *Madras Agricultural Journal*, **87**(7-9): 517-518.

Mishra, S.N., Paikaray, R.K. and Mishra, K.N. (1999). Effect of lime, organic and inorganic nutrients on wheat (*Tritium aestivum*)- soybean (*Glycine max*) cropping system in acid red soils. *Indian Journal of Agronomy*, **44** (1): 26-29.

Motsara, M.R. (2002). Available nitrogen, phosphorous and potassium status of Indian Soils as depicted by soil fertility maps. *Fertiliser News*, **47**(8): 15-21.

NAAS, (1996). *Agricultural Scientists' perceptions on plant nutrient needs, supply, efficiency and policy issues, 2000-2025:* National Academy of Agricultural Sciences, New Delhi. pp.38.

Nagarajan, R., Manicham, T.S. and Kothandaraman, C.V. (1986). Utilization of coirpith as manure for groundnut. *Madras Agricultural Journal*, **73**(8): 447-449.

Nambiar, K.K.M. and Abrol, I.P. (1989). Long-term fertiliser experiments in India: An overview. *Fertiliser News*. **34**(4): 11-20 & 26.

Narkhede, T.N., Wadile, S.C. and Attarde, D.R. (1997). Response of macro and micronutrients in combination with organic matter on yield of sesame (*S. indicum*). *Indian Journal of Dryland Agricultural Research and Development*, **12**(2): 128-129.

Narkhede, T.N., Wadile, S.C., Attarde, D.R. and Suryavanshi, R.T. (2001). Integrated nutrient management in rainfed sesame (*S. indicum* L.) in assured rainfall zone. *Sesame and Safflower Newsletter*, **16**: 57-59.

NRCG, (1988). Annual Report, National Research Centre for Groundnut, Junagadh, Gujarat.

Pannase, S.K., Sharma, R.S. and Pratibha Singh, (2001). Economy in soybean (*Glycine max* L. Merill) cultivation through zero tillage and biofertilisers. *Journal of Oilseeds Research*, **18**(1):85-86.

Panneerselvan, A., Christopher, L. and Balasubramanian, N. (2000). Soil available phosphorous and its uptake by soybean (*Glycine max* (L.) Merrill) as influenced by organic manures, inorganic fertilisers and weed management practices. *Indian Journal of Agricultural Research*. **34**(1) 9-16..

Pasricha, N.S. and Aulakh, M.S. (1986). Role of sulphur in the nutrition of groundnut. *Fertiliser News*, **31**(9): 17-21.

Pasricha, N.S., Aulakh, M.S., Bahl, G.S. and Baddesha, H.S. (1987). *Nutritional Requirements of Oilseed and Pulse Crops in Punjab (1975-86)*. Punjab Agricultural University, *Ludhiana Research Bulletin*, **15** : 92.

Pasricha, N.S., Aulakh, M.S., Sahota, N.S. and Baddesha, G.S. (1988). Fertilizer research in oilseed crops. *Fertiliser News*, **33**(9): 15-22.

Pasricha, N.S. and Tandon, H.L.S. (1993). Fertiliser management in oilseeds. In H.L.S. Tandon (Ed). *Fertiliser Management in Commercial Crops*. Fertiliser Development and Consultation Organisation, New Delhi.

Patel, J.R. and Shelke, V.B. (2000). Effect of organic and inorganic fertilisers on yield of mustard. *Journal of Maharashtra Agricultural Universities*, **25**(1): 70-71.

Patel, S.R. and Thakur, D.S. (1997). Influence of phosphorous and Rhizobium on yield attributes, yield and nutrient uptake of groundnut (*Arachis hypogaea* L.). *Journal of Oilseeds Research*, **14** (2): 189-193.

Patel, S.R. and Thakur, D.S. (1998). Effect of phosphorus and bacterial inoculation on yield and quality of groundnut. *Indian Journal of Agricultural sciences*, **68**(2): 119-120.

Pattar, P.S., Nadagouda, V.B., Salakinkop, S.R., Kannar,V.C. and Gaddi, A.V. (1999). Effect of organic manures and fertiliser levels on nutrient uptake, soil nutrient status and yield of groundnut (*Arachis hypogaea* L.). *Journal of Oilseeds Research*, **16**(1): 123-127.

Ponniah, C., Prabhakaran, J. and Jehangir, K.S. (1991). Significance of biodigested slurry and rhizobium on groundnut in acid soil. Paper presented at National Seminar on 'Integrated Input Management for Efficient Crop Production". TNAU, Coimbatore 22-25 February, 1994.

Prabhakaran, J. and Sivasubramanian, K. (1991). Response of soybean (*Glycine max*) to rhizobial inoculation using different sources of calcium and pelleting agents. *Indian Journal of Agronomy*, **36**(4): 617-618.

Ramamurthy V. and Shivashankar, K. (1995). Effect of organic matter and phosphorus on quality and nutrient uptake by soybean. *Journal of Oilseeds Research*, **12**(1): 87-90.

Ramamurthy, V. and Shivashankar, K. (1996). Effect of organic matter and phosphoreus on growth and yield of soybean (*Glycine max*). *Indian Journal of Agronomy*, **41**(1): 65-68.

Ramanjaneyulu, G.V., Ramanarao, S.V. and Hegde, D.M. (2000). *A decade of frontline demonstrations in oilseeds : An Overview*. Directorate of Oilseeds Research, Hyderabad. P-53 xii

Reddy, B.N. and Sudhakarababu, S.N. (1996a). Production potential, land utilization and economics of fertiliser management in summer sunflower (*Helianthus annuus* L.) based crop sequences. *Indian Journal of Agricultural Sciences*, **66**(1): 16-19.

Reddy, B.N. and Sudhakarababu, S.N. (1996b). A critical review of plant nutrient supply needs, efficiency and policy issues for oilseed crops for the year 2000-2005. *In* J.S. Kanwar and J.C. Katyal (Eds). *Plant Nutrient Needs, Supply, Efficiency and Policy Issues: 2000-2025*. National Academy of Agricultural Sciences, New Delhi. pp-200-216.

Reddy, M.S.S., Reddy, S.R., Reddy, M.G., and Reddy, M.G.R.K. (1990). Residual effect of organic matter and fertiliser applied to groundnut (*Arachis hypogaea*) and maize (*Zea mays*) on succeeding groundnut in groundnut-maize-groundnut sequence. *Indian Journal of Agronomy*, **36**: 298-300.

Roy, R.N. (1992). Integrated plant nutrition system - an overview. *In* H.L.S. Tandon (Ed). *Fertilisers, Organic Manures, Recyclable Wastes and Biofertilisers*, Fertiliser Development and Consultation Organization, New Delhi, pp.1-11.

Sagare, B.N., Guhe, Y.S. and Deshmukh, A.R. (1992). Biological yield and uptake of nutrients by peanut as influenced by enriched FYM products. *Journal of Maharashtra Agricultural Universities*, **17**(1): 100-102.

Saini, J.S., Sahota, T.S. and Dhillon, A.S. (1989). Agronomy of rapeseed-mustard and their place in new and emerging cropping systems. *Journal of Oilseeds Research*, **6**(1): 260-268.

Sardana, V. and Sidhu, M.S. (1994). Response, relative efficiency and optimum dose of N and P for Indian rape (*B. campestris* Var. *toria*) + Swede rape (*B. napus*) intercropping systems with or without organic manures. *Annals of Agricultural Research*, **20**(1): 122-124.

Sarkar, A., Yadav, R.L., Gangwar, B. and Bhatia, P.C. (1999). *Crop Residues in India*. Technical Bulletin, PDCSR, Modipuram, Meerut, p. 34.

Sawarkar, S.D. (1997). Effect of inorganic and biofertilisers (*Azospirillum*) on the yield of rainfed niger. *Journal of Oilseeds Research*, **14**(2): 330-332.

Shakthawat, R.P.S. and Bansal, K.N. (1999). Effect of organic manures and nitrogen on yield, quality and economics of sunflower. *Journal of Oilseeds Research*, **16**(1): 132-134.

Shanwad, U.K. and Agasimani, C.A. (2001). Effect of INM on productivity and soil fertility in sunflower + pigeonpea intercropping system under rainfed condition. *Journal of Oilseeds Research*, **18**(2): 279-280.

Sharma, R.K., Shrivastava, U.K., Tomer, S.S., Tiwari, P.N., and Yadav, R.P. (1999). Nutritional management in soybean-mustard crop sequence. *Indian Journal of Agronomy*, **44**(3): 493-498.

Singh, G.R., Parihar, S.S., Chaure, N.K., Chaudhary, K.K. and Sharma, R.B. (1997). Integrated nutrient management in summer sesame. *Indian Journal of Agronomy*, **42**(4): 699-701.

Singh, J. and Singh, K.P. (1997). Integrated nutrient management in sunflower (*H. annuus*). *Indian Journal of Agronomy*, **42**: 370-374.

Singh, M.V. (1999). *Sulphur management for oilseed and pulse crops. Bulletin No.3, Indian Institute of Soil Science*, Bhopal, p.54.

Singh, M.V. (2001). Micronutrient status of Indian soils and crop responses to their application. *Proc. of National Seminar on Biofertilisers and micronutrients*, New Delhi.

Subba Rao, I.V. (1994). Integrated nutrient management for sustainable productivity of oilseeds in India. *In* M.V.R. Prasad (Eds). *Sustainability in Oilseeds*, Indian Society of Oilseeds Research, DOR, Hyderabad, pp.264-270.

Subba Rao, A., Sammi Reddy, K. and Takkar, P.N. (1995). Phosphorus management - A key to boost productivity of soybean-wheat cropping system on swell-shrink soils. *Fertiliser News*, **40**(12): 87-95.

Sukhija, P.S., Single, P.L., Dhillon, K.S., Pasricha, N.S., and Munshi, S.K. (1985). Influence of urea application and manuring on biochemical composition of *Brassica juncea* seeds. *Journal of Research*, PAU., **23**(4): 607-614.

Swindale, L.D. (1982). In Managing Soil Resources, Plenary Session papers. *Trans. 12th International Congress on Soil Science*, New Delhi, pp. 67-100.

Takkar, P.N. and Nayyar, V.K. (1989). Crop response to micronutrient applications. *Proc. FAI-NR Seminar, Jaipur*, pp.96-123.

Talashikar, S.C., Dosani, A.A.K., Mehta, V.B. and Pawar, A.G. (1997). Integrated use of fertilisers and poultry manure to groundnut crop. *Journal of Maharashtra Agricultural Universities*, **22**(1): 205-207.

Tandon, H.L.S. (1990). *Fertiliser recommendations for oilseed crops*. A guide book Fertiliser Development and Consultation Organization, New Delhi.

Tandon, H.L.S. (1992). Fertilisers and their integration with organics and biofertilisers. *In* H.L.S. Tandon (Ed). *Fertilisers, Organic Manures, Recyclable Wastes and Biofertilisers*, Fertiliser Development and Consultation Organization, New Delhi, pp.12-35.

Tandon, H.L.S. and Messick, D.L. (2002). *Practical Sulphur Guide*. The Sulphur Institute, Washington DC, p.20.

Tandon, H.L.S. and Sekhon, G.S. (1988). *Potassium Research and Agricultural Production in India*. Fertiliser Development and Consultation Organization, New Delhi. p.144.

Tiwari, K.P., Namdeo, K.N. and Lal, J.P. (1993). Response of soybean to different nitrogen levels with and without Brady-*rhizobium* inoculation. *Journal of Oilseeds Research*, **10** (2): 299-301.

Tiwari, K.P., Namdeo, K.N., Tomar, R.K.S. and Raghu, J.S. (1995). Effect of macro and micronutrients in combination with organic manures on the production of sesame (*Sesamum indicum*). *Indian Journal of Agronomy*, **40**(1): 134-136.

Tomer, N.K., Chahal, R.S. and Khanna, S.S. (1980). Balanced nutrition - a key to raise yield of mustard under dryland. *Fertiliser News*, **25**(10): 31-33.

Umamaheshwari, P., Padmalata,V. and Murali Rao, M. (2001). Evaluation of different sources of biofertilisers in association with inorganic and organic manures in groundnut. *Agricultural Science Digest*, **25**(4): 250-251.

Vara, J.A., Madhwadia, M.M., Patel, B.S., Patel, J.C. and Khanpara, V.D. (1994). Response of soybean (*Glycine max*) to nitrogen, phosphorous and *Rhizobium* inoculation. *Indian Journal of Agronomy*, **39**(4): 678-680.

Vijayalakshmi, K. (1984). Proc. of Workshop on alfisols in the Semi Arid Tropics, ICRISAT, Andhra Pradesh.

Vinod Kumar., Ghosh, B.C. and Bhat, V. (2000). Complementary effect of crop wastes and inorganic fertilisers on yield, nutrient uptake and residual fertility in mustard (*Brassica juncea*)-rice (*Oryza sativa*) cropping sequence. *Indian Journal of Agricultural Sciences*, **70**(2): 69-72.

Wani, S.P. and Lee. (1992). Role of biofertilisers in upland crop production. *In* H.L.S. Tandon (Ed). *Fertilisers Organic Manures, Recyclable Wastes and Biofertilisers*, Fertiliser Development and Consultation Organization, New Delhi. pp.91-112.

Wankhade, S.Z., Dabre, W.M., Lanjewar, B.K., Sontekey, P.Y. and Takzure, S.C. (1992). Effect of seed inoculation with Rhizobium culture and molybdenum on yield of groundnut (*Arachis hypogaea*). *Indian Journal of Agronomy*, **37**(2): 384-385.

❑❑❑

System Based Integrated Nutrient Management, 2012
© *B. Gangwar & V.K. Singh (eds.), pp. 155-167*
New India Publishing Agency, New Delhi (India)
e-mail : info@nipabooks.com; website : www.nipabooks.com

CHAPTER **10**

Soil Test Based Integrated Nutrient Management under Cropping Systems

B.M. SHARMA AND M.C. MEENA

The population of India is estimated to touch 1.4 billion by the year, 2025, which will need at least 300 million tonnes of food grain. This necessitates the production of additional food grain from the same land without loosing the production potential of the soil. This, in turn, requires extensive research to provide a scientific basis for enhancing and sustaining food production as well as soil productivity with minimum environmental degradation. In order to attain this it is essential that the amount of nutrients removed from the soil should be replenished through judicious use of fertilizers and manures. This needs a more comprehensive approach for fertilizer use, incorporating components like soil test, field research and economic evaluation of the results. Soil testing is one of the best scientific means for quick and reliable determination of soil fertility status. Soil test crop response study in the field provides soil test calibration between the level of soil nutrients as determined in the laboratory and the crop response to fertilizers as observed in the field for predicting the fertilizer requirements of the crop.

Objectives of soil testing.

Soil testing provides a sound information about the suitability, fertility and productivity of the soil. This enables the farmers to make the most profitable use of some of the costly inputs in farming. However, lack of information among the farmers about soil testing and their importance in management of resources necessitates that the work of soil testing should be taken extensively. Soil testing is a useful tool for making fertilizer recommendations for various crops and cropping sequences as well as reclamation of problem soils. Thus the major objective of soil testing are to identify the type of soil related problems like salinity, alkalinity, and acidity and to suggest appropriate reclamation/amelioration measures and to evaluate the fertility status in terms of available nutrients for making soil test based fertilizer recommendations.

Role of soil testing in balanced use of fertilizer nutrients.

Nutrients are removed continuously by crops from the soils and their replenishment through fertilizers and manures is essential. The ideal ratio of nutrients for optimum growth depends on crop and soil. When the added amount of nutrient through fertilizers and manures is less or more than that required by the crop, the imbalanced use thus results in poor crop growth. The application of phosphorus with nitrogen increases the nitrogen use efficiency. Some times even on NP application, decrease in yield occurs because K becomes the limiting factor. Similarly with optimum dose of NPK, the deficiency of zinc may appear. Soil test based recommendation ensures balanced use of fertilizers for higher yield and profits. Balanced fertilization includes, besides major nutrients, secondary and micronutrients whose deficiencies appear in different soils in various agro-climatic regions due to intensive cultivation with high analysis fertilizers.

Response of crop to added nutrient depends largely upon the inherent capacity of soil to supply the nutrient as per the requirement of crop. Chemical tests have long been used to estimate the nutrient availability in soil to predict the probability of obtaining a profitable response to applied nutrients. On the basis of testing, soils are rated as low, medium and high categories (Table 1), accordingly fertilizer recommendations should be given. In a low nutrient soil, crop respond remarkably to the application of that nutrient. On the other hand, in high nutrient soil, the crop may show little or no response to application of nutrient. In medium nutrient soil, the response is intermediate. Therefore, soil testing is helpful in adjusting the amount of fertilizer nutrient according to soil test values and crop demand. In our country at present following methods/approaches are used for giving fertilizer recommendations.

Table 1 : Ratings of soil test values for primary nutrients

Nutrient Rating	Low	Medium	High
Organic Carbon (%) *	< 0.5	0 – 0.75	> 0.75
Alkaline $KMnO_4$-N (kg ha^{-1})	< 280	280 – 560	> 560
Olsen's P (kg ha^{-1})	< 10	10 – 25	> 25
Amm. acetate-K (kg ha^{-1})	< 120	120 – 280	> 280

* As an index of available N

General fertilizer recommendations.

Development of these recommendations are based on multi-locational trials conducted with graded doses of N, P and K fertilizers and their economic evaluation to arrive at an optimum dose for a particular crop. The general recommendations hold good under medium soil fertility conditions. In this approach the variation in soil fertility is not taken into consideration, and hence, under high or low soil fertility conditions the applied nutrients prove often a wasteful expenditure and insufficient, respectively. In both the cases, optimum fertilizer use efficiency can not be achieved. Results of long term fertilizer experiments have shown a considerable build-up of P due to continuous application of even general recommended dose in intensive cropping systems. In few cases the build-up of available P reached a level where no more phosphate application is needed for next few crops (Nambiar, 1994; Singh *et al.*, 1998).

Fertilizer recommendations based on soil fertility ratings.

In this approach medium soil fertility is equated with general recommended dose. In case of low and high fertility categories, the fertilizer doses are increased or decreased 25 to 50 per cent of the general recommended dose as per the situation. At present, most of the fertilizer recommendations issued from soil testing laboratories in India are based on this approach. In this approach, no scientific method is used for increasing or decreasing the doses as value changed from medium to low or high category. Secondly range in medium category is quite large i.e. in case of potassium it varied between 121 to 280 kg K ha^{-1} and attract the same fertilizer dose of potassium for any value between this range which is not justified. These rating were developed in 1965 for old varieties of crops, unfortunately since then these rating are same irrespective of types of soil and crops. Moreover, crops have different requirement of nutrients which also change according to the yield levels to be produced. Therefore, these semi-quantitative fertilizer recommendations based on ratings need to be replaced by more refined recommendations.

Fertilizer recommendations based on critical limit of soil available nutrient.

Critical level concept of available nutrient was developed by Cate and Nelson (1965). Critical limit is the level of soil available nutrient above which that nutrient is not considered as a primary limiting factor. The probability of getting economic response to fertilizer application in soils having available nutrient above the critical limit is quite low, while in soils below the critical limit the probability of getting economic response is quite high. Critical limit varies, depending on the soil types, crops and varieties, soil test methods used and seasonal variations. This concept separates the soils in responsive and non-responsive groups, but it does not suggest quantification of fertilizer dose for individual situations in responsive group. Hence, this concept may be more useful for fertilizer recommendation of micronutrients. The critical limits in soil and accordingly fertilizer recommendations and given in Table 2.

Table 2 : Critical limit of micronutrients in soil and fertilizer recommendations

Micronutrient	Critical limit (mg kg^{-1})	Material	Soil application (kg ha^{-1})	Foliar application
Zinc	0.6	Zinc Sulphate	20	0.5% zinc sulphate + 0.25% lime
Copper	0.2	Copper Sulphate	15	0.1% copper sulphate + 0.05% lime
Iron	4.5	Ferrous sulphate	50	2% ferrous sulphate
Manganese	2.0	Manganese sulphate	25	2% manganese sulphate
Boron	0.5	Borex	10	0.2% solubar
Molybdenum	0.2	Sodium Molybdate	1	seed treatment @ 100 g/ha

Fertilizer recommendations based on targeted yield approach.

The principle behind this concept is the existence of highly significant linear relationship between grain yield and uptake of nutrients (NPK) by crops, which means that, for unit production of grain a definite amount of nutrient is obtained from the soil. This linear relationship forms the basis for fertilizer recommendations for targeted levels of yield production first advocated by Truog (1960). Subsequently Ramamoorthy *et al.,* (1967) established the theoretical basis and experimental proof for its applicability. In this methodology the information on parameters namely, (i) nutrient requirement (kg q^{-1} of grain / economic produce), (ii) per cent contribution from soil available nutrient, and (iii) per cent contribution from fertilizer/manure nutrient are to be derived from the soil test crop response field experiment.

Targeted yield concept is primarily based on balanced prescription of fertilizer nutrients. This procedure provides a scientific basis for balanced prescription of nutrients through fertilizers and manures after considering available nutrients in soil and their actual requirement by crop. Once the actual requirement of nutrients for specific level of production is known, the needed fertilizer dose can be calculated after considering the per cent efficiencies of soil available and fertilizer nutrients. Therefore, it provides the fertilizer dose in balanced and quantitative terms in relation to soil test values and crop requirement which is necessary to optimize the response to added fertilizers, maximize the profit and achieving the desired yield target with ± 10 % deviation. Application of fertilizers according to the need of the situation also helps in maintenance of soil fertility and ensures sustainability of crop productivity.

Soil test crop response methodology

Soil Test Crop Response (STCR) is a methodology for developing soil test based fertilizer recommendations for targeted yield of crops.

Development of basic data

The basic data are developed using the following expressions :

Without FYM

$$NR = \frac{\text{Total uptake of nutrient } \left(\text{kg ha}^{-1}\right)}{\text{Grain yield } \left(\text{q ha}^{-1}\right)}$$

$$\% \text{ CS} = \frac{\text{Total uptake of nutrient from control plot (kg ha}^{-1}\text{) without FYM}}{\text{Soil test value } \left(\text{kg ha}^{-1}\right) \text{ of control plot without FYM}} \times 100$$

$$\% \text{ CF} = \frac{\begin{array}{c}\text{Total uptake of nutrient from treated} \\ \text{plot } \left(\text{kg ha}^{-1}\right) \text{ without FYM}\end{array} - \begin{array}{c}\text{Soil test value of treated plot} \times \frac{\%\text{CS}}{100} \\ \left(\text{kg ha}^{-1}\right) \text{without FYM}\end{array}}{\text{Fertilizer nutrient applied } \left(\text{kg ha}^{-1}\right) \text{in treated plot without FYM}} \times 100$$

With FYM

$$\% \text{ CFYM} = \frac{\begin{array}{c}\text{Total uptake of nutrient from control} \\ \text{plot } \left(\text{kg ha}^{-1}\right) \text{ with FYM}\end{array} - \begin{array}{c}\text{Soil test value of control plot} \times \frac{\%\text{CS}}{100} \\ \left(\text{kg ha}^{-1}\right) \text{with FYM}\end{array}}{\text{Nutrient added } \left(\text{kg ha}^{-1}\right) \text{in control plot through FYM}} \times 100$$

$$\% \text{ CF} = \frac{\begin{bmatrix}\text{Total uptake of nutrient} \\ \text{from treated plot} \\ \left(\text{kg ha}^{-1}\right) \text{with FYM}\end{bmatrix} - \begin{bmatrix}\text{Soil test value of} & & \text{Nutrient added} & \\ \text{treated plot} \left(\text{kg ha}^{-1}\right) & \times \frac{\%\text{CS}}{100} + & \left(\text{kg ha}^{-1}\right) & \times \frac{\%\text{CF YM}}{100} \\ \text{with FYM} & & \text{through FYM} & \end{bmatrix}}{\text{Fertilizer nutrient applied } \left(\text{kg ha}^{-1}\right) \text{in treated plot with FYM}} \times 100$$

Development of fertilizer adjustment equations

The fertilizer adjustment equations are developed as :

The basic data for developing soil test based fertilizer / integrated fertilizer equations for National Capital Region Delhi are given in Table 3, 4, 5 & 6.

Without FYM

$$FD = \frac{NR \times 100}{\%CF} \times T - \frac{\%\ CS}{\%\ CF} \times STV$$

With FYM

$$FD = \frac{NR \times 100}{\%CF^*} \times T - \frac{\%\ CS}{\%\ CF^*} \times STV - \frac{\%\ CFYM}{\%\ CF^*} \times \text{Nutrient in FYM}\left(kg\ t^{-1}\right) \times FYM$$

where :

NR = Nutrient requirement in kg q^{-1} of grain production

%CS = Per cent contribution from soil available nutrient

%CF = Per cent contribution from fertilizer nutrient without FYM

%CF* = Per cent contribution from fertilizer nutrient with FYM

%CFYM = Per cent contribution from FYM nutrient

FD = Fertilizer dose (kg ha^{-1})

STV = Soil test value (kg ha^{-1})

T = Yield target (q ha^{-1})

FYM = Farmyard manure (t ha^{-1})

The merits of soil test and crop-need based balanced fertilizer recommendations based on targeted yield are:

1. It ensures the achievement of desired yield target with ± 10 % deviation under optimum management conditions.
2. Efficient use of fertilizers according to soil fertility and crop-need ensure high profit and response to applied fertilizers.
3. It ensures maintenance of soil fertility at appropriate levels in cropping system for sustainable crop production.
4. Choice of choosing appropriate yield target for fertilizer recommendations according to the availability of resources and soil fertility.
5. Identification of suitable crop rotations from the point of view of relative ability of crops and crop varieties to utilize soil and fertilizer nutrients.

Table 3 : Basic data for developing soil test based fertilizer adjustment equations for targeted yields of crops in NCR of Delhi

Parameter	N	P_2O_5	K_2O
Wheat			
NR	2.61	0.84	3.16
% CS	25.00	59.00	30.65
% CF	49.12	24.35	114.72
Rice			
NR	2.09	0.73	2.57
% CS	19.9	55.70	19.40
% CF	42.40	16.30	111.00
Mustard			
NR	4.97	1.86	5.23
% CS	29.30	44.40	27.30
% CF	67.10	29.90	84.20
Pearlmillet (*Bajra*)			
NR	3.65	1.50	4.97
% CS	20.16	54.90	29.40
% CF	52.40	26.20	127.10
Maize			
NR	2.99	1.11	3.16
% CS	23.50	51.90	23.00
% CF	45.20	23.20	115.00
Barley			
NR	1.93	0.67	3.50
% CS	33.80	52.40	41.40
% CF	53.10	22.9	158.30
Pigeonpea (*Arhar*)			
NR	7.18	1.99	3.84
% CS	37.90	56.50	20.10
% CF	131.90	32.70	121.90
Gram			
NR	4.80	0.98	4.40
% CS	35.50	43.90	29.30
% CF	90.70	19.20	144.70
Mungbean			
NR	6.42	2.21	5.40
% CS	22.10	32.20	20.40
% CF	70.90	21.10	54.10
Soybean			
NR	6.06	1.58	4.64
% CS	32.50	36.40	21.00
% CF	91.80	26.10	120.10

NR = Nutrient requirement kg q^{-1} of grain production
% CS = Per cent contribution from soil available nutrients
% CF = Per cent contribution from fertilizer nutrients

Table 4 : Soil test based fertilizer adjustment equations for targeted yields of crops in NCR of Delhi

Wheat	Mustard
$FN = 5.31\ T - 0.51 SN$	$FN = 7.41\ T - 0.44\ SN$
$FP_2O_5 = 3.45\ T - 5.55\ SP$	$FP_2O_5 = 6.22\ T - 3.41\ SP$
$FK_2O = 2.75\ T - 0.32\ SK$	$FK_2O = 6.21\ T - 0.39\ SK$
Pearl millet *(Bajra)*	Maize
$FN = 6.97\ T - 0.38\ SN$	$FN = 6.61\ T - 0.52\ SN$
$FP_2O_5 = 5.73\ T - 4.81\ SP$	$FP_2O_5 = 4.77\ T - 5.13\ SP$
$FK_2O = 3.92\ T - 0.28\ SK$	$FK_2O = 2.75\ T - 0.24\ SK$
Rice	Barley
$FN = 4.93\ T - 0.47\ SN$	$FN = 3.69\ T - 0.64\ SN$
$FP_2O_5 = 4.48\ T - 7.82\ SP$	$FP_2O_5 = 2.93\ T - 5.24\ SP$
$FK_2O = 2.31\ T - 0.21\ SK$	$FK_2O = 2.22\ T - 0.31\ SK$
Pigeonpea *(Arhar)*	Gram
$FN = 5.44\ T - 0.29\ SN$	$FN = 5.29\ T - 0.39\ SN$
$FP_2O_5 = 6.09\ T - 3.95\ SP$	$FP_2O_5 = 5.10\ T - 5.24\ SP$
$FK_2O = 3.15\ T - 0.20\ SK$	$FK_2O = 3.04\ T - 0.24\ SK$
Mungbean	Soybean
$FN = 9.06\ T - 0.31\ SN$	$FN = 6.60\ T - 0.35\ SN$
$FP_2O_5 = 10.47\ T - 3.50\ SP$	$FP_2O_5 = 6.05\ T - 3.19\ SP$
$FK_2O = 9.98\ T - 0.45\ SK$	$FK_2O = 3.86\ T - 0.21\ SK$

Where, FN, FP_2O_5 and FK_2O represent dose (kg ha^{-1}) of N, P_2O_5 and K_2O, respectively, to be applied through fertilizer, SN, SP and SK represent soil test values of available N, P and K (kg ha^{-1}), respectively, and T stands for yield target (q ha^{-1})

The yield targeting equations are valid under the following situations:

1. These should be used for similar soils occurring in a particular agro-eco region
2. Targets chosen should not be unduly high or low and should be within the range of experimental yields obtained. The maximum target should not exceed 75-80% of the highest yield achieved for that crop in the area.
3. The fertilizer N recommendations for leguminous crops should be the same as the general recommended dose for that crop in the area.
4. Adjustment equations must be used within the experimental range of soil test values and cannot be extrapolated.

Table 5 : Basic data for developing soil test based integrated fertilizer adjustment for targeted yields of crops in NCR of Delhi

Parameter	N	P_2O_5	K_2O
Wheat			
NR	2.48	0.89	3.20
% CS	26.40	57.60	32.20
% CF	64.40	32.00	133.30
% CFYM	21.10	10.40	27.20
Mustard			
NR	4.13	2.03	4.43
% CS	23.60	58.60	23.10
% CF	62.20	33.40	115.40
% CFYM	25.40	14.70	43.20
Pearlmillet (*Bajra*)			
NR	3.42	1.19	4.90
% CS	18.50	36.20	24.20
% CF	63.90	25.20	169.80
% CFYM	26.80	11.90	58.80
Maize			
NR	2.55	1.22	2.63
% CS	17.80	49.00	16.30
% CF	50.80	31.00	116.90
% CFYM	20.60	15.40	32.20
Soybean			
NR	6.06	1.58	4.64
% CS	32.50	36.40	21.00
% CF	94.30	29.50	132.50
% CFYM	27.90	18.80	29.50

where : NR = Nutrient requirement in kg q^{-1} of grain production,
% CS = Per cent contribution from soil available nutrients,
% CF = Per cent contribution from fertilizer nutrients, and
% CFYM = Per cent contribution from FYM nutrients.

5. Good and recommended agronomic practices need to be followed while raising crops.
6. Other micro and secondary nutrients should not be yield limiting.
7. For obtaining the real benefit of fertilizer application based on targeted yield approach, soil testing needs to be done as frequently as possible.

Integrated nutrient management

Green revolution in India started in late sixties with the introduction of high yielding varieties, increase in the use of high analysis fertilizers and cropping intensity making India self sufficient in food production. In this process, the use of organic manures as a source of plant nutrients declined substantially. With increase in the cost of fertilizers, agriculture is becoming more and more expensive. Simultaneously indiscriminate and imbalanced use of chemical fertilizers has the negative effective on productivity. Occurrences of multi-nutrient deficiencies and over all depletion of soil productivity under chemical fertilizer use have been widely reported. Decline in soil productivity under chemical fertilizer use gave the momentum to organic farming and integrated nutrient management.

Integrated nutrient management envisages the use of chemical fertilizers in conjunction with organic manures, legumes in cropping system, bio-fertilizers and other locally available nutrient sources for sustaining health and productivity. The combined application of organic manures and chemical fertilizers generally produces higher crop yield than when each is applied alone. This increase in crop productivity may be due to combined effect of nutrient supply, synergism and improvement in soil physical, biological and chemical properties of the soil

Farmyard manure, compost, wormi-compost, gobar gas slurry, Sewage sludge, press mud etc. are the sources of organic manures. Among the organic manures, Farmyard manure (FYM) constitute an important component of

Table 6 : Integrated soil test based fertilizer adjustment equations for targeted yields of crops in NCR of Delhi

Wheat	Mustard
FN = 3.85 T– 0.41 SN –1.64 FYM	FN = 6.64 T - 0.38 SN - 1.72 FYM
FP_2O_5 = 2.78 T - 4.12 SP - 1.72 FYM	FP_2O_5 = 6.10 T - 4.02 SP - 2.43 FYM
FK_2O = 2.04 T - 0.29 SK - 0.88 FYM	FK_2O = 3.84 T - 0.24 SK - 1.21 FYM
Pearlmillet (*Bajra*)	Maize
FN = 5.35 T - 0.29 SN - 2.23 FYM	FN = 5.02 T - 0.35 SN - 1.82 FYM
FP_2O_5 = 4.72 T - 3.29 SP - 2.48 FYM	FP_2O_5 = 3.93 T - 3.62 SP - 2.29 FYM
FK_2O = 2.88 T - 0.17 SK - 1.35 FYM	FK_2O = 2.25 T - 0.17 SK - 1.00 FYM
Soybean	
FN = 6.43 T - 0.34 SN - 1.33 FYM	
FP_2O_5 = 5.36 T - 2.83 SP - 2.92 FYM	
FK_2O = 3.50 T - 0.19 SK - 0.88 FYM	

where FN, FP_2O_5 and FK_2O represent dose (kg ha^{-1}) of N, P_2O_5 and K_2O, respectively, to be applied through fertilizer, SN, SP and SK represent soil test values of available N, P and K (kg ha^{-1}), respectively, FYM represents dose of farmyard manure (t ha^{-1}), and T stands for yield target (q ha^{-1}).

integrated nutrient management for maintaining soil fertility and crop productivity. FYM contains 0.5% N 0.2% P and 0.4% K, while compost contains 1.5% N, 0.6% P and 1.2% K, wormi-compost is more rich in NPK content. It was observed from the field experiments that fertilizer dose can be reduced by 5, 2 and 3 kg ha^{-1} of nitrogen, phosphorus and potassium, respectively, with each ton application of FYM without affecting the crop yield.

Green manures are valuable and potential source of nitrogen and organic matter. Sesbania, sunnhemp, gwar and cowpea are the main crops for green manuring. Incorporation of green manure crop before transplanting rice, partially meet the nutrient requirement of the crop. In upland crops the most efficient time of incorporation of green manure crops has been found to be 15 days before the sowing of the crop. If green manure is used then dose of fertilizer nitrogen can be reduced by 50 kg ha^{-1}. Besides being the source of nutrients, green manure influences availability of nutrients through its favourable effect on soil oxidation reduction processes, pH, cation exchange capacity, increased chelation capacity and improving physical and biological properties of the soil. The green manuring also plays an important role in reclamation of salt affected soils.

Growing legumes in rotation and incorporation of legume residues after picking of pods also significantly improve the yield of cereals crops. Use of blue green algae and Azolla helps in fixation of nitrogen from atmosphere about 30 to 40 kg ha^{-1} in rice field and helps in increasing grain yield of rice. Similarly some biofertilizers like Azotobactor in wheat and Rhizibium culture in leguminous crops helps in fixation of nitrogen about 35 to 30 kg ha^{-1}. Use of phosphorus solubilizing bacterial cultural or VAM in field helps in increasing the availability of phosphorus by solubilizing the phosphorus in soil. Therefore, fertilizer requirement of the nutrient in crop production can be reduced by application of biofertilizers.

Limitations in integrated nutrient management

1. Limited availability of dung. Large portion of dung is used for making cakes for fuel purpose.
2. Limited availability of crop residue, because they are used either for feed or for fuel.
3. Organic manures are bulky in nature thus required in large quantity, therefore, cost involved in their transport is high.
4. Green manuring crops are generally grown in summer, and if the facility of irrigation is not sufficient then it becomes difficult to grow them. Additional cost and time involved in growing them is also an important factor in its adoption.

5. Limited availability of good quality biofertilizers. There is no reliable quality control at present.

Prediction of post-harvest soil test values after maize, pearl millet and wheat

Using the data on initial soil test values, post-harvest soil fertility status and fertilizer doses applied in soil test crop response field experiments on maize, pearl millet and wheat, the following equations for predicting the post-harvest soil test values were developed.

Maize (MH-108)

PHN = 67.78 + 0.68 ISN + 0.018 FN

PHP = 2.06 + 0.86 ISP + 0.105 FP

PHK = 37.80 + 0.81 ISK + 0.076 FK

Maize (PEMH-02)

PHN = 54.65 + 0.72 ISN + 0.015 FN

PHP = 2.44 + 0.66 ISP + 0.128 FP

PHK = 38.84 + 0.82 ISK + 0.152 FK

Pearlmillet (PBH-47)

PHN = 53.16 + 0.73 ISN + 0.012 FN

PHP = 1.21 + 0.93 ISP + 0.092 FP

PHK = 45.18 + 0.75 ISK + 0.134 FK

Pearlmillet (ICMH-541)

PHN = 56.2 + 0.71 ISN + 0.021 FN

PHP = 1.7 + 0.87 ISP + 0.064 FP

PHK = 28.5 + 0.82 ISK + 0.135 FK

Wheat (PBW-343)

PHN = 52.4 + 0.68 ISN + 0.015 FN

PHP = 2.1 + 0.79 ISP + 0.072 FP

PHK = 15.6 + 0.88 ISK + 0.146 FK

The predicted values can be utilized for recommending the fertilizer doses for succeeding crop, thus eliminating the need of soil testing for each crop in the sequence. This provides the way for giving the fertilizer recommendations for whole cropping sequence based on initial soil test values.

References

Nambiar, K.K.M. (1994) Soil fertility and crop productivity under Long Term Fertilizer Use in India, ICAR, New Delhi.

Ramamoorthy, B., Narsinham, R.L. and Dinesh, R.S. (1967) Fertilizer application for specific yield targets of Sonora-64. *Indian Fmg.* **17**(5): 43-45.

Singh, D., Rana, D.S. and Kumar, K. (1998) Phosphorus removal and available P balance in a Typic Ustochrept under intensive cropping and long term fertilizer use. *J. Indian Soil Sci.* **46**(3): 398-401.

Truog, E. (1960). Fifty years of soil testing. *Trans. 7th Int. Cong. Soil Sci.* **3** : 46-57.

□□□

System Based Integrated Nutrient Management, 2012
© B. Gangwar & V.K. Singh (eds.), pp. 169-189
New India Publishing Agency, New Delhi (India)
e-mail : info@nipabooks.com; website : www.nipabooks.com

CHAPTER **11**

Integrated Nutrient Management in Soybean Based Cropping Systems

A.K. VYAS AND B.G. SHIVAKUMAR

Soybean [*Glycine max* (L.) Merrill.] is the leading oilseed crop in the world. Its area and production are far ahead of other oilseeds at the global level (Table 1). The commercial cultivation of this crop in India started in the early 1970's. Within a span of 4 decades its spread has been phenomenal and today it occupies around 9.67 million hectares with a production of about 10.22 million with an average productivity of around 1006 kg ha^{-1} (DSR, 2010) (Table 2). It is now the fourth largest among field crops after rice, wheat and maize (GOI, 2007) in India and has become the leading oil seed crop surpassing groundnut and mustard and rapeseed. It is contributing nearly 37% to the total oilseeds produced in India. The central India comprising the states of Madhya Pradesh, Maharashtra and Rajasthan account for more than 95% of the total area under this crop. Chhattisgarh, Karnataka, Andhra Pradesh, Uttar Pradesh etc account or the bulk of the rest of the area. It is predominantly a rainfed crop with only about 3% of the total area getting some assured irrigation facility. Further, it is concentrated in areas which receive lesser rainfall in its early vegetative phase. A large number of crops have been replaced by soybean to a variable extent in its march towards achieving its present position. Badal *et al.* (2000) have reported that a large number farmers switched over to this crop from sorghum, groundnut, maize, cotton etc in the state of Madhya Pradesh owing to greater yield potential even under adverse conditions (Table 3). Further, it has become an integral part of several cropping systems including both intercropping and sequential cropping systems in different parts of the country.

Table 1 : Pattern of edible oil production (million tonne) field crops in the world

Crop	Production
Soybean	35.59
Rapeseed	16.82
Sunflower	10.80
Cotton seed	6.33
Groundnut	5.57
Corn	2.19
Sesamum	0.90
Safflower	0.16

Source: FAO (2009)

Table 2 : Growth of soybean production in India over the years

Year	Area ('000 ha)	Production ('000 tonnes)	Yield (kg ha^{-1})
1971	30	10	426
1981	610	440	728
1991	2564	2602	1015
2001	6420	5270	822
2005	6900	6600	952
2009	9670	10220	1006

Source: DSR (2010)

Table 3 : Crops replaced by soybean in Madhya Pradesh

S. No	Crop replaced	% Farmers
1	Sorghum	61
2	Groundnut	22
3	Maize	18
4	Cotton	12
5	Pearlmillet	3
6	Pigeonpea	4

Soybean is one of the most energy rich crops. It contains around 40% high quality protein, 20% oil and good source of a large number vitamins and mineral (Table 4) and compares quite above other competitive crops like pulses. Thus its nutrient requirement is also very high as compared to many other similar crops. However, its capacity to fix atmospheric nitrogen with the help of *Rhizobium* bacteria has made it less dependent on the nitrogenous fertilizers. But other nutrients viz. phosphorus, potassium, sulphur and other micronutrients need to be managed scrupulously for its successful cultivation. The nutrient management becomes much more critical when soybean is cultivated in different cropping systems under varied soil conditions. With nutrient sources becoming limited and expensive gradually, their judicious management is of great importance to realize higher productivity and nutrient use efficiency. The integration of different sources of nutrients and adoption of system based nutrient management are becoming increasingly significant for sustenance of the productivity of soybean based cropping systems in the long run.

Table 4 : Approximate food value of soybean (100 g^{-1})

Constituents	Value	Remarks
Protein	43%	Soybean is a rich source of best quality plant protein, poly unsaturated fatty acid rich oil, fibre and minerals. It contains good amount of omega -3 fatty acid
Carbohydrates	21%	
Fat	19%	
Minerals	5%	
Fibre	4%	
Energy	430 Kcal	Good source of dietary energy
Phosphorus	690 mg	Soybean is a good source of phosphorus, calcium, magnesium, iron, zinc, manganese, copper etc.
Calcium	240 mg	
Magnesium	175 mg	
Iron	10 mg	
Zinc	3 mg	
Manganese	2 mg	
Copper	1 mg	
Carotene	426 mg	Soybean contains a good amount of carotene
Niacin	3 mg	
Thiamine	1 mg	
Riboflavin	1 mg	
Phyto-chemicals	Reasonable amounts	It also contains good amount of isoflavones, phytic acids, phtosterols, trypsin inhibitor.

Soybean based cropping systems in India

Soybean is suitable for any traditional cropping sequence/ companion cropping/ intercropping / mixed cropping systems due to its (i) comparatively better tolerance to shade and drought (Lawn, 1982), (ii) efficient light utilization due to lanceolate leaf type and compact canopy, (iii) does not excessively compete for soil moisture (Nathanson *et al.*, 1984; Wright *et al.*, 1988), (iv) better ability to fix atmospheric di-nitrogen and high level of tolerance to aluminum (Tanaka, 1983), (v) succeeding crop benefited from the residual nitrogen fixed by rhizobium to the tune of 30 to 60 kg ha^{-1}, (vi) higher and remunerative land equivalent ratio (LER) makes soybean a better inter-crop/companion crop/mixed crop. Intercropping of soybean with other crop(s) has been found to be appropriate with pigeonpea, sorghum, maize, sugarcane, cotton, fingermillet and plantation crops and found to be highly remunerative and biologically efficient (LER 1.25 to 1.70). Some of the prominent soybean-based inter- cropping systems for different zones are given in Table 5.

Table 5 : Soybean based remunerative cropping systems for different zones

Zone	Crop sequence	Intercropping/mixed cropping
Central (Madhya Pradesh, Bundelkhand region of U.P., Rajasthan, Gujrat, Northern and western parts of Maharashtra)	Soybean-wheat, Soybean – chickpea, Soybean-wheat-corn fodder, Soybean- potato, Soybean- garlic/potato-wheat, Soybean- rapeseed or mustard, Soybean-pigeonpea, Soybean - safflower, Soybean- sorghum Soybean - linseed	Soybean + pigeonpea, Soybean + sorghum, Soybean + groundnut, Soybean + pearl millet, Soybean + cotton, Soybean in mango/guava orchard
Southern (Karnataka, Tamil Nadu, Andhra Pradesh, Southern parts of Maharashtra)	Soybean - finger millet - peas, Soybean – oat/ cowpea/barley, Soybean - finger millet - beans, Soybean – groundnut, Soybean - safflower	Soybean + pigeonpea, Soybean + fingermillet, Soybean + sugarcane, Soybean + sorghum, Soybean + groundnut, Soybean in coconut/ mango/ guava orchard and soybean in agro-forestry
Northern Plain (Punjab, Haryana, Delhi, North-Eastern plains of U.P., Western Bihar)	Soybean –wheat, Soybean – potato, Soybean – chickpea Soybean –mustard	Soybean +pigeonpea, Soybean + corn, Soybean + sorghum, Soybean in mango/ guava orchards, Soybean in agro-forestry
Northern hill (Himachal Pradesh, Uttarkhand.)	Soybean – wheat, Soybean – pea, Soybean – lentil, Soybean – toria	Soybean + corn, Soybean + pigeonpea
North east (Assam, Meghalaya, West Bengal, Bihar, Orissa)	Soybean – paddy, Paddy – soybean	Soybean + fingermillet, Soybean + paddy, Soybean + pigeonpea, Soybean + temperate grass

Source: Billore *et al.* (2004)

It is generally recommended to rotate legumes with non-legumes. Soybean forms excellent cropping sequence with cereal for effective utilization of residual nitrogen. Further, it fits well in traditional cropping sequences on account of availability of 85 to 130 days maturity duration varieties. The remunerative sequential cropping systems for different zones are given in Table 5.

The farmers grow soybean followed by wheat in case adequate / limited irrigation is available. Under rainfed conditions, chickpea is predominantly grown as succeeding crop. The other cropping systems like soybean-linseed, soybean-sunflower, soybean-safflower and soybean-mustard are agronomically feasible and economically viable have been adopted to a limited extent. Sequential crop after soybean can enhance system efficiency. For the predominant soybean growing regions having vertisols, it has been found that after a rainfed soybean, crops like pigeonpea, sorghum or chickpea can be taken on the residual soil moisture. Among them, the highest gross yield was obtained with soybean-safflower followed by soybean-chickpea and the highest gross returns were obtained with soybean-pigeonpea followed by soybean-chickpea (Bhaskar *et al.*, 1992).

Optimization of crop yield as well as water use efficiency of rainfed soybean based cropping systems on vertisols can be achieved through judicious use of N, P and S (Sharma and Gupta, 1992). For soybean and soybean based cropping systems also, the nutrient management techniques recommended to farmers should necessarily address the criteria like (a) technically efficient, (b) socially acceptable and (c) economically viable.

Fertilizer management scenario and nutrient recommendations

Soybean is mostly grown under rainfed conditions by resource poor farmers on soils which have been both thirsty as well as hungry consequently leading to poor productivity of the crop. In general, fertilization scenario in soybean has been grim. Firstly, fertilizer did not find much favour with the rainfed farmers. Subsequently, when soybean was made component of different cropping systems under assured water availability, high analysis fertilizers like urea and di-ammonium phosphate did replace the ammonium sulphate and single super phosphate. Due to total reliance on imports, use of muriate of potash or sulphate of potash has also faced benign neglect. The farm yard manure and other organic materials have found their way to only cereal crops (Vyas, 2008).

Soybean and soybean based cropping systems are high nutrient removing systems (Table 6 and 7). A soybean–wheat cropping system producing 7.7 t ha^{-1} soybean equivalent yield was reported to remove 260 kg ha^{-1} N, 85 kg

P_2O_5 and 204 kg K_2O ha^{-1} (Tandon and Sekhon, 1987), while, the same cropping system yielding 6.5 t ha^{-1} removed 416 g zinc, 3362 g iron, 488 g manganese and 710 g copper (Takkar, 1996).

Table 6 : Approximate uptake of nutrients by soybean

Major nutrients	(kg t^{-1}*)	Secondary nutrients (kg t^{-1}*)		Micro-nutrients (g t^{-1}*)	
N	70.7	S	6.7	Zn	77
P2O5	30.9	Ca	14.0	Fe	346
K2O	57.7	Mg	7.6	Mn	83
				Cu	30

* per tonne of seeds
Source: Tandon (2002)

The build up of nutrients also depends on the soil type and nutrient management practices. Tiwari *et al.* (2002) reported the build up to N and P under long term soybean-wheat system on Vertisols at Jabalpur, while, Wani *et al.* (2007) reported overall lower nitrogen depletion in soil when it was applied as compared to control in soybean-chickpea cropping system. Vyas *et al.* (2006) observed maximum soybean equivalent yield of soybean + pigeonpea intercropping system with 75% of recommended dose of fertilizers and 5 tonnes of farm yard manure. .

The combined application of crop residues (5 t ha^{-1}), FYM (5 t ha^{-1}), zinc (5 kg ha^{-1}) along with recommend dose of fertilizers (RDF) to soybean and RDF to wheat was found to increase the yield of both soybean and wheat in soybean-wheat cropping system. It further improved the organic carbon content in soil, available status of N, P, K and Zn in soil besides improving the nodulation, economics of soybean-wheat cropping system (Shivakumar and Ahlawat, 2008).

In a site specific nutrient management experiment at Guntur, application of recommended dose of fertilizer was sufficient for cotton when soybean was intercropped with cotton and it was at par with cotton crop raised with 25% N through organic sources alongwith 100% N through fertilizers indicating that soybean a legume crop, might have transferred additional N to the cotton crop (Anon., 2005).

In a 2 year rotation involving cotton-soybean-wheat at Bhopal and cotton-soybean-chickpea at Dharwad with recommended dose of fertilizers to soybean and succeeding crops showed that showed that yield attributes of soybean showed the residual effect of zinc and farm yard manure applied to previous cotton (Table 8). Further the INM treatments to previous cotton showed the maximum yield of soybean and chickpea, while wheat was unaffected (Anon., 2005).

Table 7 : Nutrient uptake (kg ha^{-1}) pattern of soybean based cropping systems

Cropping system	Location	*Kharif*			*Rabi*		
		N	P	K	N	P	K
Soybean - wheat @40N (½ N basal + ½ N top dressed) + $80P_2O_5$ +$20K_2O$ kg ha^{-1}	Morena, M.P.	96.0	14.9	59.1	148.2	34.8	68.1
Soybean - wheat @60:30:20N:P_2O_5:K_2O kg/ha + lime @ 1.25 t ha^{-1} + FYM @ 4 t ha^{-1} + B @ 10 kg ha^{-1}	Semiliguda, Orissa	149.6	18.2	46.0	105.8	9.0	63.0
Soybean – rapeseed @ 40 kg S ha^{-1}	Semiliguda, Orissa	138.4	14.1	32.8	77.9	14.2	53.1
Soybean – rapeseed @ 1.25 t lime ha^{-1}	Semiliguda, Orissa	144.2	14.2	29.6	78.6	16.4	52.4
Soybean- gobhi sarson @ 90 kg P_2O_5 ha^{-1}	Ludhiana, Punjab	-	16.85	-	-	29.9	-

Table 8 : Recommended organic manure and fertilizer dosages in different soybean growing areas

Soybean growing zone	Recommended organic manure and fertilizer
Northern Hill Zone (HP, Uttarkhand)	10 t FYM ha^{-1} 20:80: 20: 20 N:P_2O_5: K_2O:S kg ha^{-1}
Northern Plain Zone (Punjab, Haryana, Delhi, NE Plains of UP, Western Bihar)	10 t FYM ha^{-1} + 20:60: 20: 20 N:P_2O_5: K_2O:S kg ha^{-1}
Central Zone (MP, Bundelkhand Region of UP, Rajasthan, Gujarat, Western parts of Maharashtra)	10 t FYM ha^{-1} 20:60: 40: 20 N:P_2O_5: K_2O:S kg ha^{-1}
Southern Zone (Karnataka, TN, AP, Southern parts of Maharashtra)	10 t FYM ha^{-1} 20:80: 20: 20 N:P_2O_5: K_2O:S kg ha^{-1}
North Eastern Zone (Assam, West. Bengal, Meghalaya, Eastern Bihar, Orissa, Chhattisgarh)	10 t FYM ha^{-1} 20:80: 40: 20 N:P_2O_5: K_2O:S kg ha^{-1}

In general application of 30 kg N ha^{-1} has been found optimum, however the response varies from 40-60 kg N ha^{-1} depending upon the soil type. On an average phosphorus at 26.2 kg P ha^{-1} was found normal for obtaining higher yield and net returns from soybean. The response to potassium is low due to the fact that most of the soybean growing areas are rich in potassium. However, sporadic responses are also observed to application of 30-40 kg K_2O ha^{-1}. The recommended organic manures and chemical fertilizers to soybean and soybean based cropping systems have been furnished in Table 9.

Further, the positive effect of combined application of nutrients on the productivity of crops, nutrient uptake, improvement in protein content, oil content and soil fertility have also been observed by many researchers at many places. The right quantity, right time of application, synergy with other agronomic practices with regard to nutrients like phosphorus, potassium, sulphur, zinc, boron, etc. have also been observed. In many frontline demonstrations, integrated nutrient management has been highlighted to record higher yield of crop as well as higher income (Singh *et al.*, 2007; DSR, 2009).

INM option for soybean and soybean based cropping systems

By 2025 AD, there would be a large demand of nutrients to a tune of 45 million tonnes of NPK to achieve the desired level of production of different crops to sustain the rising population. Out of which, 35 million tonnes of nutrients could be met out through fertilizers and the remaining 10 million tonnes of nutrients has to come from other sources like organic manures, crop residues, biofertilizers, etc. The overdependence on fertilizer would cause depletion of natural deposits, which are finite and the import of fertilizers or raw material used in their manufacture would cause heavy drain of foreign exchange reserves. Besides, fertilizers are also susceptible to various losses consequently causing nutrient loss as well as environmental pollution. Continuous and non-judicious use of costly fertilizers adversely affects chemical and biological properties of the soil and thus hampers soil productivity and quality. Hence, there is an urgent need for INM in crop production. The INM is the rationalization of plant nutrient management in order to upgrade the efficiency of the plant nutrient supply to the crops and the farmers income through the adequate association of local and external plant nutrient sources accessible and affordable to the farmers (Ange, 1997). This concept envisages the management systems, which make use of all possible sources of nutrients in an integrated manner to achieve the desired level of productivity. The sources used must be economically viable, socially acceptable, locally available and eco-friendly. There is a great prospect for integrated nutrient management in soybean and soybean based cropping system. The various components of IPNS are discussed independently below.

Inorganic sources

Soybean cultivated soils show widespread deficiency of N and P, deficiency of K and S in large pockets and deficiency of Zn and B in small and scattered pockets. Therefore, there is need to manage these nutrients efficiently so as to obtain sustainable production of soybean and soybean based cropping systems.

Nitrogen

The N required for plant growth can be obtained either from symbiotic N_2-fixation or from the direct uptake from soil. N_2-fixation by field grown soybean may start as early as the second week after sowing. Keyser and Fudi (1992) reported that soybean can fix N_2 to the extent of 50-300 kg ha^{-1} and capable of meeting its nitrogen requirement. Soybean crop is an efficient symbiotic fixer meeting the plant requirement to the extent of 25 to 75%. It has been observed that soybean in soybean-wheat sequence leaves behind 45 kg N for succeeding crop enabling fertilizer economy in it (Trikha, 1985).

There are numerous reports suggesting a basal starter dose of 15-25 kg N ha^{-1} placed 5-8 cm below soil surface giving initial boost to the crop and helping in harnessing better yields (Pate and Dart, 1961). Further, the N requirement at the seed development stage is crucial. It has been suggested that as the nodule senescence starts after peak flowering, the symbiotically fixed nitrogen may not be adequate to meet the N needs of developing pod, and thus cause reduction in seed yield (Mahadkar and Saraf, 1992). This fact indicates that splitting the application of recommended N or a supplemental application of nitrogen at the time of flowering/grain filling stage of the crop may help in enhancing the soybean yields. The balanced and continuous supply of nitrogen increases productivity and energy budgeting of soybean. An additional application of 20 kg nitrogen at 60 days after sowing increased the seed yield by 11% over recommended dose of fertilizers (Vyas *et al.*, 2006).

Phosphorous

Unlike other crops, soybean does not exhibit characteristic phosphorus deficiency other than stunted growth. Meeting the P requirement of soybean not only maximizes its production but also benefits the succeeding crops like wheat on two accounts i.e. the usual residual effect of P and advantage as a result of better N_2-fixation under optimum P (Sharma and Vyas, 2001). The soybean uses hardly 1/5th of the applied P, while the rest gradually becomes available to succeeding crops. It acts as a catalyst to augment availability of native and fixed P (Prasad *et. al.*, 1998). Though, soybean responds well up to 60-80 kg P_2O_5 ha^{-1}, the actual crop requirement varies from 19-43 kg P ha^{-1} (Nambiar and Ghosh, 1984) Deep placement of phosphorus below seed zone as single super phosphate at the time of sowing is found to improve the yield and quality of soybean. The optimum level for P application to soybean in soybean-wheat cropping system is worked out to be 88-94 kg ha^{-1}.

Potassium

The potassium deficient plants cannot use water and other nutrients efficiently and are less tolerant to environmental stresses, pests and diseases. They show irregular yellowing around leaflet margins, chloratic areas form a continuous irregular yellow border, appearing first on older leaves usually during late flowering and early seed filling stages of crop. An healthy crop of soybean requires about 101-120 kg K ha^{-1} (Nambiar and Ghosh, 1984). Three years studies at NRCS revealed significant response of soybean-wheat cropping system to 50 kg K_2O ha^{-1} in two splits i.e.25 kg as basal + 25 kg at flowering to both crops. Tiwari *et al.* (2002) have summarized the response of soybean under varied conditions for inclusion of K in fertilizer schedule and resorting to split application along with the effect on seed quality. This aspect needs special attention to sustain the enhanced productivity of soybean and quality seed production. The positive response of soybean to potassium has also been observed by Vyas *et al.* (2007), Billore *et al.* (2008a) and Billore *et al.* (2009).

Secondary nutrients

Sulphur is an important nutrient after N, P and K for leguminous oilseed crops like soybean. Its application significantly influences the seed yield of soybean. The requirement of sulphur depends on the soil available sulphur and cropping system. On an average 20-60 kg S ha^{-1} has been observed to be optimum for soybean depending upon the soil type and the crop productivity (Hegde and Sudhakara Babu, 2009). Soybean crop responded significantly up to 20 kg S ha^{-1} (Chandel and Saxena, 1988) at Pantnagar, up to 40 kg S ha^{-1} applied through gypsum (Aulakh *et al.*, 1990) at Ludhiana, 60 kg S ha^{-1} applied through SSP (Nambiar, 1988) and 80 kg S ha^{-1} applied through pyrites (Sharma and Gupta, 1992). For soybean, SSP and gypsum are preferred due to association of P with the former and cheap and easy availability of gypsum with the latter. Joshi and Billore (1998) reported that on Vertisols at Indore, the response of sulphur was observed up to 20 kg ha^{-1} only. The economic optimum S levels were 10.38 and 21.85 kg S ha^{-1} giving soybean yield 1415 and 1114 kg ha^{-1} for PK 472 and MACS 58. Further, the applied sulphur was found to exhibit residual effect on the succeeding crop. Application of sulphur @ 25 kg ha^{-1} to soybean resulted in 25% increase in the yield of wheat grown after soybean in Morena region of Madhya Pradesh (Shrivastava, 2001).

The response to calcium is generally observed in acidic soils. Calcium as a nutrient to soybean in acid soils in Jharkhand @ 2-4 q ha^{-1} increased the yield by 26.7% mainly due to calcium rather than increase in phosphorus availability. Further, the magnesium too is found deficient in acid soils and its application @ increased the soybean yield by 10.5- 25.6% (Mathur, 1994).

Micronutrients

Micronutrients are found deficient in soybean growing areas. The application of these has elicited good response from the soybean both in sole and cropping system. The qualitative improvements like increased oils content, protein content etc. has been observed. On an average an increase of 3.8 q ha^{-1} seed yield was observed in soybean due to zinc fertilization (Sarkar and Singh, 2003). In zinc deficient soils (<0.45 ppm) application of zinc to soybean in soybean-wheat cropping system increased the yield of wheat by 6-27% in the Morena region of Madhya Pradesh (Shrivastava, 2001). In general, application of zinc @ 5-10 kg ha^{-1} (Pal and Gangwar, 2004) recorded good response in soybean. Generally basal application of zinc as $ZnSO_4$ @ 25 kg ha^{-1} at sowing or through foliar application (0.5%) at flowering was found sufficient. The regular application of boron @ 0.5 kg ha^{-1} in swell shrink soils of India (Singh *et al.*, 2003) increased the seed yield of soybean. Recommended rates for soil application vary from 0.5 - 2.0 kg B ha^{-1}, while foliar application @ 0.2 % solution of sodium tetra-borate twice is found sufficient. At Guna, Madhya Pradesh, application of boron @ 1 kg ha^{-1} to soybean had positive residual effect on the succeeding chickpea and increased its yield by 68% (Srinivasrao *et al.*, 2009). Response to iron application in soybean to the extent of 3 q ha^{-1} in Tamil Nadu (Takkar *et al.*, 1989), 4.5 q ha^{-1} at Pantnagar (Chaturvedi and Chandel, 2005) have been observed. The copper application is found to increase the protein and oil content of soybean. Likewise, the manganese application increased the soybean yield by 3.1 q ha^{-1} (Takkar *et al.*, 1989). At Pantnagar, application of 0.5 kg B ha^{-1} increased the yield of soybean by 30% over control, likewise an increase in grain yield of soybean by 1.6 q ha^{-1} and 3.5 q ha^{-1} due to application of molybdenum @ 0.8 kg ha^{-1} was also noticed at Coimbatore and Pantnagar, respectively (Saxena and Chandel, 1997). The application of FYM @ 10 t ha^{-1} alone was sufficient to compensate the requirement of different micronutrients (Zn, B and Mo) rather than going for individual application of micronutrients. The treatment was adjudged stable and sustainable for production of soybean (Billore *et al.*, 1999).

Organic sources

Soil organic carbon (SOC) stock is considered to be a single parameter to judge soil quality and health. To keep our soils healthy, it is necessary to ensure its restoration by appropriate agricultural activities. Application of organic inputs, use and incorporation of live mulches and recycling of crop residues regularly is one way to take care of soil organic carbon stocks and provide sustainability to land resources and cropping systems. The organic sources of nutrients in the country generated annually through livestock and human beings and from crop residues, compost, etc. range between 10.5-16.2 million tonnes of NPK, of which 3.9-5.7 million tonnes of the plant nutrient can be easily made available for agricultural use through systematic planning and utilization. As per the recent estimates, about 25% nutrient needs of Indian Agriculture can be met by utilizing various organic sources namely FYM (200 million tonnes), crop residue (30 million tonnes), urban/rural waste (10 million tonnes), green manuring (25 million tonnes). An all out effort is required to utilize the locally available organic resources and recycle the nutrients from them efficiently. The conjunctive use of organic inputs with mineral fertilization paves way for meeting the crop needs and sustainable production. At Indore, results accrued over four years revealed that the incorporation of farmyard manure or crop residues in conjunction with in inorganic fertilizers significantly enhanced the sustainability and stability over the control with respect to productivity of soybean-wheat cropping system. Integrated use of organic sources and 50% nutrients through inorganic fertilizers were more sustainable and stable as well as performed well under unfavourable environments (Billore *et al.*, 2008b).

Organic manures

The organic manures and composts have been used as sources of nutrients since time immemorial. However, in the recent years due to chemical fertilizers and clean cultivation their use in agriculture has been neglected. The use of organic manures improves soil physical, chemical and biological properties, besides improving the fertilizer use efficiency. These stimulate the proliferation of diverse group of soil microorganisms and play an important role in the maintenance of soil fertility. The organic manures are also rich in micronutrients and their use in conjunction with high analysis fertilizers help to overcome the micronutrient deficiency in the soil. Application of different organic sources at different rates reduce soil bulk density, increase total porosity, water filled pore space, aggregate stability and slaking resistance, plant available water and water retention.

Field studies conducted at Coimbatore indicate that application of FYM registered a significantly higher grain yield in soybean-sunflower system than with no FYM application. A long-term field experiment at Ranchi also indicates a significantly higher grain and straw yield with NPK + FYM + Lime under soybean - potato - wheat system. At Dharwad, application of FYM (10 t ha^{-1}) + RDF increased yields under both soybean-wheat and soybean-chickpea systems.

The beneficial effects of integrating organic residues with mineral fertilizers on soybean-wheat systems have been realized in terms of higher system productivity and improved soil environment for plant growth and soil health. Research reports with dominant soybean based cropping systems (Joshi *et al.*, 1998; Billore *et al.*, 1999) leads to the fact that integration of FYM with inorganic sources results in enhanced and sustained production. Similarly, soybean-safflower system performed better when both the crops were supplemented with FYM @ 6 t ha^{-1} with 20 kg N and 13 kg P (Vittal *et al.*, 2002). Application of 125% recommended dose of fertilizers (RDF) alongwith 10 t FYM ha^{-1} to soybean and wheat was enough for yield maximization of soybean-wheat system in clay loam soils at Jabalpur (Vishwakarm *et al.*, 2002). The conjunctive use of soybean residues @ 5 t ha^{-1} + FYM @ 5t ha^{-1} + zinc @ 5 kg ha^{-1} applied to soybean was most productive and remunerative in succeeding wheat (Billore *et al.*, 2005). In a long term experiment carried out at Almora, in soybean-wheat cropping system, partial factor productivity decreased with time in the unfertilized and the inorganic fertilizer management treatments, whereas they increased with N+FYM and NPK+FYM treatments for both the crops. The latter treatments also recorded higher soil organic carbon content and total nitrogen content. (Ranjan Bhattachary *et al.*, 2008).

Crop residues

Crop residues can also be used successfully for supplying nutrient requirement of soybean. The annual production of more than 350 million tonnes of crop residues in India, offer a great opportunity for their use as source of nutrients in many soybean growing areas reducing the dependence on chemical fertilizers. In a study at Delhi, inclusion of crop residues increased the grain yield of soybean significantly as compared to control. However, the combined application of fertilizer and crop residue did not show substantial improvement over fertilizer alone (Anon. 2008).

Biofertilizers

The use of efficient strains of *Rhizobium* for biological nitrogen fixation, phosphate solubilizing bacteria (PSB) for solubilizing fixed phosphorus in the soil, vesicular arbuscular mycorrhizae (VAM) for mobilizing the nutrient from the deeper layers in the rhizosphere and plant growth promoting rhizobacteria (PGPR) to modify and enhance the rhizosphere quality will greatly aid in meeting the nutrient demand of soybean and help in reducing substantially the application of nutrients in the form of organic manures or chemical fertilizers. Careful planning and adoption of these biofertilizers constitute an important component of INM strategy.

Biological nitrogen fixers

Biological nitrogen fixation (BNF) is an important mechanism found in soybean. It offers an economically attractive and ecologically sound means of reducing external inputs and improving the quality and quantity of internal resources of N. In soybean, N fixation is through a symbiotic association between the bacteria of the genus *Bradyrhizobium* and soybean crop. It requires selection of efficient strains, improved seed treatment or soil application techniques and complementary agro-techniques to enhance their activity. However, the problems associated with use of *Bradyrhizobium japonicum* culture viz. non-awareness about importance of culture among farmers, non-availability of potent culture in time, marketing of sub-standard and out dated culture limit the efficacy. Besides, high soil temperature during summer months may minimize the number of active cells of *Bradyrhizobium* in soil, making it necessary to resort to seed inoculation each year. Educating the farmers about proper use of *Bradyrhizobium* culture, which is a low cost input and economizes fertilizer use, is important.

Application of *Rhizobium* inoculation with micro-nutrients results in higher response. In an AICRP on soybean trial, there was significant impact of supplementing micronutrients with *Rhizobium* (Bj) on number of nodules, nodules biomass and seed yield. At Delhi combined application of Bj + B, at Pantnagar Bj + Zn, at Sehore Bj + B+ Zn + Mo, at Kota Bj +B + Zn+ Mo and at Ludhiana Bj + Zn were observed give very good yield (DSR, 2009). At Sehore and Pantnagar these treatments further increased the N content in seed indicating the improvement in protein content. Further, co-inoculation of two strains of *Rhizobium* was also found to increase the productivity of soybean. In one such study at Delhi, grain yield of soybean (PK-416) with co inoculation of *Bradyrhizobium japonicum* and *Rhizobacterium sp.* under rainfed conditions was substantially higher than other treatments. In another study the use of *Rhizobium* seed treatment in soybean was found to leave significant residual effect on the succeeding crop of mustard in soybean-mustard cropping system (Sanjeev Kumar and Shivani, 2002) indicating the potential benefits of *Rhizobium* inoculation in soybean crop in cropping systems. Co-inoculation of *Bradyrhizobium* with *Pseudomonas striata* and *Glomus lamellosum* in conjunction 60 kg P_2O_5 ha^{-1} recorded significantly higher grain yield of soybean (Vyas *et al.*, 2005).

Phosphate solubilizers

Phosphorus is essential for proper symbiosis and effective nitrogen fixation. Often a large portion of applied phosphorus gets fixed and become unavailable to crop plants. The heterotropic microorganisms viz. bacteria- *Bacillus megaterium, B. circulans, B. subtilis, Pseudomonas striata* etc., fungi- *Aspergillus awamori, Pencillium digitatum, Trichoderma* etc. have the ability to solubilize inorganic sources through excretion of organic acids like citric, glutamic, lactic, oxalic, glyoxalic, maleic, fumaric tartaric and ketoglutaric acid and make available the fixed phosphorus in the soil. This greatly reduces the need for application of phosphorus fertilizer in larger amount for every crop. The grain yield of soybean crop was significantly increased by 240 kg ha^{-1} due to rock phosphate plus *P. striata* treatment whereas with 80 kg P_2O_5 as superphosphate increase was hardly 100 kg ha^{-1}. A significant increase in P uptake by grain due to inoculants was obtained in both treatments with and without phosphatic fertilizers. *P. striata* showed maximum benefit on this aspect and superior to super phosphate application. N uptake was also favourably influenced due to inoculants by rendering greater phosphate availability to crop. Inoculation with phosphate solubilizing bacteria (PSB) in the presence of rock phosphate and FYM increased available P content. In addition, co-inoculation of *Rhizobium* with phosphate solubilizing bacteria (PSB) has been reported to enhance the P availability and yield. Shrivastava (2001) reported that the co-inoculation of *Rhizobium* and phosphate solubilizing bacteria (PSB) resulted in 25% saving of N and P in soybean.

Phosphate mobilizers

The symbiotic association between plant roots and fungal mycelia is termed as mycorrhizae. They are commonly called vesicular arbuscular mycorrhizae (VAM). These fungi belonging to the genera *Endogone, Glomus, Entrophosphora, Gigaspora, Acaulospora, Scutellospora* are obligate parasites and are mainly responsible for mobilization of phosphorus for plants. VAM possess special structure like vesicles and arbuscules within the root cortical cells. The arbuscules help in the transfer of nutrients from the fungus to the root system and the vesicles, which are sac like structure store P as phospho-lipids. VAM has shown to help plants acquire mineral nutrients from soil especially immobile elements such as P and Zn and mobile elements such as S, Ca, K, Fe, Mg, Mn, etc. In soils, where such elements may be deficient or otherwise less available, VAM increase efficiency of mineral uptake resulting in enhanced plant growth. Further more, mycorrhizal fungi can reduce plant response to other soil stresses such as high salt levels, toxicities, associated with mined soils or landfills, heavy metals. VAM have in some cases reduced the disease incidence due to some morphological or physiological change in the plant. They also alter soil texture by increasing the extent of soil particle aggregation and thus stability. In field trails with VAM (*Glomus fasiculatum*) indicated about 26.5% yield increase over control in soybean. The beneficial effect of mycorrhizal symbiosis is achieved at moderate rather than at higher P level though the efficiency and edaphic optima of fungi vary greatly with different levels of P.

Plant growth promoting rhizobacteria (PGPR)

Soil bacteria that aggressively colonize the rhizosphere and exert beneficial effects on plant growth are termed plant growth promoting rhizobacteria (PGPR). They belong to several genera *viz., Actinoplanes, Agrobacterium, Alcaligenes, Arthrobacter, Bacillus, Pseudomonas, Erwinia, Rhozobium, Bradyrhizobium,* etc. The nature of the growth promotion by the rhizobacteria involves the production of hormones, enhanced mineral and water uptake, improve plant growth by colonizing the root system and preventing the establishment of or suppressing deleterious rhizosphere microorganism (DRMO) or minor pathogens on the roots. *P. putida* stimulates root growth, shoot height and P uptake by shoots and roots. *P. fluorescens* also induced increased root hair formation. Co-inoculation of PGPR *viz Pseudomonas* and *B. japonicum* increased soybean plant root and shoot weight, grain yield, plant vigor, nodulation and N_2-fixation. The effect of PGPR on soybean nodulation is through increased plant vigour, rather than directly by N_2-fixing symbiosis.

Integrated nutrient management steps

With different sources of nutrients available, their careful integration facilitating easy adoption to increase nutrient use efficiency, increased productivity and monetary returns in synergy with other agronomic practices is necessary for realizing maximum benefits of INM practices. The nature of integration and management practices depend on different alternatives available, cropping systems and farmers' interest. The generalized steps for adoption of INM for soybean and soybean based cropping systems are:

- Summer cultivation after *rabi* crop.
- Application of organic manure 5-10 t ha^{-1} $year^{-1}$ preferably or at least every third year to soybean.
- Seed treatment with *Bradyrhizobium japonicum* @ 5 g kg^{-1} seed, PSB @ 5 g kg^{-1} seed, and soil application of VAM.
- Effective weed control in system to avoid nutrient drain and maintaining soil fertility.
- Application of recommended fertilizer levels to the crops in the system preferably based on soil test values.
- Need based application of micronutrients.
- Reverting the crop residues/wastes wherever possible to soil after the harvest of each crop.

Future thrust/concerns

In recent times the increasing scarcity of fertilizers and their escalating prices coupled with decreasing partial factor productivity of inorganic nutrients have forced many farmers to think of alternatives available and integration of different sources of nutrients to increase productivity, monetary returns and ensure sustainability of soybean and soybean based cropping systems in many areas. The INM hold a great promise in this direction. The following issues need to be emphasized to make the best use of INM concept.

- Development of land map for suitability of soybean crop and crop sequences based on soil site and agro-ecological zone is to be prepared for appropriate planning at macro level to achieve higher productivity from the system.
- Identification of reasons for decline in soil health and quality under soybean based cropping systems and measures to over come them.
- Harnessing synergies among different nutrients and nutrient sources.
- Evolve measures to increase input use efficiency to make the production system more cost effective and energy efficient.
- Development of management schedule for nutrients, moisture and thermal stresses based on integrated approach as all these are closely interrelated (Reddy *et al.*, 2010).
- Creating awareness among the policy makers, farmers and all other concerned about the inm possibilities.

References

Ange, A.L. (1997). Integrated plant nutrition system – An overview. In. Proceedings of the National Workshop on IPNS (Eds. Kumar, V., Govil, B.P. and Kaore, S.V.). 10-12 March 1997, held at Bhubaneswar, IFFCO, New Delhi. pp.16-33.

Anonymous, (2005). *Annual Report*. Central Institute of Cotton Research, Nagpur.

Anonymous, (2008). *Annual Report 2007*. Division of Agronomy, Indian Agricultural Research Institute, New Delhi.

Aulakh *et. al.* (1990). P-S Interrelationships for Soybean on P and S deficient soils. *Soil Science*, **150**: 705-709.

Badal, P.S., Billore, S.D. and Joshi, O.P. (2000). Socio Economic Impact of Soybean cultivation in M.P. *ICAB Project*, Division of Agricultural Economics, Indian Agricultural Research Institute, New Delhi.

Bhaskar, K.S., Balkar, S.Y., Pattiwar, V.V. and There, D.N. (1992). Efficient soybean-based cropping systems for the black cotton soils. *Indian Farming*, April:3-6.

Billore, S.D., Joshi, O.P. and Ramesh, A. (1999). Production potential, economic feasibility and energy indices of soybean based cropping system. In: *Proceedings of VI WSRC held at Chicago*, USA from 4-7 August, pp. 581.

Billore, S.D., Ramesh, A., Vyas, A.K., Joshi, O.P. and Pandya, N. (2008a). Evaluation of potassium uptake and utilization efficiency in soybean genotypes. *Soybean Research* **6**:31-36.

Billore, S.D., Vyas, A.K. and Joshi, O.P. (2004). Soybean agronomy. In. Singh, N.B., Chauhan, G.S., Vyas, A.K. and Joshi, O.P. (Eds). Soybean Production and Improvement in India. National Research Centre for Soybean (ICAR), Indore, India. pp.106-125.

Billore, S.D., Vyas, A.K., Ramesh, A., Joshi, O.P. and Khan, I.R. (2008b). Sustainability of soybean (*Glycine max*) – wheat (*Triticum aestivum*) cropping system under integrated nutrient management. system. *Indian Journal of Agricultural Sciences* **78**(4):358-361.

Billore, S.S., Ramesh, A., Vyas, A.K., Joshi, O.P. and Pandya, N. (2009). Potassium use efficiencies and economic optimization as influenced by levels of potassium and soybean (*Glycine max*) genotypes under staggered planting. *Indian Journal of Agricultural Science* **79**(7):510-514.

Billore, S.S., Vyas, A.K. and Joshi, O.P. (2005). Effect of integrated nutrient management on productivity, energy use efficiency and economic of soybean (*Glycine max*) – wheat (*Triticum aestivum*) cropping system. *Indian Journal of Agricultural Science* **75**(10):644-646.

Chandel, A.S. and Saxena, S.C. (1988). Technology for raising soybean productivity in UP. *Indian Farming*, **88**: 10-12.

Chaturvedi, S. and Chandel, A.S. (2005). *Indian Journal of Agronomy* **43**:731-733.

DSR, (2009). Director's Report and Summary Tables of Experiments (2008-09). All India Coordinated Research Project on Soybean, Directorate of Soybean Research, Indore.

DSR, (2010). Director's Report and Summary Tables of Experiments (2009-10). All India Coordinated Research Project on Soybean, Directorate of Soybean Research, Indore.

FAO, (2009). Food and Agriculture Organization, FAOSTAT, Rome.

GOI, (2007). Agricultural Statistics at a Glance. Directorate of Economics & Statistics, Department of Agriculture & Co-operation, Ministry of Agriculture, Govt. of Indian, New Delhi.

Hegde, D.M. and Sudhakara Babu, (2009). Improving yield and quality of oilseeds through secondary and micronutrients. *Indian Journal of Fertilizers* **5**(4):29-36, 39-46 & 49-55.

Joshi, O.P., Billore, S.D. and Bundela, V.P.S. (1998). Fertilizer management in soybean-chick pea crop sequence. *Indian Journal of Oilseeds Research*, **15**(1): 118-123.

Keyser h.h. and fudi li. (1992). Potential for increasing biological nitrogen fixation in soybean. *Plant and soil*, **141**:119-135.

Lawn, R.J. (1982). Response of four grain legumes to water stress in south eastern Queensland: Physiological response mechanism. *Australian Journal of Agricultural Research*, **33** : 481-486.

Mahadkar, U.V. and Saraf, C.S. (1992). Nitrogen nutrition of pulses – A review. *Journal of Maharashtra Agricultural Universities*, **17**(1): 30-37.

Mathur, B.S. (1994). Acid soils of Bihar – Their characteristics and Management, SSAC(BAU), Ranchi, *Research Bulletin No.* **1** : 44.

Mishra, S.N., Paikary, R.K. and Mishra, K.N. (1999a). *Indian Journal of Agronomy* **44**(1): 26-29.

Mishra, S.N., Paikary, R.K. and Mishra, K.N. (1999b). *Indian Journal of Agronomy* **44**(1): 56-58.

Nambiar, K.K.M. (1988). Crop responses to S and S-balance in intensive cropping system. In: *Proceeding of TSI-FAI Symposium on Sulphur in Indian Agriculture.* S III/3/ 1-16.

Nambiar, K.K.M. and Ghosh, A.B. (1984). Highlights of research of a long-term fertilizer experiment in India (1971-82). *Technical Bulletin 1*, Long-term fertilizer experiment projects, IARI. pp.100.

Nathanson, K., Lawn, R.J., Jassun, D.L.M. de, and Byth, D.E. (1984). Growth, nodulation and nitrogen accumulation by soybean in saturated soil culture. *Field Crop Research*, **8**: 73-92.

Pal, S.S. and Gangwar, B. (2004). Nutrient management in oilseeds based cropping systems. *Fertilizer News*, **49**(2): 37-38 and 41-45.

Pate, J.S. and Dart, P.J. (1961). Nodulation studies in Legume. *Plant and Soil*, **15**(4): 329-346.

Prasad, R., Rai, R.K., Sharma, S.N., Singh, S., Shivay, Y.S. and Idnani, L.K. (1998). Nutrient management. In : *Fifty Years of Agronomic Research in India*, Yadav RL, Singh Punjab, Prasad R and Ahlawat IPS (eds.), Indian Society of Agronomy, New Delhi. pp. 51-86.

Ranjan Bhattachary, Kundu, S., Ved Prakash, and Gupta, H.S. (2008). Sustainability under combined application of mineral and organic fertilizers in a rainfed soybean-wheat systems of the Indian Himalayas. *European Journal of Agronomy* **28**:33-46.

Reddy, K.S., Mohanty, M., Rao, D.L.N., Rao, A.S., Blamey, F.P.C., Dalar, R.C., Dixit, C.S.K.., Pandey, M. and Menzies, N.W. (2010). Development of farmers' participatory integrated nutrient management technology using the Mother – Baby Trial approach. World Congress of Soil Science, Soil Solutions for a Changing World, 1 – 6 August 2010, Brisbane, Australia.

Sanjeev Kumar and Shivani, (2002). Extended Summaries Vol. 1. 2nd *International Agronomy Congress*, 26-30 November 2002, New Delhi. pp.460-461.

Sarkar, A.K. and Singh, S. (2003). *Fertilizer News* **48**(4);47-54.

Saxena, S.C. and Chandel, A.S. (1997). *Indian Journal of Agronomy* **42**(2):329-332.

Sharma, K.R., Shrivastava, U.K. Tiwari, P.N. and Yadav, R.P. (1999). *Indian Journal of Agronomy* **44**(3):493-498.

Sharma, R.A. and Gupta, R.K. (1992). Response of rainfed soybean *(Glycine max)* - Safflower (*Carthamus tinctorius*) sequence to nitrogen and sulphur fertilization in vertisols. *Indian Journal of Agricultural Sciences*, **62**: 529-534.

Sharma, S.C. and Vyas, A.K. (2001). Residual effect of phosphorus fertilization and farmyard manure on productivity of succeeding wheat (*Triticum aestivum*) after soybean (*Glycine max*). *Indian Journal of Agronomy* **46** (3) : 416-420.

Shivakumar, B.G. and Ahlawat, I.P.S. (2008). Integrated nutrient management in soybean (*Glycine max*)-wheat (*Triticum aestivum*) cropping system. *Indian Journal of Agronomy* **53**(4):273-278.

Shrivastava, U.K. (2001). System Based Low Cost Production Technologies for Small Farmers (Eds. Gangwar, B. and Yadav, R.L.). Project Directorate of Cropping Systems Research, Modipuram, Uttar Pradesh. 179p.

Singh, D.K., Gautam, U.S. and Singh, R.K. (2007). Study on yield gap and level ot demonstrated crop production technology in Sagar District. *Indian Research Journal of Extension Education* 7(2&3):94-95.

Singh, J., Kolar, J.S. and Deol, J.S. (2000). *Indian Journal of Agronomy* **45** (1) : 31-36.

Singh, M.V., Patel, K.P. and Ramani, V.P. (2003). Crop responses to secondary and micronutrients in swell-shrink soils. *Fertilizer News*, **48** (4) : 63-68.

Srinivasarao, Ch, Wani, S.P., Sahrawat and Rajasekharrao, B. (2009). Nutrient management strategies in participatory watershed in semi arid tropical India. *Indian Journal of Fertilizes* **5** (12):113-116-119,125-128.

Takkar, P.N. (1996). *Journal of Indian Society of Soil Science* **44** : 563-581.

Takkar, P.N., Chibba, I.M. , and Mehta, S.K. (1989). Twenty Years of Coordinated Research on Micronutrients in Soils and Plants 1967-1987, IISS, Bhopal. 314p.

Tanaka, A. (1983). The physiology of soybean yield improvement. In : *Proceedings of Symposium on "Soybean in Tropical and Sub-Tropical System"*, Tsukuba, Japan, pp. 77-80.

Tandon, H.L.S. (2002). Nutrient Management Recommendations for Pulses and Oilseeds. Fertiliser Development and Consultation Organization, New Delhi.

Tandon, H.L.S. and Sekhon, G.S. (1987). Potassium Research Agricultural Production in India. FDCO, New Delhi.

Tiwari, A., Dwivedi, A.K. and Dikshit, P.R. (2002). *Journal of Indian Society of Soil Science* **50** : 442-447.

Trikha, R.N. (1985). The potential of soybean in Indian cropping system. In: *Soybean in Tropical and Subtropical Cropping System.* Proceedings of Symposium, Taukuba, Japan. The Asian Vegetable Research and Development Centre, Shanua, Taiwan, China. pp. 77-80.

Vishwakarma, S.K., Tiwari, K.K., Sharma, R.S. and Rathi, G.S. (2002). Extended Summaries Vol. 1. 2nd *International Agronomy Congress*, 26-30 November 2002, New Delhi. pp.1056-1057.

Vittal, K.P.R., Marutisankar, Singh, H.P. and Samra, J.S. (2002). Sustainability of practices of dry land agriculture: Methodology and Assessment, All India Coordinated Research Project for Dryland Agriculture, CRIDA, ICAR, Hyderabad. pp.100.

Vyas, A.K. (2008). Integrated nutrient management in soybean and soybean cropping systems. Improved Soybean Production Technology, KVK Kasturba, Indore.

Vyas, A.K., Billore, S.D. and Joshi, O.P. (2006). Productivity and economics of integrated nutrient management in soybean (*Glycine max*) + pigeonpea (*Cajanus cajan*) intercropping system. *Indian Journal of Agricultural Sciences* **76** (1):7-10.

Vyas, A.K., Billore, S.D., Joshi, O.P. and Pachlania, N.K. (2007). Influence of balanced nutrition on productivity of soybean. *Soybean Research* **5** : 21-25.

Vyas, A.K., Billore, S.D., Joshi, O.P. and Pachlania, N.K. (2006). Productivity of soybean (*Glycine max*) genotypes as influenced by nitrogen and sulphur nutrition. *Indian Journal of Agricultural Sciences* **76** (4):272-273.

Vyas, A.K., Billore, S.D., Joshi, O.P., Pandya, N. and Pachlania, N.K. (2005). Effect of lone dual inoculation of AM and PSB with phosphorus application on symbiotic attributes and soybean yield. PROM Review, Prom society, Udaipur.

Vyas, A.K., Billore, S.S., Chauhan, G.S. and Pandya, N. (2009). Agro-economic analysis of soybean based cropping systems. *Indian Journal of Fertilizers* **4** (5):41-42 & 45-51.

Wani, S.P., Srininvasarao, Ch., Rego, T.J., Pardhasaradhi, G. and Somnath Roy, (2007). *Indian Journal of Dryland Agriculture Development* (In press).

Wright, G.C., Smith, C.J. and Nelson, I.B. (1988). Growth and yield of soybean under wet soil culture and conventional furrow irrigation in south-eastern Australia. *Irrigation Science*, **9** : 127-142.

□□□

System Based Integrated Nutrient Management, 2012
© *B. Gangwar & V.K. Singh (eds.), pp. 191-205*
New India Publishing Agency, New Delhi (India)
e-mail : info@nipabooks.com; website : www.nipabooks.com

CHAPTER **12**

Site-Specific Nutrient Management for Sustained Higher Crop Productivity

V.K. SINGH

The apparent reasons for undesirable productivity under intensively cultivated areas are depletion of native nutrient reserves, emergence of multi-nutrient deficiencies, and consequent decline in factor productivity of applied nutrient (Dwivedi *et al.*, 2006; Yadav, 1998; Yadav, 2003). Surveys conducted at PDFSR in different agro-climatic regions of Indo- Gangetic Plain indicated that in order to achieve previously attained yield levels, farmers have started applying greater doses of N than the recommended ones (Dwivedi *et al.*, 2001; Sharma, 2003). Such indiscriminate use of N is likely to further worsen the nutrient imbalance in soil-plant systems, besides increasing the pest incidence, cost of production and environmental problems (Dwivedi *et al.*, 2001). The annual productivity of rice - wheat could hardly exceed 9 t ha^{-1} in the multi-location experiments on yield maximization (Dwivedi and Singh, 2011). On the other hand, long-term experiments (Singh and Mishra, 2010) and other studies indicate that crop productivity can be sustained with balanced fertilization, taking into account the emerging deficiencies of secondary and micronutrients (Tiwari *et al.*, 2006)

Nutrient management is a major component of a soil and crop management system. It becomes even more important in intensive high yielding cropping systems because of the higher investments at stake. The SSNM is applying those concepts to areas within a field that are known to require different management options from the field average. Site specific nutrient management is a concept that can be applied to any field and any crop. Yet, it is most often thought of in relation to use of computer and satellite equipment, and does not require large farming operation. The technology tools certainly expand the capabilities for using site-specific management. This is really a "repackaging" of management concepts that have been developed and promoted for many years.

The SSNM is basically a systematic approach to apply sound agronomic management to small areas of a field that can be identified as needing special treatment. The component of site-specific management may not be new, but now we have the capability with new technology to use them more effectively. Site-specific management includes practices that have been previously associated with maximum economic yield (MEY) management, best management practices (BMPs) as well as general agronomic principles. The systematic implementation of these practices into site-specific systems is probably our best opportunity to develop a truly sustainable agriculture system. Over all, site-specific nutrient management (SSNM) is a systematic approach of nutrient management that takes into account all nutrient deficiencies involving nutrient supply not only to meet crop demands but also to improve soil fertility levels, which finally ensures higher nutrient use efficiency, higher crop productivity and higher economic returns (Dobermann, *et al.,* 2004). In foregoing chapter, the yield gap analysis between farmer's field and genetic potential and SSNM approaches for attaining maximum productivity and profitability are being discussed for making future strategies for planners, development agencies and researchers.

Yield gaps analysis for achieving high yield targets

Crop yields and farmers profit can be increased in a sustainable and environmentally sound fashion if we have a better understanding of yield gaps. To isolate the most important constraints to achieving optimal yield and profit, yield gaps are analyzed stepwise by estimating the yield potential, the attainable yield, and the actual yield (Fig. 1).

The yield potential (Yp) is the theoretical maximum yield of a crop in any given season determined solely by climate and germplasm. Water and nutrients are at optimal levels and yield-reducing factors such as pests and diseases are absent. Yp is commonly estimated using plant growth models and could fluctuate from year to year (±10%) because of climate.

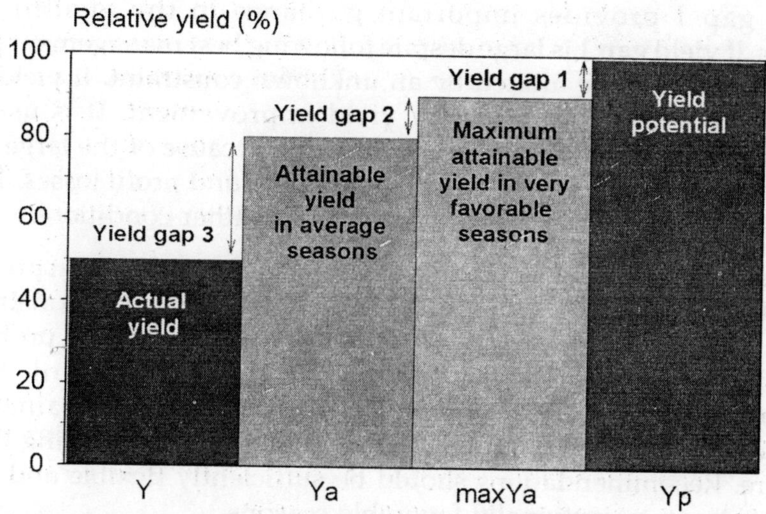

Fig. 1 : Example for the effect of nutrient and crop management on actual yield (Y), attainable yield (Ya), and maximum attainable yield (maxYa) in relation to the yield potential (Yp).

The attainable yield (Ya) is defined as the yield achieved in farmers' fields with best management practices including water, pest, and general crop management where nutrients are not limiting. The attainable yield varies – like the yield potential – from season to season and year to year depending on climate. The optimal economic yield is often linked to the attainable yield. The maximum attainable yield (maxYa) in any given season could be close to the yield potential, if management is excellent and weather conditions are very favorable. The actual yield (Y) in farmers' fields is often lower than the attainable yield due to constraints like water availability, pests and diseases, and poor crop and nutrient management practices.

Actual, attainable, and potential yield can be used to identify exploitable yield gaps (Fig. 1). The management objective of farmers should be to minimize yield gap 3, the difference between attainable and actual yield (Ya-Y). To narrow this yield gap, farmers need to evaluate promising new technologies (e.g., planting density, nutrient management) that offer improvements in yield and/or productivity against current practices. Larger yield increases can be achieved when several constraints (e.g. pests and disease problems and inappropriate nutrient management) are overcome simultaneously.

Yield gap 2 is largely determined by factors that are difficult or impossible to control including the variation in climatic conditions. Best management practices such as the use of a leaf color chart (LCC) for fine tuning N management increase the likelihood of keeping yield gap 2 small.

Yield gap 1 provides important guidance in the identification of constraints. If yield gap 1 is large despite following best management practices, attainable yield must be limited by an unknown constraint. If yield gap 1 is small, there is no further room for yield improvement. It is usually not economical to aim at fully reducing yield gap 1 because of the large amounts of inputs required and the high risk of crop failure and profit losses. This yield gap is smaller in seasons with very favorable weather conditions.

Farmers need to understand the effect of specific practices on productivity and profitability and the synergy achievable when several constraints are overcome simultaneously, for example when pest or disease problems are alleviated through more appropriate nutrient management. Fertilizer recommendations can then be developed based on the attainable yield (= yield target) to achieve high yield and profit while minimizing the risk of crop failure. Recommendations should be sufficiently flexible and robust to achieve maxYa in exceptionally favorable seasons.

Background of SSNM

Existing fertilizer recommendations for crop often consist of one predetermined rate of nitrogen (N), phosphorus (P), and potassium (K) for vast areas of crop production. Such recommendations assume that the need of a crop for nutrients is constant over time and over large areas. But the growth and needs of a crop for supplemental nutrients can vary greatly among fields, seasons, and years as a result of differences in crop-growing conditions, crop and soil management, and climate. Hence, the management of nutrients for crop requires a new approach, which enables adjustments in applying N, P, and K to accommodate the field-specific needs of the crop for supplemental nutrients.

In the SSNM approach, the plant's need for fertilizer N, P, or K is determined from the gap between the crop demand for sufficient nutrient to achieve a yield target and the supply of the nutrient from indigenous sources, including soil, crop residues, manures, and irrigation water. This SSNM approach enables:

1. Dynamic adjustments in fertilizer N, P, and K management to accommodate field- and season-specific conditions.
2. Effective use of indigenous nutrients originating from sources other than fertilizer.
3. Efficient fertilizer N management through the use of the leaf color chart (LCC), which helps ensure that N is applied at the time and in the amount needed by the rice crop.

4. Use of the omission plot technique to determine the requirements for fertilizer P and K.

Here yield targets are sets based on grain yield obtained in nutrient omission plots which is used for estimating fertilizer N, P, and K requirements.

The SSNM terminology

Attainable yield: The average grain yield of crop in farmers' fields with good management practices and without nutrient limitation to yield.

Nutrient-limited yield: The grain yield for a crop not fertilized with the nutrient of interest but with good management and ample supply of all other nutrients from indigenous sources or fertilizer. The nutrient limited yield is an indirect measurement of the soil indigenous nutrient supply.

Indigenous nutrient supply: The amount of a particular nutrient from all sources except mineral fertilizer (i.e., soil, crop residues, irrigation water, manure) available to the crop during a cropping season.

Nutrient omission plot: A plot not fertilized with the nutrient of interest while all other nutrients are applied in sufficient amounts. This plot used to measure grain yield and plant biomass as an indicator of the effective indigenous nutrient supply.

Yield response (to fertilizer): Measured by the difference in grain yield from a fully fertilized plot and a nutrient omission plot. The yield response largely determines the total requirement for fertilizer N, P, and K to meet the crop's nutrient demand for a high yield at maximum economic return.

What is the site-specific nutrient management approach?

Site-specific nutrient management (SSNM) enables farmers to optimally supply their crop with essential nutrients. With SSNM, plant-essential nutrients are supplied as and when required to ensure the 'feeding' of the crop to optimally meet its nutrient needs.

The site-specific nutrient management (SSNM) approach emphasizes 'feeding' crop with nutrients as and when needed. SSNM strives to enable farmers to dynamically adjust fertilizer use to optimally fill the deficit between the nutrient needs of a high-yielding crop and the nutrient supply from naturally occurring indigenous sources such as soil, organic amendments, crop residues, manures, and irrigation water. The SSNM approach does not specifically aim to either reduce or increase fertilizer use. Instead, it aims to apply nutrients at optimal rates and times to achieve high yield and high efficiency of nutrient use by the crop, leading to high net profit per unit of fertilizer invested. The SSNM approach involves three steps:

Step 1: Establish an attainable yield target

Crop yields are location and season specific – depending upon climate, cultivar, and crop management. The yield target for a given location and season is the estimated grain yield attainable with farmers' crop management when N, P, and K constraints are overcome. The amount of nutrients taken up by a crop is directly related to yield. The yield target therefore indicates the total amount of nutrients that must be taken up by the crop.

Step 2: Effectively use existing nutrients

The SSNM approach promotes the optimal use of existing (indigenous) nutrients coming from the soil, organic amendments, crop residue, manure, and irrigation water. The uptake of a nutrient from indigenous sources can be estimated from the *nutrient-limited yield*, which is the grain yield for a crop not fertilized with the nutrient of interest but fertilized with other nutrients to ensure they do not limit yield.

Step 3: Apply fertilizer to fill the deficit between crop needs and indigenous supply

Fertilizer N, P, and K are applied to supplement the nutrients from indigenous sources and achieve the yield target. The quantity of required fertilizer is determined by the deficit between the crop's total needs for nutrients (as determined by the yield target) and the supply of these nutrients from indigenous sources (as determined by the nutrient-limited yield). The required fertilizer N is distributed in several applications during the crop growing season to best feed the crop's need for supplemental N. Fertilizer P and K are applied in sufficient amounts to overcome deficiencies and maintain soil fertility. In the SSNM approach, fertilizers are applied using the following principles to achieve high yield and high efficiency of plant use. Here an example of rice crop is given:

1. Apply only a moderate amount of fertilizer N to young rice within the 14 days after transplanting (DAT) or 21 days after sowing (DAS), when the need of the crop for supplemental N is small.
2. Apply fertilizer N after 14 DAT or 21 DAS based on the crop's need for supplemental N, as determined by leaf N status. The leaf color chart (LCC) is a tool that could be used for assessing leaf N status and the crop's need for N.
3. Apply all fertilizer P near transplanting or sowing.
4. Apply fertilizer K twice – 50% near transplanting or sowing and 50% at early panicle initiation. When fertilizer K rates are relatively low (for example, =30 kg K_2O ha^{-1}), all fertilizer K can be applied near transplanting or sowing.

Methodology for computation of nutrient doses under SSNM

SSNM aimed at dynamic field-specific management of N, P, and K fertilizers to optimize the supply and crop demand for nutrients. The crop's need for fertilizer N, P, or K was determined from the gap between the crop demand for sufficient nutrient to achieve a yield target and the nutrient supply from indigenous sources. Initially a modification of the QUEFTS model (Janssen *et al.*, 1990) was used to predict the amount of fertilizer N, P, and K required for a specific yield target. In current SSNM version grain yield targets and grain yield in nutrient omission plots can now be directly used in a simplified manner by estimating fertilizer N, P, and K requirements as follows:

a. Establish an attainable yield target for farmers' fields

The yield target in the SSNM approach is directly used to calculate fertilizer rates for farmers' fields. It must be reasonably attainable by farmers. A yield target higher than one attainable by farmers would lead to recommendations of more fertilizer than required for high use efficiency and profit. A yield target below that realistically attainable by farmers could result in sub-optimal yield and profit. Grain yield from a fully fertilized plot with no nutrient limitations and good management (for example, the NPK plot or NPK plus micronutrient plot in the nutrient omission plot technique) can be used to estimate the yield target.

b. Estimate fertilizer N required and formulate dynamic N management

The difference between the yield target and N-limited yield (i.e., yield with no N fertilizer and no limitation of other nutrients) provides an estimate of anticipated crop response to fertilizer N. The estimated yield response to fertilizer N and a targeted efficiency for fertilizer N use are used to approximate the total requirement of the crop for fertilizer N, which is dynamically apportioned among multiple times of application to best match the crop's need for N.

c. Estimate field-specific nutrient-limited yields for P and K

Nutrient-limited yields are determined by the nutrient omission plot technique. The K-limited yield is determined in a K omission plot receiving no K fertilizer but sufficient N and P to ensure they do not limit yield. The P-limited yield is determined in a plot receiving no P fertilizer but sufficient supply of other nutrients.

d. Determine fertilizer P and K rates

The crop's need for fertilizer P is based on a comparison of the yield target and P-limited yield, whereas the crop's need for fertilizer K is based on a comparison of the yield target and K-limited yield. The SSNM approach advocates sufficient use of fertilizer P and K to both overcome P and K deficiencies and maintain soil P and K fertility.

Effect of SSNM on crop productivity

The SSNM is basically a systematic approach to apply sound agronomic management to small areas that can be identified as needing special treatment. Site-specific nutrient management (SSNM), considering indigenous nutrient supply of the soil and productivity targets is a strategy that may provide sustained high yields on one hand, and assure restoration of soil fertility on the other. On-station and on-farm experiments under AICRP_IFS indicated SSNM as a promising agro-technique to attain high productivity goals when nutrients are applied for high yield targets considering all deficient nutrients under rice- wheat system (Table 1).

Table 1: Grain yield response to SSNM and state recommended fertilizer doses over farmers' nutrient management practice

Treatment	Rice			Wheat			Rice-wheat system		
	Yield	Response		Yield	Response		Yield	Response	
	(kg ha^{-1})	kg ha^{-1}	%	(kg ha^{-1})	kg ha^{-1}	%	(kg ha^{-1})	kg ha^{-1}	%
				Sabour					
SSNM	8228	3267	65.90	5175	1924	59.2	13403	5191	63.2
SR	6032	1071	21.60	4551	1300	40.0	10583	2371	28.9
FP	4961	-	-	3251	-	-	8212	-	-
				Palampur					
SSNM	5285	1145	27.66	3415	1264	58.8	8700	2409	38.3
SR	4698	558	13.48	2991	840	39.1	7681	1398	22.2
FP	4140	-	-	2151	-	-	6291	-	-
				Ranchi					
SSNM	6758	2563	61.10	4046	1468	56.9	10804	4031	59.5
SR	5960	1765	42.10	3396	18	31.7	9356	2583	38.1
FP	4195	-	-	2578	-	-	6773	-	-
				R.S. Pura					
SSNM	8401	1708	25.52	4640	1347	40.9	13041	3055	30.6
SR	7384	691	10.32	4074	781	23.7	11458	1472	14.7
FP	6693	-	-	3293	-	-	9986	-	-
				Ludhiana					
SSNM	10431	1296	14.19	6024	391	6.9	16455	1687	11.4
SR	9808	673	7.37	5792	159	2.8	15600	832	5.6
FP	9135	-	-	5633	-	-	14768	-	-
				Faizabad					
SSNM	8279	3078	59.18	4428	1750	65.4	12707	4828	61.3
SR	6132	931	17.90	3419	741	27.7	9551	1672	21.2
FP	5201	-	-	2678	-	-	7879	-	-

Contd...

				Kanpur					
SSNM	9231	2341	33.98	5691	1153	25.4	14912	3494	30.6
SR	8283	1393	20.22	5265	727	16.2	13548	2120	18.6
FP	6890	-	-	4538	-	-	11428	-	-
				Modipuram					
SSNM	10184	3155	49.9	6101	1552	34.1	16285	4707	40.7
SR	7731	702	9.99	5411	862	19.0	13142	1564	13.5
FP	7029	-	-	4549	-	-	11578	-	-
				Varanasi					
SSNM	7028	1004	16.67	4193	807	23.8	12464	1932	18.3
SR	6527	503	8.35	3853	467	13.8	11610	1078	10.2
FP	6024	-	-	3386	-	-	10532	-	-
Mean over location									
SSNM	8203	1951	35.6	4857	1295	41.3	12787	3295	38.1
SR	6951	921	16.8	4306	744	23.8	11040	1549	18.3
FP	6030	-	-	3562	-	-	9492	-	-
CD at 5%	599	-	-	248	-	-	710	-	-

SSNM= Site- specific nutrient management; FP= Farmers nutrient management practice
Source: Singh *et al.*, 2008

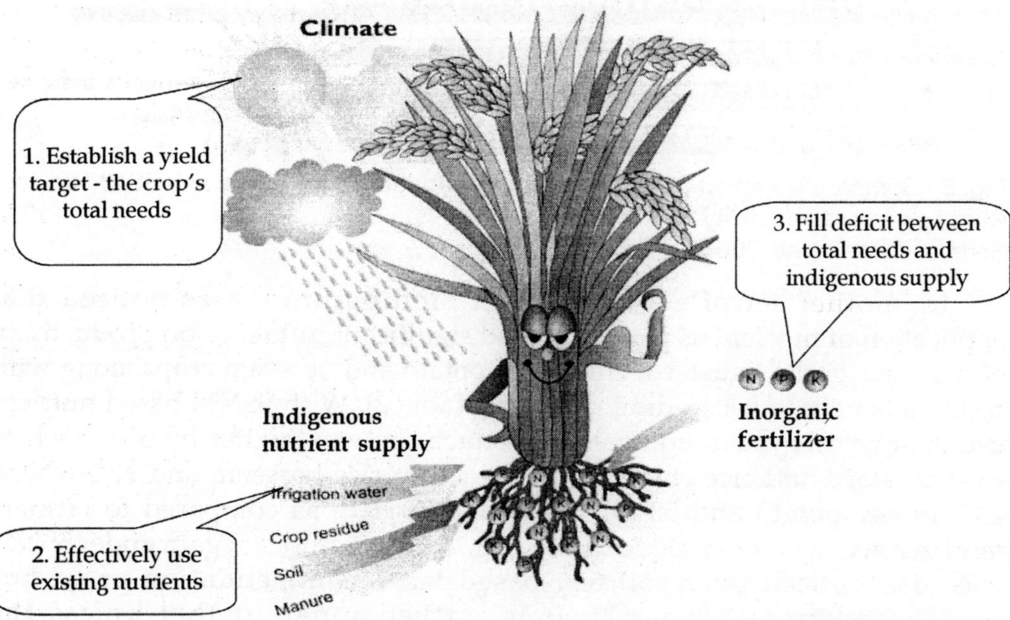

Fig. 2 : Site- specific nutrient management under rice crop (Dobberman *et. al.*, 2004)

Studies conducted under AICRP_IFS indicated that grain yields of 15-17 (t ha^{-1} yr^{-1}) are attainable in rice-wheat cropping system and yield gain under SSNM was apparently higher as compared to FP at all the locations (Fig. 3).

Among the different locations highest annual grain productivity of >16 t ha^{-1} at Modipuram and Ludhiana, 14-16 t ha^{-1} at Kanpur, 12-14 t ha^{-1} Faizabad, Varanasi, Pantnagar, Sabour, R.S. Pura, 10-12 t ha^{-1} at Ranchi and 8-10 t ha^{-1} at Palampur was recorded. Average annual grain productivity was 12.8 t ha^{-1} out of which the contribution of rice was 60% and that of wheat 40%. Again, averaged over locations, the SSNM brought about a yield advantage of 3.3 t grain ha^{-1} or 38% over farmers' nutrient management practices (Fig. 3).

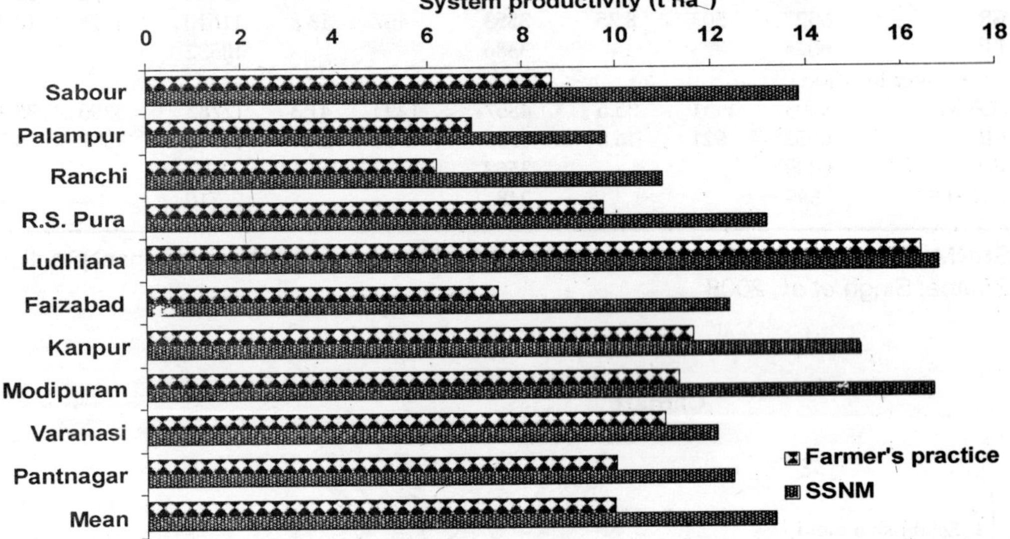

Fig. 3 : Comparative effect of farmers nutrient management practice and SSNM on rice-wheat system productivity
Source: Tiwari *et. al.*, 2006

In another set of experiments at Modipuram it was noticed that application of nutrient as per SSNM had significant influence on productivity of mustard, garlic, mustard, chick pea, potato and berseem crops along with residual benefit to succeeding rice crop (Table 2). With SSNM based nutrient use the overall system equivalent productivity was higher by 35% each in rice- mustard and rice-chickpea, 28% each in rice-berseem and rice-wheat, 43% in rice-potato and 58% in rice-garlic system as compared to farmers fertilizer management practice (FFP) (Fig. 4). The significant gain under SSNM was also noticed over soil test based lab recommendation or ad-hoc recommendations (Table 2). It is further apparent that unless the recommendations emerging from state soil-testing labs (STLR) have a yield target-specificity, the yield gains with STLR may not be different from local *ad hoc* recommendations, and there may also be a constant threat of long-term deterioration in soil fertility due to greater drain of native nutrient reserves.

Table 2 : Effect of nutrient management options on productivity (kg ha^{-1}) of different rice based cropping systems

Nutrient management options	Mustard	Chickpea	Garlic clover	Berseem green (fodder)	Potato tuber	Wheat
Farmers fertilizer practice (FFP)	1688	1970	4512	75050	17900	5029
State recommendation (SR)	2090	2188	6575	85101	22600	5610
Soil Test lab recommendation (STLR)	2105	2210	6640	80029	21800	5658
Site- specific nutrient management (SSNM)	2312	2652	7534	92219	27500	6255
CD<0.05	126	214	512	-	2010	416
Residual rice yield						
Farmers fertilizer practice (FFP)	6745	6839	7070	7161	7248	6790
State recommendation (SR)	7276	7903	7329	8071	8115	7237
Soil Test lab recommendation (STLR)	7771	8042	8469	8386	8310	7027
Site- specific nutrient management (SSNM)	9041	9250	9025	9338	9542	8836
CD<0.05	772	823	664	938	734	543

Source: Singh *et al.*, 2010

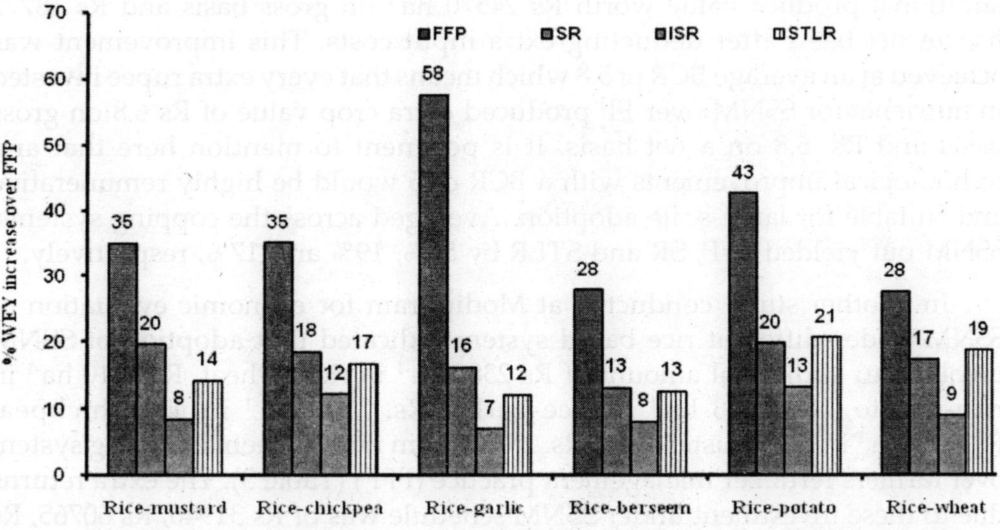

Fig. 4 : Per cent increase in system productivity (Rice equivalent, REY) in SSNM treatment over other nutrient management option under different cropping systems.
Source: Singh *et al.*, 2010

Economics of SSNM vs other nutrient management options

Economics is the most important factor which governs the technological adoption at farmer's field. Studies conducted under AICRP_IFS during 2003-06 under rice- wheat system reveals that the SSNM practice involved an additional expenditure ranging from Rs. 1090 to Rs. 5875 ha^{-1} (average Rs. 3361 ha^{-1}) over the farmer's practice in rice cultivation (Table 3). This additional expenditure generated extra produce (grain + straw) worth Rs. 14340 ha^{-1}, on an average within a range of Rs. 6625 to Rs. 21561 ha^{-1}. After deducting the additional costs, the resulting net returns worked out at Rs.10980 ha^{-1} with a BCR of 4.6.

Similarly, moving from farmers practice (FP) to SSNM involved an additional fertilizer expenditure of Rs. 300 to 2955 ha^{-1} with an average of Rs.1438 ha^{-1} for wheat production (Table 3). The generally lower additional investment needed in case of wheat as compared to rice is in fact due to that cost incurred for sulphur and micro nutrients (applied to rice only) have been debited to rice. Since wheat has also benefited from residual effect of these nutrients applied to rice, the net returns have been affected proportionately. The additional net return under SSNM over FP ranged from Rs. 2277 at Ludhiana to Rs. 13503 at Sabour. As expected, the improvements in wheat were associated with higher BCR (Rs rupees^{-1} investment) than the improvements in rice productivity because of high additional input cost debited to rice for S and minor nutrients applied. The cumulative effect of SSNM under RWCS involved an additional expenditure of Rs. 4798 ha^{-1} and resulted additional produce value worth Rs 24570 ha^{-1} on gross basis and Rs 18772 ha^{-1} on net basis after deducting extra input costs. This improvement was achieved at an average BCR of 5.8 which means that every extra rupee invested in nutrients for SSNM over FP produced extra crop value of Rs 6.8 on gross basis and Rs. 5.8 on a net basis. It is pertinent to mention here that any technological improvements with a BCR of 5 would be highly remunerative and suitable for large-scale adoption. Averaged across the copping systems, SSNM out yielded FFP, SR and STLR by 50%, 19% and 17%, respectively.

In another study conducted at Modipuram for economic evaluation of SSNM under different rice based system indicated that adoption of SSNM involves an additional amount of Rs 2388 ha^{-1} in rice- wheat, Rs 1210 ha^{-1} in rice-potato, Rs. 4488 ha^{-1} in rice-garlic, Rs. 3224 ha^{-1} in rice-chickpea, Rs 3110 ha^{-1} in rice-mustard and Rs. 2876 ha^{-1} in rice-berseem cropping system over farmers fertilizer management practice (FFP) (Table 3). The extra returns due to these investment under SSNM schedule was of Rs 31946, Rs 60765, Rs 166478, Rs 33034, Rs 31904 and Rs 28163 ha^{-1}, respectively under different cropping system. Which clearly indicate that an additional investment of 1Rs over FFP gave additional Rs 13.34 in rice- wheat, Rs 50.2 in rice-potato,

Rs. 37.1 in rice-garlic, Rs. 10.2 in rice-chickpea, Rs 10.3 in rice-mustard and Rs. 9.8 in rice-berseem cropping system.

Table 3: Changes in economic returns while shifting from farmers' nutrient management practice to SSNM in rice- wheat cropping system

Location	Crop	SSNM vs Farmers' practice			
		Extra cost of fertilizer (Rs ha^{-1})	Value of extra produce (Rs ha^{-1})	Net return (Rs ha)^{1}	Net return (Rs./Re^{-1} extra invested in nutrients)
Sabour	Rice	2755	21561	18806	6.8
	Wheat	1695	15198	13503	8.0
	System	4450	36759	23309	7.3
Palampur	Rice	3048	7556	4508	1.5
	Wheat	1418	9984	8566	6.0
	System	4465	17539	13074	2.9
Ranchi	Rice	3098	16914	13817	4.5
	Wheat	1665	11595	9930	6.0
	System	4763	28510	23747	5.0
R.S. Pura	Rice	5875	11271	5396	0.9
	Wheat	2955	10639	7684	2.6
	System	8830	21911	13081	1.5
Ludhiana	Rice	2940	8552	5612	1.9
	Wheat	810	3087	2277	2.8
	System	3750	11639	7889	2.1
Faizabad	Rice	4190	20313	16123	3.8
	Wheat	1860	13823	11963	6.4
	System	6050	34137	28087	4.6
Kanpur	Rice	3760	15449	11689	3.1
	Wheat	1635	9107	7472	4.6
	System	5395	24556	19161	3.6
Modipuram	Rice	1090	20822	19732	18.1
	Wheat	300	12259	11959	39.9
	System	1390	33080	31690	22.8
Varanasi	Rice	3490	6625	3135	0.9
	Wheat	600	6373	5773	9.6
	System	4090	12998	8908	2.2

Source: Singh *et. al.*, 2008

Table 4 : Extra cost/extra return due to fertilization (Rs. ha^{-1}) over farmers' fertilizer practice

Nutrient management options	Rice-wheat	Rice-potato	Rice-garlic	Rice-chickpea	Rice-mustard	Rice-berseem
			Extra cost			
State recommendation (SR)	285	-1840	1418	739	345	831
Soil Test lab recommendation (STLR)	128	-1510	1611	642	529	662
Site- specific nutrient management (SSNM)	2388	1210	4488	3224	3110	2876
			Extra return			
State recommendation (SR)	10985	29518	104196	14204	12610	13547
Soil Test lab recommendation (STLR)	8541	27950	118861	15539	17210	14260
Site- specific nutrient management (SSNM)	31946	60765	166478	33034	31904	28163

Source: Singh *et al.*, 2010

Conclusion

The site specific nutrient management strategy described in this chapter has demonstrated promising agronomic and economic potential and provided a strong conceptual and scientific basis for the simplified and refined SSNM approach. On station experiments clearly reveal that crop productivity and farm profitability can be improved up to significant level by adopting SSNM along with sustained soil health. Adoption of SSNM in larger area in future would be needed to feed increasing population. Site and crop specific balanced fertilization in addition to maintaining food security will help sustaining soil health and environment with improved nutrient use efficiency. An integration of modern tools like GIS based mapping may become useful for its wider applicability under larger area of domain.

References

Dobermann, A., Witt, C. and Dawe, D. (2004). Increasing productivity of intensive rice systems through site-specific nutrient management. Science Publishers and IRRI. pp. 410.

Dwivedi, B.S. and Singh, V.K. (2011). Balanced fertilization for maximizing the rice-wheat system productivity under Typic Ustochrept soils of western Indo-Gangetic Plain. *Journal of Farming Systems Research and Development* (In press).

Dwivedi, B. S., D. Singh, Chonkar, P.K., Sahoo, R.N. Sharma, S. K and Tiwari, K. N. (2006). Soil Fertility evaluation-a potential tool for balanced use of fertilizer, pp.1-66, IARI New Delhi-PPI-PPIC, India programme, Gurgaon, India.

Dwivedi, B.S., Shukla, A.K., Singh, V.K., Yadav, R.L. (2001). Results of participatory diagnosis of constraints and opportunities (PDCO) based trials from the state of

Uttar Pradesh. In: A. Subba Rao, S. Srivastava (Eds), *Development of farmers' resource-based integrated plant nutrient supply systems: experience of a FAO-ICAR-IFFCO collaborative project and AICRP on soil test crop response correlation* Bhopal: Indian Institute of Soil Science. pp. 50-75.

Sharma, S.K. (2003). Characterization and mapping of rice- wheat system: its changes and constraints to system sustainability. *NATP (PSR 4.1) Final report,* (2003-04), New Delhi

Singh, V. K., Tiwari, R. Gill, M. S., Sharma, S. K. Tiwari K. N., Shukla, A. K and Mishra, P.P. (2008). Economic viability of site-specific nutrient management in rice-wheat cropping. *Better Crops-India,* **2** (1): 16-19.

Singh,V.K., Majumdar, K., Singh, M.P. Raj- Kumar and Gangwar, B. (2010). Maximizing productivity and profit through site- specific nutrient management in rice- based cropping systems. *Better Crops International.*

Tiwari, K.N., Sharma, S.K., Singh, V.K., Dwivedi, B.S. and Shukla, Arvind K. (2006). Site-specific nutrient management for increasing crop productivity in India: Results with rice-wheat and rice-rice system. p-92. PDCSR Modipuram and PPIC India Programme Gurgaon.

Yadav, R.L. (2003). Assessing on-farm efficiency and economics of fertilizer n, p and k in rice-wheat systems in India. *Field Crops Research,* **81** : 39-51.

Yadav, R.L. (1998). Factor productivity trends in a rice-wheat cropping system under long-term use of chemical fertilizers. *Experimental Agriculture,* **34** : 1-18.

❏❏❏

System Based Integrated Nutrient Management, 2012
© *B. Gangwar & V.K. Singh (eds.), pp. 207-221*
New India Publishing Agency, New Delhi (India)
e-mail : info@nipabooks.com; website : www.nipabooks.com

CHAPTER **13**

Nutrient Management in Vegetable Based Cropping Systems

A.K. PANDEY

India is a leading vegetable producing country in the world. Presently it occupies 7.981 million hectare area with the annual production 129.077 million tones (2008-09). The country being belessed with the unique gift of nature of diverse climate and distinct seasons, make it possible to grow an array of vegetables number exceeding more than hundred types. However, potato being the staple food and easy to mix in several preparations ranks first (26.6%) in total production of vetetables followed by onion (10.5%) and other important Solanaceous vegetables like tomato (8.6%), brinjal (8.0%). Cauliflower and cabbage are most preferred winter vegetables and their total share in the country's vegetable production is 5.1 and 5.3%, respectively. Other important vegetables, which are primarily grown in the country, are okra, onion, vegetable peas and a good range of cucurbits (Fig. 1)

There is miss-match in area production and productivity of vegetables among various states of the country. The West Bengal ranks first in the total production of vegetables in the country and its production during the year 2008-09 was in tune of more than 22.704 million tones, contributing the more than one-fifth of total country's vegetable production. Uttar Pradesh, Bihar and Orissa in north India and Maharashtra, Tamil Nadu and Karanataka in south India are the other leading vegetable producing states in the country. In last one decade, there has been considerable progress in enchancing the productivity of vegetables which is presently 16.2 tonnes hectare^{-1}. Incessant growth of urbanization, ceaseless fragmentation of land holdings and depleting natural resources are the major challenges before the expansion of any agricultural commodity either cereals or vegetables. Hence, our attention must be focused on vertical expansion strengthened with the boon of the technology instead of horizontal expansion just by increasing the crop area. To ensure the nutrition security of the burgeoning population of the country, it is estimated that up to 2020, the country's vegetable demand would be around 155 million tones (Rai *et al.*, 2004; Pandey *et al.*, 2009). To achieve this target, it is *sine qua non* to integrate the various technologies right from production to post-harvest.

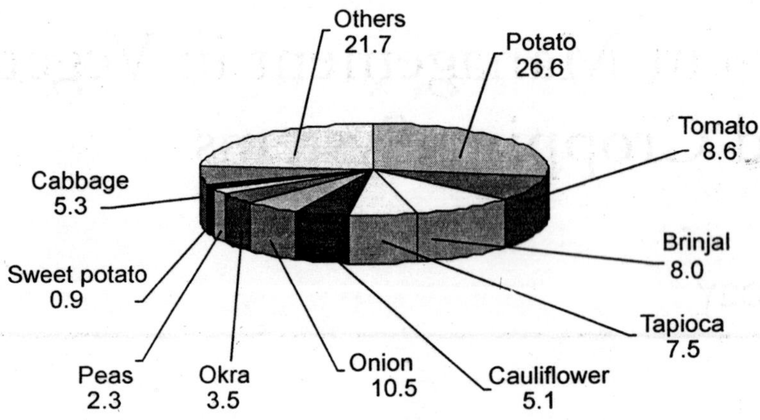

Fig.1 : Production share of major vegetables in India

Nutrient management in vegetable crops

The quantity of nutrients which the farmer needs to apply depends on the yield potential of the cultivar, the level of available plant nutrients already in the soil, and growth conditions. Since vegetative and reproductive stages overlap in this group of crops, they need a continuous and steady supply of nutrients throughout their life span. It is necessary to adopt appropriate nutrient management practices which help to supply nutrients in quantities adequate to just meet crop demand and minimize losses, thereby increasing the nutrient use efficiency. Such practices will be environmentally friendly, and lead to sustainability in vegetable production.

Various approaches for judicious use of nutrients

Split applications and fertigation

Application of N in four splits at 30-day intervals has been recommended by Singh *et al.* (1988) to achieve maximum yields and profits in chilli production. Subhani *et al.* (1990) obtained the highest yield of chili when both N and K were applied in four splits at planting, 30, 60 and 90 days after transplanting (DAT). Jaime *et al.* (1987) compared furrow and trickle irrigation in the N nutrition of eggplant. He observed that 180 kg N ha^{-1} applied in trickle irrigation could yield (16.8 kg $plant^{-1}$) more than that obtained with 360 kg N ha^{-1} under furrow irrigation (16.2 kg $plant^{-1}$). An application of 50% of total NPK in trickle irrigation resulted in a higher nutrient uptake in tomato than when the same NPK was applied before planting (Dangler and Locascio, 1990). Goyal *et al.* (1985) reported that fertigated pepper, tomato and eggplant receiving 9, 18 or 30 g urea $plant^{-1}$ all had higher yields than plants receiving a side application of urea (15 + 15 g $plant^{-1}$ at planting and first harvest). Fertigation with acidifying N fertilizers such as urea may sometimes have an adverse effect on growth and productivity, if Al toxicity is induced by soil acidity below the water emitters (Haynes, 1988). In such areas, methods of countering or preventing soil acidification below trickle emitters need to be developed if fertigation with acidifying N fertilizers is to be practiced routinely. Mulching, especially when an overhead irrigation system was used, considerably increased the total recovery of applied nitrogen in tomato (Sweeny *et al.*, 1987). Further, slow-release fertilizers hold great promise for the production of solanaceous vegetables such as eggplant and tomato. Gezerel and Donmez (1988) compared slow-release fertilizer (Plantacote) and conventional fertilizers at 100:80:90:30 kg ha^{-1} NPK Mg. They found that slow-release fertilizers produced 92 metric tonnes ha^{-1} of tomato, compared to only 42 metric tonnes ha^{-1} when ordinary commercial fertilizers were used.

Integrated management of nutrients

The basic concept underlying the principles of integrated nutrient management is the maintenance, and possible improvement, of soil fertility for sustaining crop productivity on a long-term basis. Sustained productivity may be achieved through the combined use of various sources of nutrients, and by managing these scientifically for optimum growth, yield and quality of different crops, in a way adapted to local agro-ecological conditions. In vegetable production in Asian countries, farmers have been using organic manures for centuries, together in recent decades with chemical fertilizers, to meet the nutrient demands of crops. Integrated nutrient use has assumed great significance in recent years in vegetable production, for two reasons. Firstly, the need for continued increases in per hectare yields of vegetables requires that applications of nutrients increase. Not enough chemical fertilizer is available in many developing countries in Asia, to meet crop nutrient requirements. Secondly, the results of a large number of experiments on manures and fertilizers conducted in several countries reveal that neither chemical fertilizers alone, nor organic sources used exclusively, can sustain the productivity of soils under highly intensive cropping systems (Singh and Yadav, 1992). Subbiah *et al.* (1985) obtained higher yields of tomato and eggplant with combined use of FYM and fertilizers. For eggplant, applications of 100 kg N ha^{-1}, half in urea (50%) and half in poultry manure (50%), resulted in higher yields (45.8 mt ha^{-1}) than the same level of nitrogen applied in urea alone (37.8 mt ha^{-1}). The integrated use of urea and poultry manure also resulted in a higher nutrient uptake (Jose *et al.*, 1988). Jablonska (1990) reported that the combined use of rye straw and nitrogen resulted in higher yields of tomato, eggplant and pepper than either N fertilizer or FYM used alone. Hosmani (1993) also reported higher yields of chili with integrated use of chemical and organic fertilizers than with the use of either of these separately.

Use of micronutrients in vegetable crops- an urgent compulsion for yield and quality

Micronutrients are required only in very small quantities. Their concentrations in plant tissues are only a small proportion of the concentrations of macronutrients (Brady and Weil, 1999). Micronutrients play many complex roles in plant nutrition, but most of them are used in the functioning of a number of enzyme systems (Brady and Weil, 1999). However, there is considerable variation in the specific functions of the various micronutrients in plants and in microbial growth processes. For example, copper, iron, and molybdenum are an essential part of the complex reactions which make up photosynthesis and many other metabolic processes. Zinc and manganese function in many plant enzyme systems as bridges. They connect the enzyme with the substrate upon which it is meant to act. Since most micronutrients are relatively immobile in the plant, they are not readily transferred from older leaves to younger ones. Therefore the concentration of the nutrient tends to be lowest in younger leaves. Symptoms of deficiency are most pronounced in young leaves which develop after the supply of the nutrient has run low. Sources of the seven micronutrients vary markedly from one area to another. Organic matter is an important secondary source of some trace elements. Copper, in particular, is often tightly held by organic matter, to the extent that its availability can be very low in organic peat soils (Histosols). Soil profiles of uncultivated soil tend to have higher concentrations of micronutrients in the surface soil, much of them presumably in the organic matter. There is a certain connection between levels of soil organic matter and the content of copper, molybdenum, and zinc. Although the elements present in the organic matter are not always readily available to plants, their release through decomposition is undoubtedly an important fertility factor. Animal manure is a good source of micronutrients, most of them being present in organic forms.

It is difficult to diagnose a micronutrient deficiency from the visible symptoms alone. In fact, sometimes there may be no symptom at all except a reduction in yield. This type of latent or hidden deficiency may be suspected if crops do not respond to applied fertilizer (Chang, 1999 ; Osotsapar, 1999). All nutrients must be present in optimum quantities for the best yields. If one nutrient is lacking, it negates the value of all the others. To make diagnosis even more difficult, crops often suffer from multiple micronutrient deficiencies, not just one. Moreover, the symptoms of micronutrient deficiencies often mimic those of diseases, especially virus diseases. Farmers need laboratory tests to be sure of a diagnosis, and these are expensive. However, farms in a region with the same kind of soil tend to share the same plant nutrient problems. There is no need to test on a farm-by-farm basis, although this is ideal if farmers can afford it. How can farmers protect their crops from micronutrient deficiencies? Certainly, they should never apply a good dressing of assorted micronutrients "just in case". Toxicity problems from a surplus of micronutrients in the soil can be even more damaging than a deficit, and can be more difficult to correct. Farmers should apply only the nutrient needed, in the quantity needed by the crop. Application rates for micronutrients are very low. A typical boron application, for example, is one kilogram hectare^{-1}. Critical levels in leaves of most micronutrients are only 10-20 mg kg^{-1}. However, a lack of this very small amount is likely to cause serious damage to the plant. For this reason, if there are micronutrient deficiencies, it is important that they are diagnosed and treated without delay. In short-term crops such as vegetables, by the time symptoms are marked enough to diagnose deficiency, it may be too late to save most of the crop.

Description of symptoms of micronutrients in vegetable crops and their correction

A. Iron

The symptoms of iron deficiency are yellowing or chlorosis of the interveinal areas of the emerging leaf. Later the entire leaf turns yellow, and finally turns white. If the deficiency is severe, the entire plant becomes chlorotic and dies. Iron deficiency can easily be mistaken for nitrogen deficiency. Lower, nitrogen deficiency affects the older leaves first, while iron deficiency affects the emerging leaves first

1. *Tomato*: Young leaves become yellow with chlorosis. In plants grown in soil, iron deficiency is corrected by three soil applications of 1% EDTA iron solution. The pH of the soil should be adjusted to 5.5-6.5.

2. *Chinese Leek*: The symptoms of iron deficiency are yellowing or chlorosis of the leaves. Severe deficiency causes the entire plant to become chlorotic and die. The symptoms are similar to those of nitrogen efficiency, but

affect the young leaves first. In contrast, nitrogen deficiency affects older leaves first. Soil conditions likely to produce symptoms of iron deficiency in Chinese leek are the same as those for rice, ginger and grape. Well-drained (aerobic) soils with a pH higher than 6.5 are likely to be deficient in available iron. The severity of the problem increases if the pH rises above 7.8 in calcareous soils.

3. *Water Spinach* : The symptoms of iron deficiency in this crop are yellowing or chlorosis of the interveinal areas of the emerging leaf. Later, the entire leaf turns yellow, and finally white. If the deficiency is severe, the entire plant becomes chlorotic and dies. Iron deficiency can easily be mistaken for nitrogen deficiency. However, nitrogen deficiency affects older leaves first, while iron deficiency affects first the emerging leaves. Well-drained soils with a pH of more than 7.0 are likely to be deficient in available iron. The severity of the problem increases with an increase of the pH to 7.8 or more in calcareous soils.

4. *Cucumber and Melon* : The symptoms are the same as those found in plants with iron deficiency grown in the open field. For cucumber or melon grown in soil, iron deficiency is corrected by three applications of 1% EDTA iron solution. The soil pH should be adjusted to 5.5-6.5.

5. *Ginger*: The symptoms of iron deficiency in ginger are yellowing or chlorosis of the interveinal areas of the emerging leaf. Later, the entire leaf turns yellow, and finally white. If the deficiency is severe, the entire plant becomes chlorotic and dies. Iron deficiency can easily be mistaken for nitrogen deficiency. However, nitrogen deficiency affects older leaves first, while iron deficiency first affects younger leaves. Iron deficiency is a common micronutrient disorder of ginger growing in neutral, calcareous or alkaline aerobic soils. The severity of the disorder increases with a higher pH. Iron deficiency may also sometimes be observed on upland acid soils. Other soil conditions are the same as for paddy rice. Well-drained soils with a pH greater than 6.5 are likely to be deficient in available iron. The severity of the problem increases with an increase of the pH to 7.8 or more in calcareous soils.

B. Manganese

Manganese (Mn) deficiency is common in leached tropical soils, particularly in calcareous soils derived from limestone. Legumes are again particularly sensitive. The main symptom of Mn deficiency is chlorosis or yellowing between the veins of new leaves. Mn deficiency also limits legume production on black calcareous soils in Asian countries. Foliar applications of 0.5% manganese sulfate solution are often recommended, applied once every ten days until the manganese deficiency has improved.

1. *Tomato* : Manganese deficiency is widespread in tomato plants. It manifests itself first as a lightening of the green color in the interveinal area of the leaves, which gradually turns to yellow .The deficiency can be corrected by applying a foliar spray of 0.3% manganese sulfate (Mn SO_4) solution. The pH of the soil, and any nutrient solution used, should be adjusted to 5.5-6.5.
2. *Egg plant* : Manganese deficiency is common in soils with a high pH, and in highly oxidized soil conditions, (dry soil with a low level of organic matter). It is also found in soils where the parent material has high lime content. The critical level of exchangeable manganese in the soil is 2.3 mg kg^{-1}, using 1M ammonium acetate (pH 7.0) extraction. The upper leaves should be tested, since it is these that tend to show the most marked symptoms. High levels of calcium, copper, iron, zinc and phosphate may prevent manganese absorption and transportation. Calcium in particular has a strong effect in preventing manganese absorption. The pH of the soil should be lowered, and the water-holding capacity improved, by applying organic matter to increase the humus content of the soil. Fertilizer applications may include the application of manganese materials to the soil, or a foliar spray of manganese solution
3. *Watermelon*: Usually the interveinal area of 4-8 leaves turns yellow before the plants begin to flower. Interveinal chlorosis and white spots appear after the plants begin bearing fruit. Deficiency symptoms spread from lower leaves to upper ones over time. The condition is usually found in soils with a high pH, and a level of easily reductive manganese of less than 100 mg Mn kg^{-1} (extracted by 2% hydroquinone + 1M ammonium acetate) The condition can be remedied by applying manganese carbonate or manganese sulfate at a rate of 500 kg ha^{-1}, or manganese oxide (MnO) at a rate of 200 kg ha^{-1}, as a basal application.

C. Boron

Boron (B) deficiency is more common in volcanic soils, acidic soils derived from igneous rocks and in calcareous soils. In annual crops, the symptoms vary from one species to another. Boron deficiency may not show any visible symptoms in the seed, but the yield may fall by as much as 50%. In peanut, soybean, papaya, and citrus, boron deficiency often results in an empty space within the seed known as "hollow heart", "woody fruit", or "stone fruit". A common result of boron deficiency in all crops as an interruption in flowering and fruiting.

1. *Cauliflower* : A common result of boron deficiency in cauliflower is an interruption in flowering and an empty space known as "hollow heart". Yields are poor, and the cauliflower heads are deformed or a discolored

brown color. Boron deficiency is common in high-rainfall areas with high temperatures (more than 33°C), on acidic soils with a coarse texture. Soil boron is highly soluble, and easily leaches out from such soils. Boron deficiency is common in calcareous soils (with a pH higher than 7.5), and soils with a very low soil boron content (less than 0.5 mg kg^{-1} hot-water-soluble B). The condition is also often found in sandy soils and acidic soils (pH lower than 5), soils with a very low organic matter content (lower than 0.75%), and soils derived from acid igneous rocks (granite). Boron deficiency may also be encountered in soils which have been given excessive applications of nitrogen and potassium fertilizers. Levels of hot-water-soluble boron (HWS-B) in low or deficient soils are less than 0.5 mg kg^{-1}. In soils with medium and adequate levels, the boron content is 0.5 - 2 mg kg^{-1}. Soils with high or excessive levels contain more than 2 mg kg^{-1}. Thus, soils with a level of boron (HWS-B) of 0.5 or less can probably not supply enough boron to support normal plant growth and yields. In most crops, a level of boron in plant tissue of 15-100 mg kg^{-1} is considered adequate for normal growth. The problem of boron deficiency can be easily solved by applying boron, usually in the form of borax, a white crystalline salt. The quantities needed for the crop are very small. As little as 5-10 kg ha^{-1} can correct boron deficiency of cauliflower. Alternatively, a foliar spray of 0.4% of borax solution can applied repeatedly every 10 days until the deficiency is corrected. For calcareous soils, sulfur should be applied at a rate of 2 mt ha^{-1}, to reduce the soil pH to 6.0-7.0.

2. *Pepper*: Pepper plants with boron deficiency are stunted or dwarfed. The leaves are small, and often become twisted and discolored. The severity of boron deficiency symptoms depends on other factors beside the boron content of the soil. High soil moisture and the application of too much lime may contribute to boron deficiency in crops. The condition can be corrected by a soil application of 10 kg ha^{-1} of borax. The pH of the soil, and of any nutrient solution used, should be adjusted to 5.5-6.5.
3. *Cucumber*: Cucumber plants with boron deficiency have abnormal shoots. The apical growing points are stunted, and eventually die. Cracks may appear in the stem, and the fruit are often deformed. The condition can be corrected by the soil application of 10 kg ha^{-1} of borax. The pH of the soil and nutrient solution should be adjusted to 5.5-6.5.
4. *Tomato*: Tomato plants with boron deficiency have stunted growth and are dwarfed. The leaves are twisted and small in size, and may have a variegated appearance. The young shoots may wither and die .As well as the level of boron in the soil, the severity of boron deficiency symptoms depends also on other factors such as high soil moisture and over-liming

of the soil. The condition can be corrected by 10 kg ha^{-1} of borax, applied to the soil. The pH of the soil and nutrition solution should be djusted to 5.5-6.5.

5. *Melon*: Melon plants suffering from boron deficiency become stunted or dwarfed. The young leaves of new shoots are smaller than normal, and curled back. They are often discolored with yellow mottles. As well as the level of boron in the soil or nutrient solution, the severity of symptoms in boron-deficient plants also depends on other factors. The most important are high soil moisture content and over-liming of the soil. The condition can be corrected by the application of 10 kg ha^{-1} of borax applied to the soil. The pH of the soil and nutrient solution should be adjusted to 5.5-6.5.

D. Molybdenum

Molybdenum (Mo) concentrations in leaves and nodules show a significant correlation with the shoot dry weight and nitrogen content in peanut, soybean, green gram and black gram. The results can be used to establish critical concentrations for the diagnosis of molybdenum deficiency. The critical values for nodules are much higher than those for leaves. In the case of peanut, the relationship between seed dry matter and leaf Mo concentrations were most reliable at the pod filling stage. Often the most obvious symptoms of Mo deficiency are similar to those of uncomplicated nitrogen deficiency. In the case of legumes, depending on nitrogen fixing activities of soil microorganisms, the plants may in fact be nitrogen-deficient. Examples of Mo deficiency are "yellow-spot" of citrus, "blue-chaff" of oats, and "whiptail" of cauliflower. Depending on the crop species, the critical deficiency levels of Mo vary from 0.1 to 1 mg kg^{-1}. Foliar applications of molybdenum fertilizer are not the only way to increase the molybdenum content of the leaves. It also can also be increased by the application of manganese and magnesium, especially the former. Research results also suggest that the foliar application of manganese might help molybdenum uptake translocation, by reducing the adsorption of molybdenum to the surface, cell wall or cell membrane of leaves. This might increase the movement of molybdenum into the phloem.

Cauliflower: Leaves are concave and sometime chlorotic. The curd is small and discolored. The condition is usually found in soils with a low pH (less than 5.0). The soil pH should be maintained at about 5.5 - 6.5. The deficiency can be corrected by a basal application of 1 kg ha^{-1} sodium molybdenum.

E. Copper

The soils in which copper (Cu) deficiency occurs are usually organic soils, calcareous soils or sandy soils. The low requirement of many plants for copper is probably the reason why copper deficiencies are fairly uncommon. Copper deficiencies are seldom seen in the crops of Asian countries. In using plant analysis to diagnose copper deficiency, we should be aware that the concentration of copper in the oldest leaves is misleading as an indicator of the total copper status of the plant. It is the copper concentration in the shoot tips which is related to vegetative growth and yield. The critical copper concentration in the shoot tips of peanut ranged from 1 to 1.5 mg kg^{-1} (dry weight). The critical value for the diagnosis of copper deficiency in soybean is 2.0 mg Cu kg^{-1} (dry matter). A foliar spray of 0.1% copper sulfate solution mixed with lime is strongly recommended to remedy Cu deficiency in crops.

1. *Tomato* : Copper deficiency in tomato produces symptoms of reduced or stunted growth of shoots. Young leaves are distorted, and there is necrosis of the apical meristem. The upper leaves and stem wilt. The condition can be corrected by applying a foliar spray of 0.1% copper sulfate pentahydrate ($CuSO_4.5H_2O$) solution with lime. The pH of the soil, and any nutrient solution used, should be adjusted to 5.5-6.5.

Indian organic - a need of integrating traditional knowledge and wisdom

In India approximately 2.6 m ha^{-1} is under organic certification. Out of which cultivated area is only 1 lakh ha while remaining 2.5 million ha is wild forest area. North East account for nearly 4400 ha under certified organic farming (including under conservation). Nation steering committee on Accreditation is the apex body for accreditation and has authorized 11 certification agencies. Ministry of Agriculture has initiated a National Project on Organic Farming for promotion of organic agriculture and for ensuring effective implementation of regulatory mechanism at affordable cost. The project is operated through a National Centre of Organic Farming at Ghaziabad and six Regional Centre of Organic Farming at Bangalore, Bhubaneshwar, Hisar, Imphal, Jabalpur and Nagpur.

Demand for Indian organic products in domestic and export markets

The products available in the domestic markets in the organic quality are vegetables, fruited, tea, coffee, cereals and pulses. Major markets for organic products like in Metropolitan cities. Mumbai, Delhi, Kolkatta, Chennai, Bangalore and Hyderabad. According to an estimate domestic sales of organic products are 7.5% of the total organic production in the country. According to an estimate made by Org- Marg 2002. Further demand of domestic organic produce in year 2005-2006 and 2006-2007 will be around 1457 and 1568 tonnes, respectively (Fig. 2).

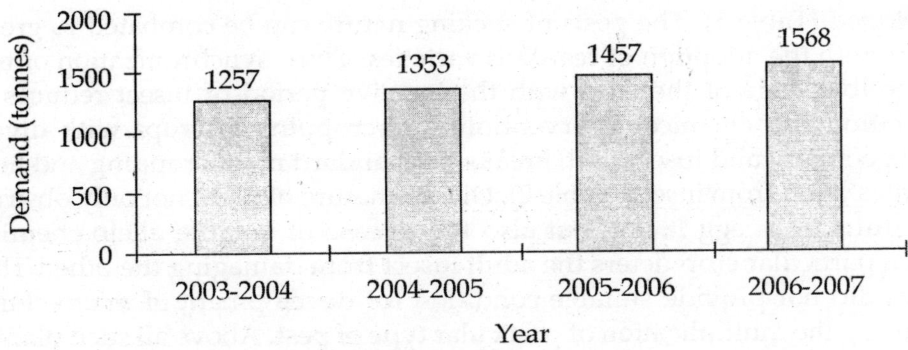

Fig. 2 : Year-wise domestic demand of organic products

International scenario of organic products

The world market for organic product is US dollar 35 billion and growing at 7-15%. USA alone accounts for 11-13 billion followed by European Union (10-11 billion US $) and Japan (350-450 million US $). Nearly 130 countries produce organic products. Total area under certified organic farming is 24 m ha^{-1} (Australia is the largest organic producer with 11.3 m ha^{-1} under organic cultivation followed by Argentina (2.8 m ha^{-1} and total (1.05 m ha^{-1}) strong growth is found especially in more countries that have an active organic sector. One of the factors that promote growth in organic markets worldwide is consumer awareness about health, environment issues and food safety. Other factors that influences fourth development of the organic farming and marketing strategies used by key players, such as retailers. Developing countries like India are expanding their organic market into developed countries and in parallel are building a domestic market. However, in order to help developing countries improve their sole opportunities in export markets. Traders need to know for general conditions that exist in the developed countries. In India export sales of organic products amount to 11,925 tonnes. These accounts for 85% of the total production and are likely to reach export sale of 21,525 tonnes during 2006-2007 (Fig. 3).

Routine agronomic practices like selection of varieties, intercultural operation, sowing/planting date, trap cropping/intercropping should be suitably modified to make uncongenial condition for pests and to enhance the activity of natural enemies. A model for integrated pest management is given in Fig. 4. Keeping in mind the diversity and intensity of pests in particular place, selection of resistant/less susceptible varieties, holds good in pest management. Plant resistance provides a built in ability to allow less pest and cut off the extra load of insecticides in the crop. Being completely safe, host plant resistance fits well with all other components. Unlike cereals, in vegetables very less number of resistant/less susceptible varieties has been developed (Table 1). The pests of sucking nature can be combated to greater extent with the adoption of resistant varieties. Thus, synchronization of most susceptible stage of the crop with the inactive period of insect reduces the infestation and chemical intervention. Intercropping of crops with diverse plant geometry and insect pest, breaks the standard monocropping and limits the infestation from insect (Table 2). Diverse nature of plant not only obstructs the adults from egg laying but also the release of volatile allelo-chemicals from a particular crop deters the adult insect from damaging the other. These effects did not provide suitable condition for development of micro-climate favouring the multiplication of particular type of pest. Above all such planting combination enhances the activity of predators and parasites also.

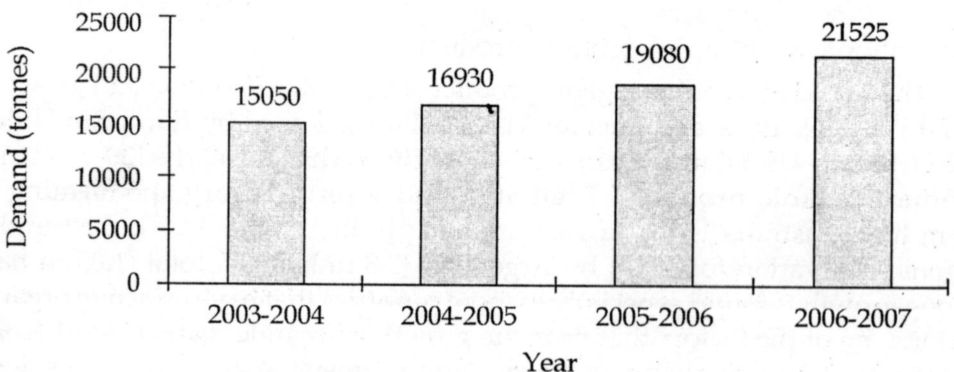

Fig. 3 : Future export demand of organic products IPM strategies for organic farming of vegetables

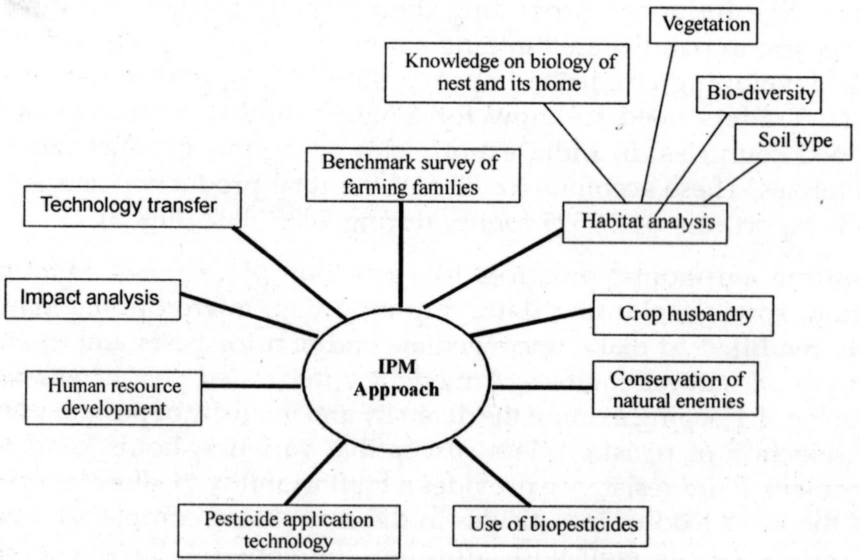

Fig. 4 : Approaches for the successful IPM

Table 1 : Less susceptible varieties of some vegetable crops

Crop	Pest	Varieties
Brinjal	Shoot and fruit borer (*Leucinodes orbonalis*), aphid, jassid, thrips whitefly	SM 17-4, PBr 129-5 Punjab, Barsati, ARV 2-C, Pusa Purple Round, Punjab Neelam Kalyanpur-2, Punjab Chamkila, Gote-2, PBR-91, GB-1, GB-6
Cabbage	Aphid (*Brevicoryne brassicae*)	All season, Red Drum Head, Sure Head, Express Mail
Cauliflower	Stem borer (*Hellula undalis*)	Early Patna, EMS-3, KW-5, KW-8, Kathmandu Local
Okra	Jassid (*Amrasca biguttula*)	IC-7194, IC-13999 New Selection, Punjab Padmini
Onion	Thrips (*Thrips tabaci*)	PBR-2, PBR-6, Arka Niketan, Pusa Ratnar, PBR-4, PBR-5, PBR-6
Round gourd	Fruitfly (*Dacus cucurbitae*)	Arka Tinda
Pumpkin	Fruitfly (*Dacus cucurbitae*)	Arka Suryamukhi
Bitter gourd	Fruitfly (*Dacus cucurbitae*)	Hissar-II

Table 2 : Intercrop combination effective in vegetable pest management

Crop combination	Target pest
Cabbage + Carrot	Diamondback moth
Broccoli + Fababean	Flea beetle
Okra + Cowpea	Yellow vein mosaic
Cabbage + French bean	Root fly
Cabbage + Tomato	Diamondback moth

Trap crop must be distinctly attractive to the pest than the main crop. It provides protection either by preventing the pest from reaching the main crop or by concentrating them in certain part of the field where they can be economically destroyed. Mustard as a trap crop along with cabbage has been successfully utilized for the control of diamondback moth, aphid and leaf webber. African marigold in tight bud stage functions as good source to trap the adults of *Helicoverpa armigera* besides it also attracts the adults of leafminer for egg laying on the leaves. Maize plants applied with bait spray trap and kill fruit fly adults when sown in combination with bitter gourd.

References

Brady, N.C., and Weil, R.R. (1999). The Nature and Properties of Soils. Prentice-Hall Inc., Upper Saddle River, New Jersey, USA. 12th edition. 881 pp.

Chang, S.S. (1999). Micronutrients in crop production of Taiwan. In: *Proceedings of International Workshop on Micronutrient in Crop Production*, held Nov. 8-13, 1999, National Taiwan University, Taipei, Taiwan ROC.

Dangler, J.M., and Locascio, S.J. (1990). External and internal blotchy ripening and fruit elemental content of trickle irrigated tomatoes as affected by N and K application rate. *Journal of the American Society of Horticultural Science* **115**: 547-549.

Gezerel, O., and F. Donmez. (1988). The effect of slow release fertilizers on the yield and fruit quality of vegetable crops growing in the Mediterranean area of Turkey. *Acta Horticulturae,* **272**: 63-69.

Goyal, M.R., Rivera, L.E. and Santiago, C.L. (1985). Nitrogen fertigation in drip irrigated peppers, tomatoes and eggplant. In: *Drip/Trickle Irrigation in Action*, Vol. 1. St. Joseph, Michigan, USA, ASAE, pp. 388-392.

Haynes, R.J. 1988. Comparison of fertigation with broadcast applications of urea-N on levels of available soil nutrients and on growth and yield of trickle irrigated peppers. *Scientia Horticulturae,* **35**: 189-198.

Hosmani, M.M. (1993). Chili Crop (*Capsicum annuum* L.). 2nd Edition. Mrs. S.M. Hosmani, Dharwad, Karnataka.

Jablonska, C.R. (1990). Straw as an organic fertilizer in cultivation of vegetables. Part II. Effect of fertilization with straw on the growth of vegetable plants. *Horticultural Abstracts,* **63**: 244.

Jaime, S., Casado, M. and Aguilar, A. (1987). Effects of trickle irrigation on the nitrogen nutrition of aubergines (Solanum melongena L.) under greenhouse cultivation. *Annales de Edafologia y Agrobiologia,* **46**: 1375-1383.

Jose, D., Shanmugavelu, K.G. and S. Thamburaj. (19880. Studies on the efficiency of organic vs. inorganic form of nitrogen in brinjal. *Indian Journal of Horticulture,* **45**: 100-103.

Osotsapar, Yongyuth. 1999. Micronutrients in crop production in Thailand, In: Proceedings of International Workshop on Micronutrient in Crop Production, held Nov. 8-13, 1999, National Taiwan University, Taipei, Taiwan ROC.

Pandey, A.K., Singh, P.M., Priti, Rai A.B. Rai, Mathura, (2009). Emeging Trends in Vegetable Export in India. International Conference on Agripreneurship and rural development. Organized by Faculty of management studies, Banaras Hindu University, Varanasi in Association with School of Agriculture and Consumer Scinces, Tannesssee State University, USA, NBRI, Lucknow and North Eastern Hill University, Tura Campus Meghalaya, India. From Dec.05-06, 2010, pp1-2.

Rai, Mathura, Singh, J. and Pandey, A.K. (2004). Vegetables: a source of nutritional security. *Indian Horticulture,* **48**(4): 14-17.

Rai, Mathura, Pandey, A.K., Singh, B. and Rai, A. B. (2009). Sabzi Niryat : vertman sthiti evam sambhavnaye. In: *Souvenir , Rajya stariya Sangosthi on Sabjion ki tikau utpadan evam vikas hetu karbnik kheti. Organized by NHRDF,* from 17-18 Noemberr, 2009. pp. 78-87.

Singh, G.B., and Yadav, D.V. (1992). Integrated plant nutrition system in sugarcane. *Fertilizer News*, **37**: 15-22.

Subbiah, K., Sundararajan, S., Muthuswami, S. and Perumal, R. (1985). Responses of tomato and brinjal to varying levels of FYM and macronutrients under different fertility status of soil. *South Indian Horticulturalist*, **33**: 198-205.

Subhani, P.M., Ravishankar, C. and Narayan, N. (1990). Effect of graded levels and time of application of N and K_2O on flowering, fruiting and yield of irrigated chili. Indian Cocoa, *Arecanut and Spices Journal*, **14**: 70-73.

Sweeny, D.M., Graetz, D.A., Locascio, S.J. and Campbell, K.L. (1987). Tomato yield and nitrogen recovery as influenced by irrigation method, nitrogen source and mulch. *Horticultural Science*, **22**: 27-29.

□□□

System Based Integrated Nutrient Management, 2012
© B. Gangwar & V.K. Singh (eds.), pp. 223-237
New India Publishing Agency, New Delhi (India)
e-mail : info@nipabooks.com; website : www.nipabooks.com

CHAPTER **14**

Integrated Nutrient Management (INM) under Protected Agriculture

AWANI KUMAR SINGH, BALRAJ SINGH AND RAKESH KUMAR

Protected cultivation of horticultural crops offers the best choice for diversification from traditional agriculture production system for a number of reasons. Production of crops under protected structures has great potential in augmenting production and quality of vegetables, flowers and in some fruit crops in main and off-season and maximizing water and nutrient use efficiency under varied agro-climatic conditions of the country. This becomes relevant to growers in India who have small land holding. They would be interested in a technology, which helps them to produce more crops each year from their land, particularly during off-season when prices are higher. The commonly protected structures are polyhouses, poly-tunnels, floating row covers, clotches, shade-net, insect-proof net and plastic-mulches (Chandra *et al.*, 2000 ; Singh *et al.* 2010). These protected structures act as physical barrier and play a key role in integrated pest management by preventing spread of insect-pests and viruses which causes severe damage to the crop (Singh *et al.*, 2003). This technology has great potential especially in peri-urban agriculture in near future, since it can be profitably used for growing high value vegetable crops like tomato, cheery-tomato, coloured peppers, parthenocarpic cucumbers, cut-flowers like rose, carnation, gerbera, lilium and fruits like strawberry, grapes etc. and for off-season cultivation of vegetables and their healthy and virus-free seedling production.

Protected cultivation is intended to mean some level of control over plant micro-climate to alleviate one or more of abiotic stresses for optimum plant growth. The microclimatic parameters being referred here are temperature, light, air-composition and the nature of rooting medium.

Advantages of protected agriculture

The protected agriculture has many advantages in comparison to cultivation in open field conditions. The major advantages are enlisted below:

- Plays a key role in INM, IPM and organic food production
- Minimizes biotic and abiotic incidences
- Enhance earliness and crop duration
- Increased yield and quality
- Production can be taken off-season as per market demands
- Save land and water for others crops
- Promote Hydro-ponics / Aero-ponics/ Aqua-ponics technology
- Can be used at roof with pot cultivation
- Helpful to breeding works and seed production
- Beneficial for marginal or weaker section farmers
- Increased employment or livelihood

Table 1 : Current status of protected agriculture in world

S. No.	Country	Area under protected agriculture
1	China	2.31 Lakh ha
2	Japan, Spain, Italy	60000 - 1. 0 Lakh ha
3	Netherlands, Israel, Turkey, Morocco	45000 – 85000 ha
4	USA, Canada, South Korea, UK,France and other EU nations	50000 – 1.0 Lakh ha
5	India	18500 + ha

Table 2 : Current status of protected agriculture in India

S. No.	State	Area under protected agriculture (ha)
1	Maharashtra	8000
2	Karnataka	3000
3	Gujarat	2500
4	NEH State	800
5	Himachal Pradesh	500
6	J&K	400
7	Uttarakhand	300

Integrated nutrient management

Integrated nutrient management (INM) refers to maintenance of soil fertility and plant nutrient supply to an optimum level for sustaining the desired crop productivity through optimization of the benefits from all possible sources of plant nutrients in an integrated manner under protected agriculture. The current trend is to explore the possibility of supplementing chemical fertilizers with organic fertilizers specially biofertilizers of microbial origin (Gupta *et al.*, 1999). Biofertilizers offer an economically and ecologically sound mean of reducing external inputs and improving the quality and quantity of internal sources. They are less expensive, eco-friendly, sustainable and long lasting and thus repeating application is not necessary. Although bio-fertilizers are not an alternate to inorganic fertilizers, they may be useful in increasing the yield and quality when combined with organic manure and inorganic fertilizers in balanced proportion.

Types of biofertilizers

1) Biological nitrogen fixers

Biological nitrogen fixation is the conversion of atmospheric nitrogen to ammonia by micro-organisms in the soil. It involves highly specialized and intricately evolved interactions between soil microorganisms and higher plants for harnessing the atmospheric nitrogen. It is a fascinating biological phenomenon, which provides low-cost nitrogen and improves crop productivity. Some of the important nitrogen fixing micro-organism are being listed here.

Free living		**Symbiotic with plants**
Aerobic	**Anaerobic**	
Azotobacter, *Beijerineckia* *Klebsiella* Cyanobacteria	*Clostridium* (some) *Desulfovibrio* Purple sulphur bacteria Purple non-sulphur bacteria Green sulphur bacteria	*Rhizobium* *Frankia* *Azospirillum*

Phosphate-solubilizing microorganisms

Phosphate solublizing microorganisms (PSM) include diverse range of soil bacteria and fungi. The bacterial species include *Pseudomonas, Micrococcus, Bacillus* etc while fungal species *are Flavobacterium, Penicillum, Fusarium,*

Aspergillus and others (Gaur *et al.*,1980).The bioconversion of insoluble to soluble P requires the presence of an array of biochemical steps/products.

Mode of action of phosphate solublizing microorganisms (PSM)

Primarily there are two schools of thought interpreting the mechanism of phosphate-solubilization by microorganism:

1. Solubilization by production of organic acids
2. Solubilization by action of phosphatase enzyme

PSM include various bacterial, fungal and actinomycetes forms which help to convert insoluble phosphate into simple and soluble forms. Species of *Pseudomonas, Micrococcus, Bacillus, Flavobacterium, Penicillum, Fusarium, Sclerotium and Aspergillus* are some of the PSM. Among these, soil bacteria belonging to the genera *Pseudomonas* and *Bacillus* are more common. The solubilization effect of PSM is generally due to the production of organic acids by these organisms. They are also known to produce amino acids, vitamins, and growth promoting substances like IAA and GA which help in better growth of plants.

Arbuscular mycorrhizal (AM) fungi

Arbuscular mycorrhizal (AM) fungi are the most ancient type of mycorrhizal symbiosis. It is a mutualistic symbiosis of co-existing partners, with the fungal partner being obligate survival on plant partner. AM fungi are the most widely distributed association in plants and are characterized by inter- and intracellular fungal growth in root cortex, forming specific fungal structures, referred to as vesicles and arbuscules. About 80% of all terrestrial plant species form this type of symbiosis and 95% of the world's present species of vascular plants belonging to families that are characteristically mycorrhizal. The hyphal branches penetrate the cortical cell walls and differentiate terminally within the cell to form dichotomously branched structures called Arbuscules. Arbuscules live for few days and followed by lipid filled storage organ called Vesicle. The *Glomus, Gigaspora, Sclerocystis, Acaulospora and endogene* are impotant genera of AM fungi.

Basic need of INM under protected agriculture

The integrated nutrient management (INM) in protected conditions require good quality irrigation water, maintenance of soil health and quality, good quality seed and cultivars, mechanization of the farm, integrated pest management, micro-irrigation system and fertigation unit along with good agricultural practices. Protected agriculture cannot be promoted without INM and IPM.

Significance of INM for protected agriculture

In protected agriculture, integrated nutrient management (INM) is of immense significance. Because under protected technology, cultivated vegetable or flower crops are sensitive to various macro- and micro-nutrient deficiencies and results in poor quality produce with reduced yields. So in order to get high quality produce with high yield, INM should be practiced under protected conditions. Deficiency symptoms in different crops under protected agriculture due to various nutrients are described below:

Table 3 : Loss of yield and quality form nutrient deficiency of horticultural crop and their control

S. No.	Crops	Problems	Cause	Yield Loss (%)	Control Measure
1	Tomato, Chili Capsicum, and Brinjal	Blossom end -rot	Ca, Deficiency, Higher dose of N	25-30%	Calcium chloride ($CaCl_2$)@5kg ha^{-1}
2	Tomato	Fruits cracking	Boron deficiency	12-15%	Borax .3 - .4% spray
3	Cauliflower	Buttoning	N deficiency	20-25%	50-60 kg ha^{-1}N
		Browning	Boron deficiency	10-15%	10-15 kg ha^{-1}borax
		Whip-tail	Molybdenum deficiency	2-5%	1.5-2 kg ha^{-1} sodium ammonium molybdenum
4	Cabbage	Scorch on leaves	Potassium deficiency	2-3%	10-12 kg Potassium chlorite
5	Cucumber	Pillow	Low Calcium	2-3%	Calcium chloride ($CaCl_2$) @ 5 kg ha^{-1}
6	Carrot	Cavity spot	Calcium deficiency	4-5%	Calcium chloride (Cacl2) @ 6 kg ha^{-1}
		Breakdown of cortex	Boron deficiency	3-4%	5-7 kg ha^{-1}borax
7	Sugar-beet	Tiphooking	Calcium deficiency	10-12%	Calcium chloride ($CaCl_2$) @ 3 kg ha^{-1}
8	Pea	Marsh spot	Manganese deficiency	5-7%	Magnesium chloride @ 2 kg ha^{-1}
9	Onion	Leaf cracking	Boron deficiency	4-5%	Borax 2-3 kg ha^{-1}

Integrated nutrient management (INM) in vegetable crops

The integrated nutrient management (INM) for cultivation of cucurbits in poly-tunnel requires green manure and FYM/Compost @ 200q ha^{-1} mixed in the soil at the time of field preparation. The dose of NPK @ 40, 20, 20 kg ha^{-1} three times a month through drip irrigation should be given. Integrated nutrient management (INM) combination for cultivation of capsicum in poly-house requires green manure and FYM/Compost @ 400q ha^{-1} and it is mixed in the soil at the time of field preparation. The NPK dose @ 140, 100, 120 kg ha^{-1} thrice a month through drip irrigation should be given. Calcium is applied @ 5kg ha^{-1}. However, integrated nutrient management (INM) combination for cultivation of tomato in polyhouse requires green manure and FYM/Compost @ 500q ha^{-1} and it is mixed in the soil at the time of field preparation. The dose of NPK @ 150, 120, 130 kg ha^{-1} thrice a month through drip irrigation should be given. Application of calcium @ 5 kg ha^{-1} and boron @ 3-4 sprays should be done. Integrated nutrient management (INM) combination for cultivation of cucumber in polyhouse is FYM/Compost @ 300q ha^{-1} and it is mixed in the soil at the time of field preparation. The dose of NPK @ 50, 30, 40 kg ha^{-1} through drip irrigation should be given. Application of calcium @ 3 kg ha^{-1} should be done. Integrated nutrient management (INM) for cultivation of summer squash in polyhouse require FYM/Compost @ 200q ha^{-1} and it is mixed in the soil at the time of field preparation. Dose of NPK @ 40, 20, 30 kg ha^{-1} through drip irrigation should be given.

Different structures of protected agriculture

Protected agriculture includes different structures such as poly-house, poly-tunnel, shading-net house, floating row cover, bird-proof net house, cloches, insect-proof net house, mulches and anti-hail net house. The selection of the greenhouse design should be determined by the grower's expectations, need, experience, and above all its cost-effectiveness in relation to the available market for the produce. Obviously, cost of greenhouse is very important and may outweigh all other considerations.

Polyhouse technology with INM

A polyhouse is quasi-permanent structure, covered with a transparent or translucent material, ranging from simple homemade designs to sophisticated pre-fabricated structures, wherein the environment could be modified suitably for the propagation or growing of plants. Polyhouse technology provides optimum conditions of light, temperature, humidity, CO_2 for the best growth of plant to achieve maximum yield with best quality (Ngouajio and Ernest, 2004). Materials used to construct a polyhouse frame may be wood, bamboo, and steel or even aluminum. Coverings materials generally used are glass, polythene, polycarbonate sheets or fiberglass reinforced plastic panels (FRP).

Type of polyhouses

1. Naturally ventilated or zero energy
2. Semi-controlled or Fan and Pad or evaporative cooled

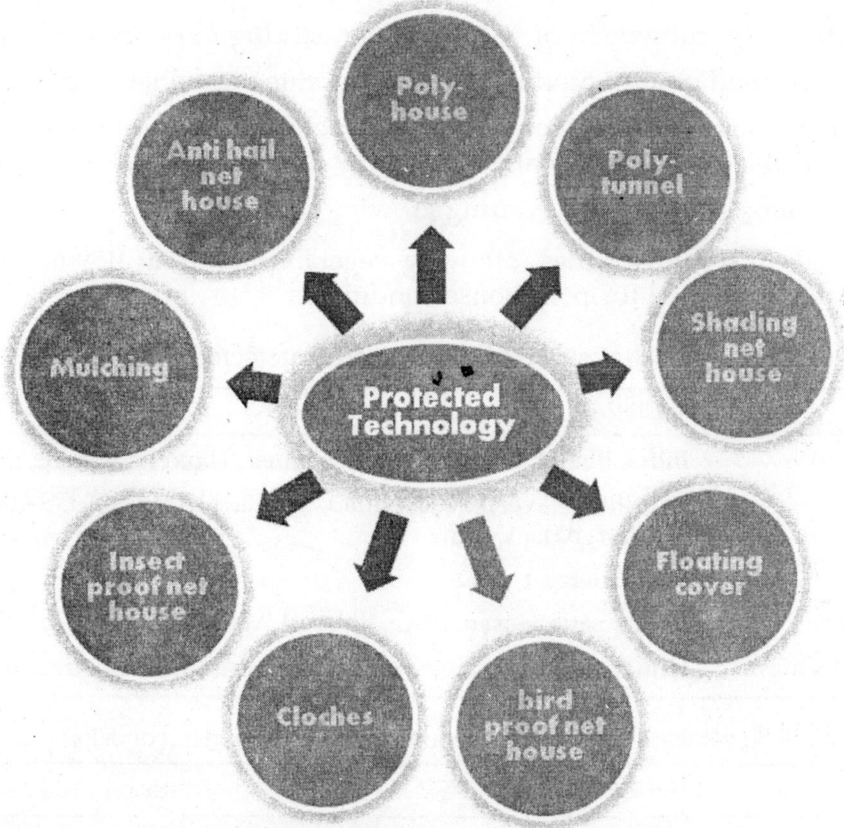

Plant environment and polyhouse climate

A plant grows best when exposed to an environment that is optimal for that particular plant species. The aerial environment for the plant growth can be specified by the following four factors:

i. Heat or temperature
ii. Light
iii. Relative humidity
iv. Carbon dioxide

While plants have precise optimum environmental conditions for best growth, most are tolerant to variations in these conditions within some limits. However, permanent damage would occur when they are exposed to conditions outside these limits. At the same time, plants are subject to attack by pests and diseases.

Objectives and utility of poly-house

The poly-houses are of much importance

- Off-season cultivation of vegetables for earning more income
- Year-round quality production of high value vegetable crops
- These permanent structures are suitable for high value off-season vegetable cultivation in any area at any time of the year
- Minimizes crop stresses during growing season

Based on several studies following vegetables and cut flower varieties can be recommended for poly-house conditions.

Table 4 : Vegetable varieties suitable for cultivation in protected conditions

Crop	Varieties
Capsicum	Indra, Bharat, Bomby Golden Summer, Tanavi , Orobelle, Swarna
Tomato	Rakshita, Naveen 2000, Nutan, Snehlata, Avinas-2, GS-600, Arka Vardan , Arka Vishal
Cherry tomato	Pusa Cherry-1, NS 2
Cucumber	Kiyan, Satis, Hasan, Priya, Poinsett, Malini, Alamir, Bronco
Summer squash	Himanshu, Seole Green, Dokato, Pusa Alankar, Australian Green

Table 5 : Cut flower varieties suitable for cultivation in Polyhouse conditions

Crop	Varieties
Rose	First Red, Grand Gala, Diplomat, Konfetti, Cora, Corvetti, Esacada, Femma, Kiss, Lambada, Laser, Movies Star, Nicole, Noblesse, Vivaldi, Soledo, Susanne
Gerbera	Monique, Diana, Anneke, Nette, Rosetta, Margarethe, Lydia, Gloria Gina, Ingrid, Pricilla, Alexis
Carnation	Dona, Pink Dona, Malaga, White Dona, Solar, Cobra , Pendy, Lorella

Poly-Tunnel technology with INM

Poly-tunnels are flexible transparent coverings that are installed over single or multiple rows of vegetables to enhance the plant growth by warming the air around the plants in the open field during winter season when the temperature is below 8°C. Plastic low tunnels are often used to promote the growth of plants during the period of winter season. Use of low plastic tunnels help in reducing the impact of frost and low temperature to a great extent in winter and high temperature during summer (Singh and Asrey, 2007).

Low tunnels are supported above the plants by using hoops of GI wire and a clear or transparent plastic of 20-30 micron is covered/stretched over the hoops and the sides are secured by placing in soil. Plastic is vented or slitted during the growing season as the temperature increase within the tunnels. The farmers can grow different varieties of summer squash (round-fruited, long-fruited), which is a emerging crop along with cultivation of netted muskmelon varieties in place of traditional varieties. Bitter gourd and round melon are two other crops with increasing demand and which usually fetches very high price during off-season and can be grown successfully by using the plastic low tunnel technology. This technology is highly suitable and profitable for the farmers living in northern plains of India.

Poly-tunnel is known as temporary or mobile micro-polyhouse or row-cover and classified in the three types:

1. Low poly-tunnel
2. High poly-tunnel
3. Trench poly-tunnel

Objectives and Utility of Poly-tunnels

Poly-tunnels are of same importance as poly-house and are used for

- Off-season cultivation of vegetables for earning more income
- Walk in tunnels are only erected over the crop during the peak winter months of December to mid- February and thereafter the structure is removed from the crop.
- Since, the plastic is used only for two months (December to mid-February) therefore, life of the plastic can be 8-10 years.
- These temporary plastic structures are suitable for off-season vegetable cultivation in northern-plains and low hills.

Shade net

Shade nets are perforated plastic materials used to cut down the solar radiation and prevent scorching or wilting of leaves caused by marked temperature increases within the leaf tissue from strong sunlight. These nets are available in different shading intensities ranging from 25% to 75%. Leafy vegetables and ornamental greens are recommended to be grown under shade nets whose growth rates are significantly enhanced compared to unshaded plants when sunlight is strong.

Floating row covers

They literally "float" or lie directly over the crop, whether direct seeded or transplanted. Materials include perforated plastic, spunbonded polyester and spunbonded polypropylene; they do not impede seedling emergence or subsequent growth of the crop. To secure the cover against wind, all the edges are buried or weighted down with sandbags, rocks, or other materials. For most crops, floating row covers require no support because they are lightweight. There are two types of these covers: perforated polyethylene and spun-bonded polyester or polypropylene which is available in several weights (rather than in thickness).

Drip-fertigation

Applying fertilizers through irrigation water, particularly through drip system, termed as fertigation, provides the most effective way of supplying nutrients to the plant roots zone or application of water and nutrients in the form of discrete droplets directly to the root zone of the crop through a network of pipelines is known as drip fertigation.

Type of drip-fertigaion

1. Pressurized drip fertigation system
2. Low-pressure drip fertigation system

Advantages of pressurize drip fertigation

- Water saving by 30-70 %
- High WUE (water losses are minimized)
- Uniform supply of the water to all the plants
- High fertilizer use efficiency (FUE), fertilizer saving (20-40 %), uniform supply and as per the crop growth stage
- Uniform germination and crop growth
- Higher yield and better quality
- Labour saving in weeding, irrigation and fertilizer application
- Energy saving
- Field sanitation - less weed growth, pests and diseases incidence
- Cultivation in undulated lands
- Cultivation in mulches, greenhouses, low tunnels, net-houses and walk-in-tunnels etc.
- Suitable to automation and simple to irrigate the field

Common formula used for drip fertigation

A. Conc. of element (ppm) =

$$\frac{\text{Quantity of fertilizer in water } (g\,m^{-3}) \times \text{\% of element infertilizer}}{100}$$

B. Quantity of fertilizer $= \dfrac{100 \times \text{concentration of elements (ppm)}}{\text{\% of element in fertilizer}}$

Table 6 : Recommended concentration (ppm) of nutrient solution for INM through drip-fertigation under protected condition

S. No.	Nutrients	Nursery	Tomato	Capsicum	Cucumber	Flowers
1	N	25	150	120	150	150
2	P	25	50	50	80	30
3	K	25	200	120	150	160
4	Ca	-	127	127	40	90
5	Mg	-	48	48	48	35
6	S	-	66	66	66	40
7	Fe	-	2.8	2.8	2.8	1.30
8	B	-	0.7	0.7	0.7	0.60
9	Cu	-	0.2	0.2	0.2	0.30
10	Mn	-	0.8	0.8	0.8	0.10
11	Zn	-	0.2	0.2	0.2	0.25
12	Mo	-	0.05	0.05	0.05	0.05

Fertilizers suitable for drip fertigation

The fertilizers which are to be applied through drip irrigation should be 100 % water soluble, dissolved quickly, have high nutrient content with no toxic elements, low-price and should be easily available.

- Urea (46% N)
- Urea Phosphate (17:44:0),
- Mix fertilizer of NPK (20:20:20)
- Ammonium Sulphate (21% N, 24%S)
- Murate of Potash (60% K_2O)
- Potassium Nitrate (13% N, 34% K_2O)
- Ammonium Nitrate (34% N)
- Calcium Nitrate (15% N, 19.5% Ca)
- Calcium chloride

- Mono Ammonium Phosphate (12%N, 61%P_2O_5)
- Phosphoric acid (54%P_2O_5)
- Other water soluble specialty fertilizers

INM for mulching technology

Mulching is the practice of covering soil around plants with an organic or synthetic material to make conditions more favorable for plant growth, development and crop production. The change in soil environment could be favorably controlled by selection of an appropriate mulching material. Bio-mulches such as leaves, straw, saw dust, peat moss, compost and gravel have been used for centuries to control weeds, hold moisture and conserve soil. More recently, paper, petroleum and plastics mulches have been tested for soil environment control. The most versatile of all the mulches at present is, however, plastics mulch which is manufactured in different film colours, thickness and widths. These plastics film mulches are generally made from low density polyethylene or linear low density polyethylene. By a proper selection of plastics mulch composition, colour and thickness, it is possible to precisely control the soil environment. Use of plastic mulches accelerates plant growth by increasing soil temperature, establishing soil moisture, weed control and production of quality produce and reduction in leaching of fertilizer (Das *et al.*, 2007; Pande *et al.*, 2005). Black is the most common colour used for mulching. Transparent mulch film increases the soil temperature. The optimum thickness of polyethylene should be around 40-50 micron.

Advantages of mulching

- Improves micro climate around the plant root zone
- Minimizes weed incidence
- Minimizes soil erosion and conserve moisture
- Save 70-75% Water .
- Increase 30-40% yield
- Continuous uptake of nutrient
- Decompose organic matter

Colour of plastic mulch

Colour of plastic mulch films have been investigated for their impact on crops. Each colour film reflects the wave-length that gives it the colour we perceive.

- Yellow has been shown to attract insects so using yellow-mulched rows to attract insects to a location where they can be killed

- Black plastic has been the standard plastic used worldwide
- Silver-reflective plastic mulch associated with higher reflectance cause insect dis-orientation and repels aphids
- Blue attracts thrips but has been very effective in greenhouse

There also appears to be some reduction in disease pressure with crops grown on specific colours.

Frames and cladding material

Wood, bamboo, GI pipe, angle iron and aluminium are common material used for the framework of a polyhouse. Durability of the structure depends on the type of the material used. Green house covering materials generally used are glass, polythene, polycarbonate sheets or fiberglass reinforced plastic panels (FRP). As the purpose of a greenhouse covering is to allow sunlight to pass through it so that the energy is retained inside. Glass was the main covering material in the early greenhouses. A brief description of greenhouse covering materials is given below:

Glass: A clean, transparent glass provides the maximum light transmittance to the extent of 90%. However, being heavier in weight, it requires elaborate structure for adequate support. It is brittle and can break with minimum shock or vibrations resulting in high maintenance costs.

Acrylic: This material has long service life, good light transmittance (80%), moderate impact resistance, but prone to scratches. It has a high coefficient of expansion and contraction. Being inflammable and costly, it is not a preferred material.

Polycarbonate: It is available in single or double wall sheets of different thickness. A new polycarbonate sheet has good light transmittance of about 78%, but reduces with age. It has excellent impact resistance and low inflammability. High cost limits its use on large scale.

Fiberglass reinforced plastic panels (FRP): These plastics consist of polyester resins, glass fibers stabilizers etc. It has initial light transmittance of about 80% and has high impact resistance with a service life ranging from 6-12 years. Good quality FRP materials for greenhouse coverings are not quite assured.

Polyethylene: A clear, new polyethylene sheet has about 88% light transmittance. Its higher strength and low cost have made it most popular replacement to glass. An ultra-violet (UV) stabilized plastic sheet can have a service life of 3 years. These sheets are generally available in 7 and 9 meter widths with 200 micron (0.2 mm) thickness.

Constrains for success of INM with protected agriculture

There are certain constraints for success of integrated nutrient management in protected agriculture such as infestations of soil & air borne diseases, poor quality of soil health and water and traditional methods of agricultural practices.

Conclusion

Keeping in view the increasing demand of off-season and high value vegetables, fruits and flowers in metro-politan cities of the country, there is urgent need for diversification from the traditional agriculture by production of high value horticultural crops under different protected structures for increasing their productivity and quality for getting high returns along with use of integrated sources of nutrients for sustained yields. Nursery raising under protected cultivation can be adopted as an agri-entrepreneur in major vegetable growing areas of the country by unemployed youths who are graduate in agriculture or post graduate in horticulture. Low cost protected technology like plastic low tunnels or walk in tunnels can be used for off-season vegetable cultivation for getting high returns from off-season produce. Similarly insect-proof net houses can be used on a large scale for safe vegetable cultivation to minimize use of pesticides in vegetable cultivation and virus-free crop production in large number of vegetables.

Protected cultivation of horticultural crops is reasoned to be the next logical step to open field agriculture. One million hectare under protected vegetable cultivation can easily increase the production of vegetables by about 40 million tonnes. This will also permit the creation of about one million additional jobs in rural sector. The field of protected cultivation is still evolving. Methods of more efficient input application are being devised. Better plant protection measures are being developed. More eco-friendly plastics are being produced by making the processes more efficient. As a result, protected cultivation could bring prosperity to vegetables growers in India and ensure nutritional security for everyone.

References

Chandra, P., Sirohi, P.S., Behera, T.K. and Singh, A.K. (2000). Cultivating Vegetable in Polyhouse. *Indian Horticulture*, **45** (3): 17-32.

Das, B., Nath, V., Jana, B.R.. Dey, P., Pramanik, K.K. and Kishore, D.K. (2007). Performance of strawberry cultivars grown on different mulching materials under sud-humid subtropical plateau conditions of Eastern India. *Indian Journal of Horticulture*, **64** (2): 136-143.

Gaur, A.C., Ostwal, K.P. and Mathur, R.S. (1980). Save superphosphate by using phosphobacteria. *Kheti*, **32**: 23-25.

Gupta, N.S., Sadavarte, K.T., Mahorkar, V.K., Jadhao, B.J. and Dorak, S.V. (1999). Effect of graded levels of nitrogen and bioinoculants on growth and yield of marigold (*Tagetes erecta* L.). *Journal Soils and Crops*, **9** (1): 80-83.

Ngouajio, M. and Ernest, J. (2004). Light transmission through coloured polythene mulches affects weed population. *Horticulture Science*, **39** (6):1302-1304.

Pande, K.K., Dirmi, D.C. and Kamboj, P. (2005). Effect of different mulches on growth, yield and quality attributes of apple. *Indian Journal of Horticulture*, **62** (2): 145-147.

Singh, A.K., Singh, A.K., Gupta, M.J. and Shrivastav, R. (2003). Effect of polyhouse on insect-pest incidence, fruit quality and production of vegetable and fruit crops. *Proceeding of the National Symposium on Frontier areas of Entomological Research* at IARI, *New Delhi, India*. pp: 123-30.

Singh, R. and Asrey, R. (2007). Cultivating strawberry: The Plasticulture Way. *Indian Horticulture*, **52** (2): 6-7.

Singh, B., Singh, A.K. and Tomar, B. (2010). In peri-urban areas protected cultivation Technology to bring prosperity. *Indian Horticulture*, **55** (4): 31-33.

□□□

System Based Integrated Nutrient Management, 2012
© *B. Gangwar & V.K. Singh (eds.), pp. 239-248*
New India Publishing Agency, New Delhi (India)
e-mail : info@nipabooks.com; website : www.nipabooks.com

CHAPTER **15**

Integrated Nutrient Management in Horticulture Based Farming Systems

AKATH SINGH

Indian horticulture has shown impressive impact on production, productivity and profitability. Currently, the total production of horticultural produce is 214.7 million tones from the 9% of total cultivated area in India, which contribute 30.4% to GDP of agriculture (Singh, 2010). In world perspective, India is second largest producer of fruits (68.5 mt) from 6.10 m ha area and contributes 11.2% in global fruit production. Vegetable crops occupying 8.0 m ha have the production of 129.1 mt (Table 1). However; India is the largest producer, consumer and exporter of spices and spice products in the world and produce more than 50 spices. The spice production in India is the order of 4.14 mt from an area of about 2.63 m ha. India is also the leading producer of plantation crops in the world and contributes 22.34% in coconut, 25% in cashewnut and 55% in arcanut. Commercial floriculture sector has recorded fast pace of growth during the last decade. During the last decade area under floriculture has expended to 1, 67, 000 ha with production of 9,87,000 MT of loose flower and 4.8 million cut flowers (Singh, 2010). The over all percent share of fruits and vegetables to their total production and contribution of different states for production of fruits and vegetables are repicted in Fig. 1 & 2. From the above fact, it is indicated that the production of horticulture produce has increased manifold but the gap in demand and supply continues simultaneously and is hardly sufficient and meets only 46% of national demand. Hence, there is a strong need to increase the production and productivity through crop diversification.

Table 1 : Horticulture scenario in India (2008-09)

Crops	Area (000 ha)	Production (000 MT)	Productivity (Mt/ha)
Vegetables	7981	129077	16.2
Fruit	6101	68466	11.2
Plantation crop	3217	11336	3.5
Spices	2629	4145	1.6
Flowers	167	987	5.9
Medicinal & aromatic	430	430	1.0
Nuts	136	173	1.3
Total	20661	214716	10.4

(Source: NHB, 2009)

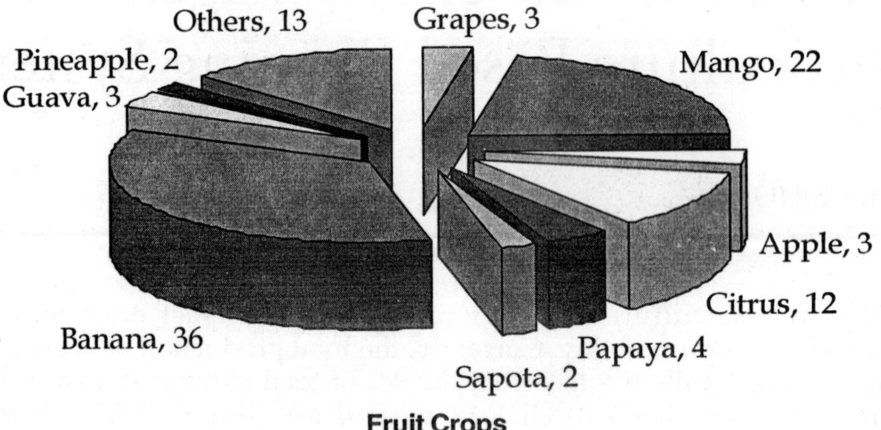

Fruit Crops

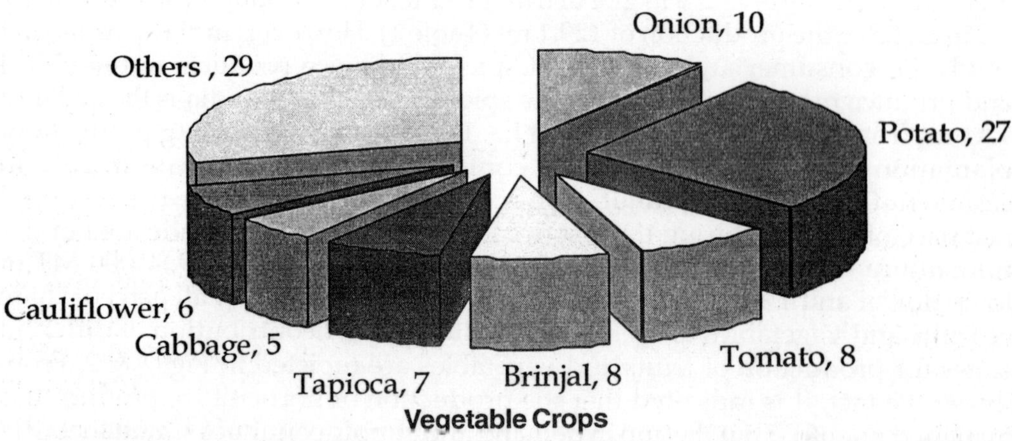

Vegetable Crops

Fig. 1 : Percent share of major fruits and vegetables in total production in India
Source : NHB, 2008-2009.

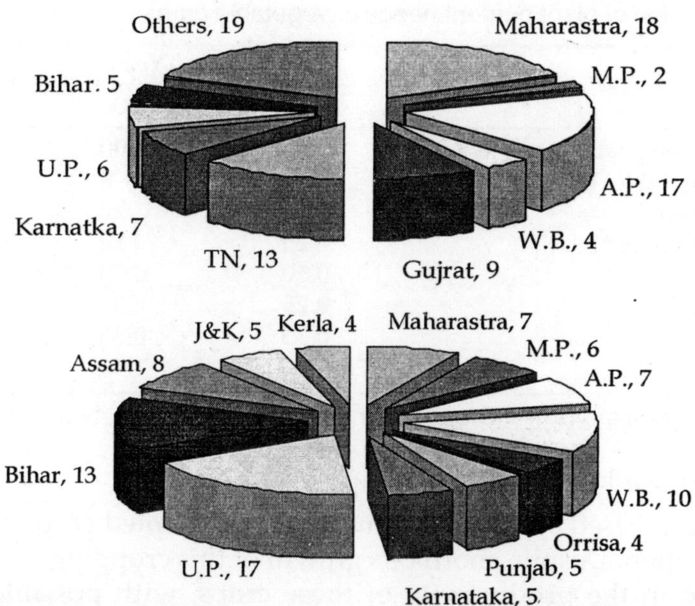

Fig. 2 : Percent share of fruits and vegetables in different states to total in fruits and vegetables production in India
Source : NHB, 2009.

On an average, a 35 t/ha crop of vegetables remove 151, 57 and 195 kg N, P_2O_5 and K_2O, respectively (Table 2 & 3). On the other hand, there is an appalling gap between potential yields and actual yields harvested by farmers. For most vegetable crops, yields realized are less than 50 percent of the potential. This task calls for better crop husbandry including the use of optimum rates of fertilizers.

Table 2 : Nutrient removal by some vegetable crops

Crops	Yield, t ha^{-1}	Nutrient Removal, kg ha^{-1}		
		N	P_2O_5	K_2O
Beans	15	130	40	160
Cabbage	70	370	85	480
Carrot	30	125	55	200
Cauliflower	50	250	100	350
Cucumber	40	70	50	120
Okra	20	60	25	90
Onion	35	120	50	160
Radish	20	120	60	120
Tomato	50	140	65	190
Peas	20	125	35	80
Mean	35	151	57	195
Nutrient ratio		2.65	1.00	3.42

Table 3 : Projection of plant nutrient needs of vegetable crops

Particulars	Year		
	1995	2000	2025
Requirement of vegetables (mt)	90	100	170
Area under vegetables (m ha)	5	6	8
Productivity (t ha^{-1})	14	20	30
Farm yard manure (mt)	125	150	200
Nitrogen (mt)	0.50	0.70	1.00
Phosphorus (mt)	0.25	0.35	0.50
Potash (mt)	0.25	0.35	0.50
Total NPK (mt)	1.00	1.40	2.00

Source: B.S. Prabhakar, Better Crops International/Vol. 10, No. 2, November 1996

Horticulture based farming system approach

Cereal crop based system is the most widely adopted cropping system in most of the region. Despite enormous growth of this cropping system, reports of stagnation in the productivity of these crops, with possible decline in production in future are seen. Important issues emerging as a threat to these systems are over mining of nutrients from soil, disturbed soil aggregates, decreasing response to nutrients, declining ground water table, build up of diseases/pests, low input use efficiency. But the challenges before us are to produce more from shrinking land and declining water in the scenario of climate change. Horticulture sector besides, improving income through increased productivity, employment, nutritional security, produces higher biomass resulting in efficient utilization of natural resources. Indian farmer cannot take risk with small land holding size. The need is to produce more from less and less resources.

Both ruminants and non-ruminants provide manure for the maintenance and improvement of soil fertility. Soil fertility depletion is a major constraint to Indian agriculture. Even when inorganic fertilizers are applied, crop yields may not be maintained under continuous cultivation on nutrient-poor soils with a low buffering capacity. The use of only mineral fertilizers can decrease soil pH and base-saturation and increase aluminium toxicity. Organic materials applied in bulk can improve soil texture, promote better absorption of moisture, reduce run-off and prevent crusting of the soil surface. Even small quantities of organic materials can bring about marked improvements in the cation exchange capacity of soils. Manure is also valuable in reversing the deterioration in soil structure in sodic soils, characterized by high contents of exchangeable sodium and low permeability. In integrated production systems different type of manure i.e. pig manure, poultry manure, vermicompost, other farm wastage, biomass recycled is drained and the clear effluent is applied as fertilizer to the crops. This is the unique feature of any farming system where a by product of one system acts as main input for the other in terms of food, feed, fibre, nutrient etc. Even under arid conditions, the application of farm manure can play an important role in stabilizing or increasing crop yields, improving the utilization of inorganic fertilizer and enhancing soil fertility. It is relevant to mention here the role of organic manure crops (green manure, cover crops, intercrops), an important component of horticulture based system which maintain soil productivity. In recent years a decline in organic matter levels is threatening the rice– wheat systems on the Indo-Gangetic Plain of India. Many species have been used for green manure and include sun-hemp (*Crotalaria juncea*), mungbean (*Phaseolus aureus*), cowpea (*Vigna unguiculata*), guar (*Cyanopsis tetragonoloba*), *Sesbania rostrata* and berseem which can substitute for up to 60–120 kg of fertiliser nitrogen/ hectare and enhance the availability of other nutrients. Contributions of NPK from some prominent species under integrated farming system (IFS) are given in Table 4.

Table 4 : Nutrient accumulation through leaf biomass of different hedge species under integrated farming system

Species	Leafy matter (q ha^{-1})	Nutrient accumulation (kg ha^{-1})		
		N	P	K
C. cajan	7.82	25.73	5.24	11.18
Crotolaria tetragona	22.98	79.74	11.03	37.46
Desmodium rensonii	5.31	19.28	2.55	8.28
Flemengia macrophylla	6.46	20.87	2.91	8.14
Indigofera. tinctoria	16.99	65.58	13.76	27.69
Tephrosia candida	15.30	54.62	4.90	25.55

(Source: Bhatt and Bujarbaruah, 2007)

The predominant land use systems in horticulture based farming systems in the country are as follows :

1. Agrihorticulture (Crops + fruits)
2. Hortipastoral (pasture +fruit plants)
3. Agrihortipastoral (crop +pasture + fruit plants)
4. Hortisilvipastoral (fruit plants +tree +pasture)
5. Agrihortisilvipastoral (crop +fruit +trees +pasture)
6. Multi tier or multistoried cropping systems
7. Fruit based system (fruits+ inter/cover crops + vermiculture)
8. Vegetable based farming system

Based on the past research and experiences the multi-storied horticulture based system are given in Table 5.

Table 5 : Suggested multistoried systems for different agro-climatic conditions

Agro-climatic conditions	Cropping system
Arid	Ber+Karonda+Khejri+Lasoda+Phalsa
Semi-arid	Aonla+Guava+Black gram
	Aonla+Custard apple+Legumes
Sub-tropical	Aonla+Guava+Black gram
	Mango+Guava+Papaya
Tropical	Mango+Citrus+Tapioca
Coastal	Coconut+Black pepper+Banana
NEH	Areca nut+ Black pepper +Pineapple
	Areca nut+ Banana +Pineapple+ Turmeric/Ginger

Fertilizer consumption

India is the third largest producer and consumer of fertilizers in the world after China and USA. Against 21.65 million tonnes of fertilizer nutrients (NPK) consumed during 2006-07, the nutrient consumption is 22.57 million tonnes respectively during 2007-08. The consumption of major fertilizers namely, Urea, DAP, MOP, SSP and complexes were 25.96, 7.50, 2.88, 2.29 and 6.57 million tonnes during 2007-08. The over all NPK requirement for vegetable production, which was 1.0 mt during 1995 increased upto 1.4 mt by 2000 AD. The future projection of NPK requirement for vegetable production will be about 2.00 mt by 2025 in the country (Table 4). The all India average fertilizer consumption is 116.5 kg ha^{-1} of NPK nutrients, though there is wide variation from state to state varying from 212.7 kg ha^{-1} in Punjab, 208.2 kg ha^{-1} in Andhra Pradesh, 190.9 kg ha^{-1} in Haryana to less than 5 kg ha^{-1} in states like Arunachal Pradesh and Nagaland. India is by and large self sufficient in respect of Urea and about 90% in case of DAP but other secondary and micro-nutrients have to be imported. It is imperative to focus on integrated muthatmangat for crop productivity of sustained economy.

INM components in farming systems

Among various inputs, fertilizers alone account for 20-30% of the total cost of production, moreover, the efficacy of fertilizers applied in soil is low due to various losses and fixation. At present, chemical fertilizers are not only in short supply but they are expensive too. Although these fertilizers contribute a lot in fulfilling the nutrient requirement of horticultural crop but their regular, excessive and unbalanced use may lead to health and ecological hazards, depletion of physico-chemical properties of the soil and ultimately poor crop yields. Hence, there is a strong need to increase the production and productivity of horticultural crops with judicious nutrient management and INM is regarded as key to achieving these goals. Integrated nutrient management (INM) refers to maintenance of soil fertility and plant nutrient supply to an optimum level for sustaining the desired crop productivity through optimization of the benefits from all possible sources of plant nutrients in an integrated manner. Therefore, it is a holistic approach where we first know what exactly is required by plants for optimum level of production, in what different forms, at what different timings in best possible method, and how best these forms can be integrated to obtain highest productivity levels with efficiency at economically acceptable limits in environmental friendly way.

Under farming system therefore, integrated nutrient management (INM) involves efficient and judicious supply of all major and micro components of plant nutrients including local available resources on sustainable basis. This is achieved through optimization of benefits derived from different sources of nutrient supply including chemical fertilizers, biofertilizers etc from out sources and organic manures, crop residues, green manures, cover crops/intercrops, organic mulch, biomass recycling, litter of poultry/duck/fish pond in integrated manner from the farming system and, which are desired as components of INM. Long term experiments involving integrated use of organic and inorganic fertilizers in farming system have demonstrated that 50% substitution of nitrogen fertilizers by farmyard manures alone helps in maintenance of sustainable crop productivity. However, one needs to realize that, although, crop residues alone may be harmful, but in combination with green manuring, it may improve soil fertility. Horticulture based farming system has a vast potential of organic waste resources, which include biomass of different intercrops, pruning woods, animal dung, animal urine, crop residues etc. Much of these organic wastes remain unutilized, leaving enormous scope for development of organic manures through recycling. These farm wastes can be converted in to more value added inputs with use of some beneficial micro or macro organism. For example, in a typical horticultural farming, there are various kind of crop residues which can be either used as mulch material in tree basin directly or can be further composted and used as manures. It is well established that organic mulching improves quality fruit production by improving the soil conditions as these mulches decompose with time and thus they improves physical, chemical and biological properties of soil. Besides, it minimizes weed competition, maintain basin and canopy temperature, improves nutrient uptake and after decomposition produces humus, humic acid and other organic acids, phenols, phenolic acid and easily available forms of nutrients. The improved soil conditions result in better growth and development of plant root systems and uptake of nutrients for quality fruit production.

INM in horticulture based farming system : Practices and impact

Number of studies indicated improved efficiency of fertilizer nutrients with combined use of organic manures and fertilizers. Different organic manure viz. cow manure, poultry manure, pig manure, rabbit manure, bioplus, neem cake along with spraying of Amritpani and Panchgavya registered 30-60% higher yield of ginger and turmeric as compared to 100% RDF, grown in farming system under Alfisol at Umiam Meghalaya (Sanwal *et al.*, 2007). Maximum yield and B:C ratio in sprouting broccoli was obtained by poultry manure @2.5t ha^{-1} + 50% RDF, which slightly varied with vermicompost @2.5t ha^{-1} + 50% RDF (Maurya *et al.*, 2008). INM study in tomato revealed that maximum yield (421.3 q ha^{-1}) along with a maximum B:C ratio (3.15:1) was obtained in treatment having *Azotobacter* inoculation + 100% NPK + FYM as compared to 100% NPK + FYM (282.8 q ha^{-1}) (Satesh and Sharma, 2007). Yadav *et al.*, (2006) reported that okra growth and yield was significantly influenced by poultry manure with NPK as compared to other organic sources like FYM and vermicompost with same dose of NPK.

Some classical examples of INM based results in different fruit crops can be cited. Highest fruit yield with improved quality was obtained with 25 kg FYM with 400g N-150g P-300g K plant^{-1} in Khasi mandarin, 15 kg FYM-1500g N-440g P- 600g K plant^{-1} in Naval orange and 15 kg neem cake-800g N-300g P-600g K plant^{-1} year^{-1} in sweet orange (Srivastava 2008). The study with organic manure and biofertilizer with chemical fertilizer, revealed 50kg FYM tree^{-1} + 150g Azotobacter+100g VAM+500 g N-250g P-500g K tree^{-1} year^{-1} showed maximum yield with excellent quality litchi in alluvial zone of West Bengal (Dutta et al., 2010). Athani *et al.*, (2007) reported highest yield (33.11 kg plant^{-1} and 9.19 t ha^{-1}) of guava supplied with 75 per cent RDF + 10 kg vermi-compost followed by 100% RDF and in situ vermiculture application (50 worms plant^{-1}). INM studies in papaya comprised RDF (200:250: 250 g. N:P:K plant^{-1}), FYM 50 kg plant^{-1}, vermicompost 20 kg plant^{-1}, poultry manure 20 kg plant^{-1}; rhizosphere bacteria culture 50 g plant^{-1} alone and in combination with reduced levels of RDF viz. 75 and 50%. The yield of papaya was statistically at par with 75% recommended fertilizers rate + 25% vermicompost + rhizosphere bacteria culture in comparison to control (recommended fertilizers dose). Quality parameters of the fruits were found to increase with decreasing level of chemical fertilizers (Kirad *et al.*, 2010). Application of 100% recommended dose of NPK with 10 kg FYM per plant and biofertilizers (*Azosprillum* and PSB @ 25 g plant^{-1}) were found beneficial in terms of yield and monetory returns in banana based system, which was followed by application of 50% NPK through organic (i.e. FYM + Green manure) and 50% NPK through inorganic and biofertilizers. Similarly, Hazarika and Ansari, (2010) reported that the treatment having the 100% recommended dose of NPK (P as as rock phosphate) in combination with farmyard manure and biofertilizers significantly influenced plant growth and yield of banana besides shortening the crop cycle. In strawberry nitrogen fertilizer can be substituted by half of their standard dose by combined application of *Azotobacetr, Azospirillum* and 100 ppm GA_3 without any adverse effect on growth and productivity (Singh and Singh, 2009).

Correction of micronutrient deficiencies is an important component of INM and there are several classical examples in horticultural crop which improve the quality of produce. Foliar spraying of important nutrients like B, Zn, Mo, Cu, Fe, Mn and multiplex significantly increased the concentration of N, P, K, S, Zn, Fe, Cu, Mn and B in tomato fruits, especially when used in a mixture (Bhatt and Srivastava, 2006). Heavy fruit drop and poor quality fruit is a common problem in most of mandarin cultivars. Babu *et al.* (2007) reported that spraying of Kinnow mandarin tree with Zn, Mg and Mn, alone or in combination retain maximum fruit with highest grade and quality fruits. Among the different levels of micronutrients, application of borax at 0.50% + $ZnSO_4$ at 0.25% was considered as the best which caused the maximum plant growth (171.62 cm plant height and 39.74 cm girth) and fruit yield (37.20 kg/plant) in papaya (Singh *et al.*, 2010). Various studies involving combination of oil cakes with inorganic fertilizers have shown promising results. These include: combination of 15 kg neem cake and 800g N-300g P-600g K tree^{-1} year^{-1} in sweet orange on alkaline loam soil, 7.5 kg neem cake-600g N-300g P-600g K tree^{-1} year^{-1}in khasi mandarin on acidic loam soil and 7.5 kg castor cake 400g N-150g P-300g K tree^{-1} year^{-1} in Sathgudi sweet orange on alkaline sandy loam soil in Tirupati A.P. (Srivastava, 2008).

Research gaps

- Mismatching of INM practices developed at research stations with the farmers' resources and their practices
- INM recommendations for different crops are not based on soil testing and nutrient release behaviour of the manures
- Nutrient balance/flow analysis vis-à-vis soil fertility management practices with special reference to INM at farm level needs to be worked out
- Nutrient release characteristics of farm residues in relation to their quality to develop decision support systems
- Biofertilizers were not included as component of INM in many cases
- Integrated Farming Systems (IFS) approach needs to be encouraged for sustaining livelihood in rural areas particularly for small and marginal farmers.

References

Athani, S.I., Prabhuraj, H.S., Ustad, A.I., Swamy, G.S.K., Patil, P.B., Kotikal, Y.K. (2007). Effect of organic and inorganic fertilizers on growth, leaf, major nutrient and chlorophyll content and yield of guava cv. Sardar. *Acta Horticulturae*. **735**: 451-54.

Babu, K. D., Dubey, A.K.and Yadav, D.S. (2007). Effect of micronutrients on enhancing the productivity and quality of Kinnow mandarin. *Indian Journal of Horticulture,* **64**: 353-356.

Bhatt, Lallit and Srivastava, B.K. (2006). Effect of foliar application of micronutrient on the nutritional compostion of tomato. *Indian Journal of Horticulture,* **63**: 286-288.

Bhatt, B.P. and Bujarbaruah, K.M. (2007). intensive integrated farming system: A sustainable approach of land use in eastern Himalayas. In : *Shaping Agrarian Prosperity through Integrated Intensive Farming, eds*: Prakash et al.,. ICAR Research complex for NEH Umaim, Meghalaya.

Dutta, P., Kundu, S. and Biswas, S. (2010). Integrated nutrient management in litchi cv. Bombay in new alluvial zone of West Bengal. *Indian Journal of Horticulture,* **67**(2): 181-184.

Hazarika, B.N. and Ansari, S. (2010). Effect of integrated nutrient management on growth and yield of banana cv. Jahaji. *Indian Journal of Horticulture,* **67**(2): 270-273.

Maurya, A.K., Singh, M.P., Srivastava, B.K., Singh, Y.V., Singh, D.K., Singh , S. and Singh, P.K. (2008). Effect of organic manures and inorganic fertilizers on growth characters, yield and economics of sprouting broccoli. *Indian Journal of Horticulrue,* **65**: 116-118.

NHB, (2009). National Horticulture data based 2008-2009.

Singh, Kirad K., Barche, S., Singh, D.B. (2010). Integrated nutrient management in papaya (*Carica papaya* L.) cv. Surya. *Acta Horticulturae.* **851**: 163-66.

Sanwal, S.K., Yadav, R.K. and Singh, P.K. (2007). Effect of types of organic manures on growth, yield and quality parameters of ginger. *Indian Journal of Agricultural Sciences,* **72**:67-72

Satesh Kumar and Sharma, S.K. (2007). Effect of integrated management strategies in tomato production. *Indian Journal of Horticulture,* **64**: 96-97.

Singh, Akath and Singh, J.N. (2009). Effect of bio-fertilizers and bio-regulators on growth, yield and nutrient status of strawberry. *Indian Journal of Horticulture,* **66**(2): 220-224.

Singh, D.K., Ghosh, S.K., Paul, P.K., Suresh, C.P. (2010). Effect of different micronutrients on growth, yield and quality of papaya cv. Ranchi. *Acta Horticulturae.* **851**: 261-63.

Singh, H.P. (2010). Dynamics of Indian horticulture to meet the challenges of food and nutritional security. In : *Souvenir: National Symposium on Emerging Trends in Agricultural Research, Sep. 11-12, 2010.*

Srivastava, A.K. (2008). Integrated nutrient management: An analysis of development, constraints and opportunities. In : *Souvenir: Third Indian Horticulture Congress, November 6-9, 2008, Bhubaneswar.*

Yadav, Pavan, Singh, P. and Yadav, R.L. (2006). Effect of organic manures and nitrogen levels on growth, yield and quality of okra. *Indian Journal of Horticulture,* **63**: 215-217.

□□□

System Based Integrated Nutrient Management, 2012
© *B. Gangwar & V.K. Singh (eds.), pp. 249-258*
New India Publishing Agency, New Delhi (India)
e-mail : info@nipabooks.com; website : www.nipabooks.com

CHAPTER **16**

Soil Chemical and Biological Properties under Integrated Nutrient Management

M.A. KHAN AND N.C. UPADHAYAY

Agriculture in India is one of the most important sectors of its economy. It provides livelihood to almost two thirds of the work force in the country and accounts for 18% of India's GDP. About 43 % of India's geographical area is used for agricultural activity and large number of production systems are in practice in different parts of the country. In pre-independence era (before the 1950s), agriculture in India was a system of harnessing nature for the sustenance of human beings, similar to the presently defined organic farming. Some countries have moved away from inorganic to organic farming yet it is not possible for India to depend entirely upon this system of farming due to large population. Total availability of plant nutrients from organic sources is projected to be 7.75 million tonnes by 2025 (Tandon, 1997) against the total requirement of 37.50 million tonnes (Katyal, 2001) for 1504 million population (Sekhon, 1997). Thus nutrients supply through organic sources alone would result in deficiency of about 30 million tonnes of nutrients. Hence, it is imperative to use both organic and inorganic sources of plant nutrients enhance food production for the increasing population (Mahajan *et al.*, 2007). In India the contribution of organic sources was 80-100 per cent in 1949-50 and reduced to about 32 per cent in 1993-94 and 20-25 per cent in 2004-05. So, it is imperative to enhance organic sources to meet total nutrient consumption. Following independence, rapid population growth in India placed great pressure on land and on these traditional farming systems; huge demands for food grains led to increased use of fertilizers and pesticides to boost production. Still a negative balance of about 8 million tones of NPK is foreseen in 2020, even if we continue to use chemical fertilizers, maintaining present growth rates of production and consumption. The most optimistic estimates at present, show that only about 25-30 per cent nutrient needs of Indian agriculture can be met by utilizing various organic sources. Many of the gains in production during the last 4-5 decades resulted from the "Green Revolution," a campaign of technological interventions in agriculture widely adopted by farmers in developing countries. Expansion of irrigation to cover rain fed areas, popularization of hybrids/transgenic varieties of crops, and use of synthetic chemical fertilizers and pesticides were the major technologies promoted. This paid rich dividends in India, quadrupling food grain production from 50 million metric tons in 1950-51 to 211 million metric tons in 2001-02, and enabling India to become self-sufficient in food grain. Now a second green revolution is also in the offing, to boost agricultural production and meet the estimated requirement of 337 million metric tons.

Integrated nutrient farming is a form of agriculture which includes the use of synthetic fertilizers and pesticides, plant growth regulators, livestock feed additives, genetically modified organisms and. organic material etc. As far as possible, integrated nutrient management, farmers rely on all available inputs such as inorganic fertilizers, synthetic pesticides, crop rotation, green manure, compost, biological pest control, and mechanical cultivation to maintain soil productivity and control pests. Soil is a dynamic, living matrix that is an essential part of the terrestrial ecosystem and constitute a medium for plant growth. It is a critical resource not only to agricultural production and food security but also to the maintenance of most life processes. Soils contain enormous number of diverse living organisms assembled in complex and varied communities. Soil biodiversity reflects the variability among living organisms in the soil- ranging from the myriad of invisible microbes, bacteria and fungi to the more familiar macro-fauna such as earthworms and termites. Plant roots can also be considered as soil organisms in view of their symbiotic relationships and interactions with other soil components. These diverse organisms interact with one another and with the various plants and animals in the ecosystem forming a complex web of biological activity. Environmental factors, such as temperature, moisture and acidity, as well as anthropogenic actions, in particular, agricultural and forestry management practices, affect to different extents soil biological communities and their functions. Soil organic matter, created through decomposition and recycling of plant and animal residues, has many important roles in organic systems, including supplying the necessary elements for plant growth. The first step in formation of soil organic matter is fixation of carbon dioxide (CO2) by plants. Capturing the energy of sunlight and efficiently recycling it through various forms of different soil organic matter is, therefore, a basic goal associated with organic production systems. The soil organic matter or carbon (C) inputs improve soil physical properties, such as aggregate stability, and provide food, habitat, and shelter for billions of soil organisms. Increased aggregate stability, improved soil structure, and surface protection provided by crop residues, manure or compost, and cover crops reduce soil erosion losses and increase water-holding capacity and aeration. Maintaining soil organic matter content at levels that are consistent with the natural characteristics of the soil (i.e., loamy soils will generally have higher organic matter than sandy soils) helps soil biological activity and the healthy microbial and macrofaunal populations that are required for efficient nutrient cycling. These populations include bacteria, fungi, actionmycetes, nematodes, and earthworms. Crop rotations (required for all organic operations) are crucial for organic systems because the legume crops (e.g., alfalfa and red clover) provide nitrogen (N) and also help recycle nutrients, such as phosphorus (P) and potassium (K). Including crops with deep root systems in the rotation helps extract nutrients from lower soil depths and return them to the surface when the vegetation dies. Crop residues also provide the carbonaceous biomass upon which soil microfauna (e.g., earthworms and beetles) and microorganisms depend on for survival. Over the last few years, the concepts of Integrated Plant Nutrient Management (IPNM) and Integrated Soil Management (ISM) have been gaining acceptance, moving away from a more sectoral and inputs-driven approach. IPNM advocates the careful management of nutrient stocks and flows in a way that leads to profitable and sustained production. ISM emphasises the management of nutrient flows, but also highlights other important aspects of the soil complex, such as maintaining organic matter content, soil structure, moisture and biodiversity.

The concluded reason was an increased soil microbe community in the manure fields, providing a healthier, more arable soil system. Increased biodiversity, especially from soil microbes such as mycorhizzae, have been proposed as an explanation for the high yields experienced by some organic plots, especially in light of the differences seen in a twenty one years comparison of organic and control fields. The bulk density of soil was less in organic farms which indicates better soil aggregation and soil physical conditions. Improvement in soil organic matter decreased the bulk density by dilution of the denser fraction of the soil.There was a slight increase in soil pH and electrical conductivity in organic farms compared to conventional farms. On an average there was 29.7% increase in organic carbon of soil in organic farms (1.22%) compared to the conventional farms (0.94%) which is a good indicator of soil quality as it works as a sink for all nutrients and known for improving all soil physical and biological properties of soil. Regular organic additions (manures and root biomass) had the positive effect on soil organic matter. Dehydrogenase, alkaline phosphatase and microbial biomass carbon were higher in organic soils by 52.3%, 28.4% and 34.4% respectively compared to the conventional farms. This establishes higher microbial activity in organically amended soils which is essential for nutrient transformations and increased availability of these nutrients to the plants. Increased nutrient availability in organic manure treatment could also be due to increased dehydrogenase and phosphatase activity. In general, increase in microbial biomass carbon in organic manure amended soils was due to increased availability of substrate-C that stimulates microbial growth, but a direct effect from microorganisms added through the compost is also possible. In organically managed soils, both macronutrients (N, P and K) and micronutrients (Zn, Cu, Fe, Mn) were available in larger quantities compared to the conventional soils, (Armand *et. al* 2008). It is well documented that there is a significant positive correlation between organic matter and micronutrient cation availability. Similar increase in soil quality by the addition of manures in organic farming was reported from the long term organic experiments in Switzerland (Ramesh et.al 2010). Treatments receiving organic amendments (Vermicompost in Maize-Potato-Onion or vermicompost+crop residue in Rice -Potato-Wheat & Sesamum-Potato-Moong) showed higher stock of C as well as C sequestration rate in all the cropping system.Incorporation of crop residues or application of secondary and micronutrients over 100% NPK gave more C stocks than 100% NPK alone as evaluated from long term experiment a CPRI Campus Modipuram (Fig. 1).

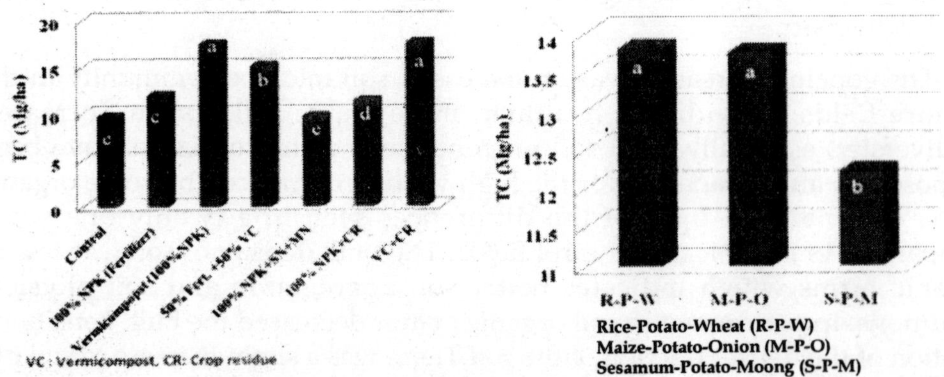

Fig. 1 : Impact nutrient management and cropping system on carbon stock in Inceptisol (bars with different lower case letters are statistically significant)

Integrated approach (NPK +FYM) increased soil organic carbon content of the soil under long term experiments at different locations in India. Table.1. Swarup (2002).

Table 1 : Effect of soil fertility management on soil organic contentr (SOC) g./Kg in long-term experiment at different locations in India.

Location	Control	N.P.K	N.P.K+FYM	Time yrs
Banglore	4.8	5.9	8.4	10
Barrackpore	4.1	5.0	5.4	24
Bhuvneswar	3.7	5.7	8.1	21
Coimbatore	4.3	4.9	6.2	23
Delhi	4.4	5.5	6.7	25
Hyderabad	4.6	5.3	8.0	23
Jabalpur	5.3	6.0	8.8	25
Ludhiana	2.5	3.3	3.8	25
Palampur	7.3	10.0	12.0	22
Pantnagar	5.0	8.3	15.0	24
Ranchi	3.0	3.5	4.8	23

The practical applications and operational methodologies in organic farming ,especially in potato based cropping systems are meager due to lack of comprehensive research in this field. Absence of package of practices prescriptions for organic farming for potato based cropping system hinder the implementation and promotion of this sustainable alternative production system. Many methods and techniques of organic agriculture have originated from various traditional farming system all over the world, where there is non-use of chemical inputs. The most commonly used organic sources of plant nutrients for potato based cropping systems are farm yard manure, vermicompost ,green manures, crop residues, oil cakes and biofertilizers etc. During the "organic transition" period, defined under US national organic standards as the first 3 years after switching from conventional to organic management, crop yields were lower than in conventional systems. Yields may eventually increase but the expectation of initially lower yields can be "a strong deterrent to those farmers who may wish to make such a change". Initially lower yields on organic farms have been attributed to the negative effects of conventional practices on the soil microorganisms that mineralize soil organic matter, or that control soil-borne pests. Gradual improvements in "soil quality" could be a possible reason to explain increasing yields with years of organic management. For many soils the form and availability of soil N is central to the conversion phenomenon. Organic farming has been described as a complex system where initially productivity get decreased but subsequently improves gradually with every passing year under organic management. Normally switching over from inorganic to organic farming required 3 to 5 years to get stabilized production. In the present study organic and inorganic systems were compared for three consecutive years (2003-2006) using four potato based crop rotations, (viz; rice-potato-wheat, maize (cob)-potato-wheat, sorghum(fodder)-potato-chickpea and sesamum-potato-mungbean), to determine which rotation was associated with the lowest risk during transition period .It was observed that the system productivity of all the four potato based cropping systems were higher under inorganic management option as compared to organic one. In the case of organic management option, the system productivity increased consistently over the years. In first year, the system productivity of different cropping systems, was 51.6-60% lower than the inorganic system. The productivity increased 17-24.4% and32-54.74% in second and third year respectively over first year. In general, the increase in system productivity was higher in systems where legume was one of the component crop. This increase in crop productivity over the years during transition phase of organic farming is attributed to gradual improvement in soil properties, such as organic carbon (10-15%),microbial biomass carbon (22-35%),mineral nitrogen (14-21%),aggregate analysis and decrease in bulk density (1.58 to1.43 g/cm$^{3)}$.The decrease in bulk density under organic system may partially result from a dilution effect caused by mixing of added organic material with denser mineral fraction of soils. Organic manure application increases the relative number of small pores (i.e.<30um diameter) in the soil ,especially for coarse textured soils. Manure used as a source of organic matter (OM) in organic farming improves soil quality as well as the traditional non-nutrient value, like soil tilth, water holding capacity, soil structure and aggregate stability ,water infiltration rates ,soil diversity and activity.(Khan etal 2007) A transition period of at least 2 years is required for annual crops before the produce may be certified as organically grown. The purpose of this study was to evaluate the effects of three organic amendments on the soil properties during transition to organic production. The organic amendments were composted farmyard manure (FYMC), vermicompost and lantana (*Lantana* spp. L.) compost applied to soil at four application rates (60 kg N ha^{-1}, 90 kg N ha^{-1}, 120 kg N ha^{-1} and 150 kg N ha^{-1}). The grain yield of wheat in all the treatments involving organic amendments was markedly lower (36–65% and 23–54% less in the first and second year of transition, respectively) than with the mineral fertilizer treatment. For the organic treatments applied at equivalent N rates, grain yield was higher for FYMC treatment, closely followed by vermicompost. In the first year of transition, protein content of wheat grain was higher (85.9 g kg^{-1}) for mineral fertilizer treatment, whereas, in the second year, there were no significant differences among the mineral fertilizer treatment and the highest application rate (150 kg N ha^{-1}) of three organic amendments. The grain P and K contents were, however, significantly higher for the treatments involving organic amendments than their mineral fertilizer counterpart in both years. Application of organic amendments, irrespective of source and rate, greatly lowered bulk density (1.14–1.25 Mg m^{-3}) and enhanced pH (6.0–6.5) and oxidizable organic carbon (13–18.8 g kg^{-1}) of soil compared with mineral fertilizer treatment after a 2-year transition period. Mineral fertilized plots, however, had higher levels of available N and P than plots with organic amendments. All the treatments involving organic amendments, particularly at higher application rates, enhanced soil microbial activities of dehydrogenase, â-glucosidase, urease and phosphatase compared with the mineral fertilizer and unamended check treatments. We conclude that the application rate of 120 kg N ha^{-1} and 150 kg N ha^{-1} of all the three sources of organic amendments improved soil properties. There was, however, a 23–65% reduction in wheat yield during the 2 years of transition to organic production_(Gopinath,et.al 2008). Organically managed surface soils also supported significantly greater microbial biomass (159.4% more), microbial carbon as a percent of total carbon (66.2% greater), readily mineralizable carbon (25.5% more), and microbial carbon to mineralizable carbon ratio (86.0% greater) These indicate larger pools of total, labile, and microbial biomass C and a higher proportion of soil total and labile C as microbial biomass. Use of FYM alone or or in combination with chemical fertilizers lead to higher numbers of microbes and enhanced microbial respiration then use of chemical fertilizer alone. All measures of microbial activity were significantly greater in the organically farmed soils, including microbial respiration (33.3% more), dehydrogenase (112.3% more), acid phosphatase (98.9% more), and alkaline phosphatase (121.5% more). Research results of INM established that besides serving as a source of plant nutrients to crop, the organic manures improve soil aeration, permeability, aggregation, water holding capacity and biological properties of soils and enhances the fertilizer use efficiency. The organic manure also provides essential micro nutrients like S, Mg, Cu, Mn and Fe etc. (Roy and Dudal, 1993). The results of 16 years long field experiment on rice-wheat system at Faizabad, Uttar Pradesh revealed that substitution of 25-50 per cent N through FYM along with 50-75 per cent recommended NPK through chemical fertilizers to rice resulted in higher yield as compared to 100 per cent chemical fertilizers application alone. The results indicated that FYM application @ 12-15 t ha^{-1} to rice could substitute nearly 60 kg N ha^{-1}. The higher reduction in pH, electrical conductivity and exchangeable sodium percentage of soil was observed by using organic manures along with chemical fertilizers as compared to fertilizers alone. The combination of chemical fertilizers with FYM registered more increase in available N and P content of the soil against the chemical fertilizers alone.

The more build up in available N and P is attributed to solubilization of nutrients from their native sources during decomposition and mineralization of organic manures (Yadav and Kumar, 2002). Depletion rate of DTPA-extractable micronutrient cations *viz.* Zn, Cu, Mn and Fe was found higher in inorganic fertilizers added as compared to the organics. Chelating action of organic compounds released during decomposition of organic sources increased the availability of micronutrients by preventing their fixation, oxidation, precipitation and leaching (Yadav and Kumar, 2000). The beneficial effects of balanced and integrated nutrient management on soil health in terms of physical, chemical and biological attributes and overall crop productivity have very well been demonstrated by the Long Term Ferti izer Experiments at Palampur,Himachal Pradesh (Sharma *et al.*, 2005). The conjunctive use of chemical fertilizers and organic manure (NPK+FYM) enhanced organic carbon, soil available nutrients, soil aggregation, water infiltrability, microbial biomass-C and microbial population compared to use of chemical fertilizers alone (Table 2).

Table 2 : Effect of chemical fertilizers, lime and FYM on soil microbial population and biomass-C

Treatment	Soil microbial parameters				
	Bacteria (CFU×10^5 g^{-1} soil)	Fungi (CFU×10^4 g^{-1} soil)	Actinomycetes (CFU×10^3 g^{-1} soil)	Azotobactor (CFU×10^3 g^{-1} soil)	Microbial biomass - C (mg/kg^{-1})
50% NPK	24.56	4.39	52.40	55.16	230
100% NPK	11.49	7.83	20.30	17.50	316
150% NPK	6.94	8.06	5.75	14.61	279
100% NPK + HW	9.21	5.75	12.10	68.53	305
100% NPK + Zn	18.23	2.61	8.86	15.93	312
100% NP	7.96	5.13	24.33	45.50	2.5
100% N	4.55	6.98	46.63	11.86	190
100% NPK + FYM	28.08	2.78	73.20	226.16	410
100% NPK (-S)	13.76	5.04	34.63	14.60	207
100% NPK + Lime	18.88	0.45	87.86	92.66	350
Control	9.53	4.45	19.76	127.83	257
CD *(P=0.05)*	0.787	2.993	2.993	7.964	19

The organically farmed soils had a significantly lower qCO_2 metabolic quotient, indicating that the microbial biomass in the organically farmed soils was 94.7% more efficient or under less stress than in the conventionally farmed soils (Sebastiana Melero *et al* 2006) The patterns of rate and amount of nutrients released from various organic sources and their goodness of fit with the nutrient requirement of the crops at different growth stages need to be worked out. This information could be used in evolving appropriate nutrient management schedule, so as to ensure optimal nutrient supply to the crop at active physiological stages having peak nutrient demands. Allelopathic effects of various plant species need to be tapped, particularly for weed and pest management. Integrated use of organic inputs such as farm yard manure and chemical fertilizers is recommended to maintain soil productivity under continuous cultivation.

References

Ademir S.F. Araújo 1,*, Luiz F.C. Leite 2, Valdinar B. Santos 3 and Romero F.V. Carneiro 1 (2009). Soil Microbial Activity in Conventional and Organic Agricultural Systems. *Sustainability*, **1** : 268-276;

Armand Wowo Koné, Jérôme Ebagnerin Tondoh, France Bernhard-Reversat, Gladys Loranger-Merciris), Didier Brunet Yao Tano (2008). Changes in soil biological quality under legume- and maize-based farming systems in a humid savanna zone of Côte d'Ivoire *Biotechnol. Agron. Soc. Environ,* **12** (2), 147-155

Gouri, P.Y.S.M. (2005). National Programme for organic production. *Bulletin of Indian Society of Soil Science. No.22* (K.P. Singh, G. Narayanasamy, RK. Rattan and N.N.Goswami ed.), 61-64

Gopinath K.A., Supradip Saha, B.L. Mina, Harit Pande, S. Kundu and H.S. Gupta. Influence of organic amendments on growth, yield and quality of wheat and on soil properties during transition to organic production Nutrient Cycling in Agroecosystems, **82** (1) : 51-60.

Katyal, J.C. (2001). Fertilizer use situation in India. *Journal of Indian Society of Soil Science,* **49** (4) : 570-582.

Khan, M.A.; Shukla Arvind K.; Upasdhayay N.C.; Singh, O.P.; Singh, B.P.; Lal, S.S. and Pandey, S.K. (2007). Changes in productivity of potato based cropping system and soil productivity during transition period of organic farming. International on organic farming and renewal sources of energy for sustainable agriculture Sep 19-21.2007 at M.PU.A.T Udaipur pp 47.

Mahajan, A., Gupta, R.D. & Sharma, R. (2007). Organic farming. *Gram Vikas Jyoti* July-September Issue, pp. 8-9.

Ramesh P., N. R. Panwar, A. B. Singh, S. Ramana, Sushil Kumar Yadav, Rahul Shrivastava and A. Subba Rao (2010). Status of organic farming in India. *Current Science,* **98** : (9) : May 2010.

Roy, R.N., & Dudal, R. (1993). In *Integrated Plant Nutrition Systems*. Fertilizer and Agriculture Organization. *Fertilizer and Plant Nutrition Bulletin*, pp 1-12.

Sebastiana Melero, Juan Carlos Ruiz Porras, Juan Francisco Herencia and Engracia Madejon (2006). Chemical and biochemical properties in a silty loam soil under conventional and organic management Soil and Tillage Research, **90** (1-2) : 162-170.

Sekhon, G.S. (1997). In J.S. Kanwar and J.C. Katyal (Eds.), *Plant Nutrient Needs, Supply Efficiency and Policy Issues: 2000-2025* (pp. 78-90). New Delhi, India: National Academy of Agricultural Sciences.

Swarup, A. (2002). *Fertilizer News* **47** (12) : 59-73.

.Tandon, H.L.S. (1997). In J.S. Kanwar and J.C. Katyal (Eds.), *Plant Nutrient Needs, Supply Efficiency and Policy Issues: 2000-2025* (pp. 15-28). New Delhi, India: National Academy of Agricultural Sciences.

Yadav, D.S. and Kumar, A. (2000). Integrated nutrient management in rice-wheat cropping system under eastern Uttar Pradesh conditions. *Indian Farming*, **50** (1) : 28-30.

□□□

System Based Integrated Nutrient Management, 2012
© B. Gangwar & V.K. Singh (eds.), pp. 259-274
New India Publishing Agency, New Delhi (India)
e-mail : info@nipabooks.com; website : www.nipabooks.com

CHAPTER **17**

Resource Conservation Technologies for Improving Nutrient Use Efficiency

K.K. SINGH

Rice-wheat cropping system is the most important food production system in India. This system has enabled the country to attain food security during last three decades and provided livelihood to millions of farming families in recent times. The productivity of this system has now started showing signs of decline or stagnation at many places. There are also concerns of declining soil health and other related environmental problems. At the same time, food demand is rapidly increasing due to increasing population and income growth. During the past 30 years, agricultural production has been able to keep pace with population demand for food. This came about through significant area and yield growth. Area growth was a result of new lands being farmed and through increases in cropping intensity, from a single crop to double or even triple crops per calendar year. Area growth will be less important in its contribution to production growth in the future as more land is used for urban areas and industry. Yield growth will have to be the mainstay for providing the means for meeting future food demands unless food imports start to play a major role. Total factor productivity is declining and farmers have to apply more fertilizer to obtain the same yields. There is, therefore, a huge challenge ahead to meet future food demands without damaging the natural resource base on which agriculture depends, producing food at a cost that is affordable by the poor, and with incentives to farmers that allow them to improve their livelihoods. Water will become a major limiting factor for sustained production in the next decade. Rapidly growing urban areas and industry will compete with agriculture for good quality water. There are already reports of declining water tables in some areas leading to more costly pumping of groundwater and increased costs of production.

This chapter describes various conservation agriculture technologies to attain the goal of increasing productivity and meeting food security needs while at the same time efficiently using natural resources, including water, providing environmental benefits and improving the rural livelihoods of farmers. The major success in the last few years has been the development and deployment of conservation agriculture technologies (CATs) with farmers in the rice-wheat systems. CATs range from simple surface seeding, where wheat seed is broadcast on the non-ploughed soil; no-till with a special opener for placing seed in the soil also without ploughing; reduced tillage and bed planting. This has benefited farmers through less cost, more yields and more income. CATs are defined as any practice that will result in improvement of the efficiency of natural resources. These technologies are rapidly gaining popularity among farmers as they result in higher production at less cost with significant benefits to the environment and more efficient use of natural resources. This ultimately results in higher profits, cheaper food, and improved farmer livelihoods. Crop diversification is also easier as less land is needed to produce staple cereals, freeing up land for other crops.

Green revolution technologies

The green revolution is one of the most striking success stories of post-independence India. The success was reflected through more efficient dry matter partitioning to reproduction and therefore, higher harvesting index with significant gain in the yield potential. It is the combination of green revolution varieties and their responses to external inputs, which produced meaningful advances in agricultural productivity. More than 90% farmers have adopted semi-dwarf wheat by 1997 (Pingali, 1999). It is not easy to escape a general relationship between grain productivity and fertilizer nitrogen especially after the evolution of semi-dwarf varieties. It is estimated that irrigated lands have expanded to reach 268 m ha with 80% in developing countries and much in Asia. This expansion is now slowing down (FAO, 1998). In addition to nitrogen fertilizers and expansion of irrigation, there has been a consistent increase in the use of external inputs including pesticides. Thanks to green revolution, the higher food availability without using the extra land represents a success story in agriculture.

There were not the varieties alone which transformed the food production scenario, but the response of these varieties to external inputs brought about a major change in the food production. The gross consumption of fertilizers increased 25-fold in developing countries to reach 91 million tones in 2002, but only increased 2-fold in developed countries. The use and rates in the developing countries surpassed that in the developed countries in the early 1990s (Cassman *et al.*, 2003). The green revolution has slowed sharply, as has yield growth, since the 1980s. The slow down or even reversal has been due to water table lowering because of ever deeper tube wells, micronutrient depletion, mono-culture, reducing bio-diversity and build up of insect, diseases and weeds, development of resistance against pesticides and high concentration of pesticides or fertilizer derived nitrates and nitrites in water courses. The amelioration of above factors adds to the cost of cultivation and, therefore, a decline in the total factor productivity. With the rise in input cost, the net profit of farmers has decreased even if the productivity is increasing slightly. Each farmer, therefore, needs to maximize earnings through alternate technologies. Seen from profitability point of view, it will be important to maintain natural resources. Resource Conservation Technologies (RCT), therefore, have become a critical component to growth in agriculture. These technologies require complementary innovations through multi-disciplinary, multi-institutional and farmers' participatory approach. This is important because the livelihood of more than a billion agriculture population in developing countries will depend on technologies that raise outputs per labour-hour and per unit area at less cost (Lipton, 2004).

Description of various conservation agriculture technologies

A basket of technologies has been developed and made available to farmers. Some are based on reduced tillage for wheat including zero-tillage. Bed planting systems increase water productivity and when combined with reduced tillage in a permanent bed system provide even more savings. Laser leveling combined with these tillage systems provides additional benefits. Many of the benefits of the tillage options for wheat are lost when rice soils are traditionally puddled. System based technologies do away with puddling so that total system productivity is increased. The various technologies are described briefly below:

Zero-tillage

This is CAT where the seed is placed into the soil by a seed drill without prior land preparation. This technology is more relevant in the higher yielding, more mechanised areas of north-western India, where most land preparation is now done with four-wheel tractors. However, in order to extend the technology in other parts, equipment for 2-wheel hand tractors and bullocks is being modified. The basis for this technology is the inverted-T openers. This coulter and seeding system places the seed into a narrow slot made by the inverted-T as it is drawn through the soil by the four-wheel tractor. The coulters can be rigid or spring-loaded depending on the design and cost of the machine. This type of seed drill works very well in situations where there is little surface residue after rice harvest. This usually occurs after manual harvesting. Where combine harvesting is becoming popular, loose straw and residue creates a problem for the inverted-T opener. Farmers presently burn residues to overcome this problem of loose stubble whether they use zero till or the traditional system. This practice needs to be discouraged because of major environmental and air pollution issues. Future strategies will look at alternative machinery and techniques to overcome this problem. Leaving the straw as mulch on the soil surface has not been given much thought. However, results suggest that this may be very beneficial to early establishment and vigour of crops planted this way and for soil moisture conservation, water infiltration and erosion. Significantly fewer weeds are found under zero-tillage compared to conventional tillage. Fields with zero tillage and those with normal tillage were sprayed with weedicide, but significantly lower weed counts were found in fields with zero tillage. This difference can be explained by the nature of the weeds found in the rice-wheat cropping system. Most of the weeds affecting the wheat crop germinate during the crop season, and since the soil is disturbed less under zero-tillage, fewer weeds are exposed and germinate. Also before the weeds are able to grow and compete, the main crop is able to cover up the surface and significantly reduce weed biomass. Weed problems typically are more severe under conventional tillage than under zero tillage, at least in the near term. Earlier planting is the main reason for the additional yields obtained under zero-tillage. Zero-tilled plots could be planted in first week of November, the optimum date for planting wheat. The results of many trials suggest that the longer the farmer delays planting, the lower the yield. This finding has been confirmed in trials throughout the Indo-Gangetic plains in the past few years. In Haryana, surveys and crop cuts have shown that zero till produces 4-500 kg ha^{-1} more grain than traditional systems. This is attributed to earlier, timely planting, less weeds, better plant stands and improved fertilizer efficiency because of placement with the seed drill. Some experiments are now in their 14th year of continuous zero till and find no deleterious effects that would make them revert to the traditional system.

The major challenge before us is to organize ourselves to meet the need for continuing productivity increases. According to McCalla (1998) much of the yield increase must come in the developing countries which will depend on: (1) what mode of scientific investigation will best generate the next phase of crop yield increases across very variable economic and institutional conditions, (2) how will the technology development process best be linked to other support services and institutions in order to increase productivity in farmers' fields, and (3) how can the synergies between genetic improvement and crop and resource management research be best exploited when organizational dynamics and creating increasing separation?

A holistic approach is needed to tackle these second-generation problems and to improve the sustainability of this cropping system. However, interventions in the form of new resource conservation technologies (RCTs) must include the component of profitability, value addition, efficiency and farmers' participatory approach for their large-scale acceptance. Introduction and popularization of zero tillage in wheat is a unique example in this context.

During the past decade, the evolution and acceleration of zero tillage has been one of the few big ideas in introducing conservation agriculture. Farmers in the Indo-Gangetic Plains have now rediscovered the virtue of technologies like zero tillage and bed planting because they are profitable and add value to the system as a whole.

Advances in this context will depend on the investment in public research which is much less than what existed from 1966 to 1985. Such mode of scientific investigation that leads to more productivity with less input cost will help generating next phase of yield increases especially in areas where wheat sowings are delayed beyond admissible limits. Technologies like zero tillage and other resource conservation technologies include bed planting, direct seeded rice, double zero tillage, leaf colour charts, green manuring, water harvesting, water recycling will help stimulating economic growth especially in the light of projected environmental changes.

Sowing the crop as early in the season as possible is perhaps the most important issue. Only early sowings can help combating the effect of terminal heat. Productivity gains give a new meaning to the management technologies of rice-wheat cropping system which now demands early sowing of both rice and wheat. What really matters is the availability of technologies which allow sowings with anchored crop residues retained on the surface. The availability of natural resources will remain a pre-condition for sustainability of the cropping system. When choosing among the research paths, we should raise the productivity of land and water at less cost so that the total factor productivity of small farms rises faster than now.

Reduced tillage

The strip and rotary till drills have been developed that prepare the soil and plant the seed in one operation. This system consists of a shallow rotavator followed by a seeding system. Soil moisture was found to be critical in reduced tillage system. The rotovator fluffs up the soil, which then dries out faster than with normal land preparation. The seeding coulter does not place the seed very deep, so soil moisture must be high during seeding to ensure germination before the soil dries appreciably. The tractor can also be used with a rotavator to quickly prepare the soil and incorporate the seed after a second pass. This speeds up the planting and results in better stands with less cost than traditional methods. However, the strip and rotary till drills do a better job because the seeds are placed at a uniform depth in the single pass.

Bed planting

In bed planting systems, wheat or other crops are planted on raised beds. This practice has increased in the last decade. Farmers have given the following reasons for adopting the new system: management of irrigation water is improved, bed planting facilitates irrigation before seeding and thus provides an opportunity for weed control prior to planting, plant stands are better, weeds can be controlled mechanically between the beds early in the crop cycle, seed rates are lower, after wheat is harvested and straw is burned, the beds are reshaped for planting the succeeding crops, burning can be eliminated, herbicide dependence is reduced and hand weeding and roguing is easier as well as less lodging occurs. Two bed widths and two or three rows or wheat planted per bed were compared with conventional flat bed planting. Two rows on 70cm beds were best. Two of the major constraints on higher yields are weeds and lodging. Both can be reduced in bed planting. The major weed species affecting wheat, *Phalaris minor,* is normally controlled using the herbicides. Preliminary observations indicate that *P. minor* is less prolific on dry tops of raised beds than on the wetter soil found in conventional flat bed planting. Cultivating between the beds can also reduce weeds. Thus bed planting provides farmers with additional options for controlling weeds. Lodging is also less of a problem on raised beds. Additional light enters the canopy and strengthens the straw, and the soil around the base of the plant is drier. Reduced lodging can have a significant effect on yield. An additional advantage of bed planting becomes apparent when beds are "permanent" - that is, when they are maintained over the medium term and not broken down and re-formed for every crop. In this system, wheat is harvested and straw is left or burnt. Passing a shovel down the furrows reshapes the beds. The next crop can then be planted into the stubble in the same bed. Research in farmer fields has also shown that rice can be grown on beds making this system's feasibility in the rice-wheat pattern. Rice can be grown on beds by either transplanting seedlings or direct seeded. At the moment, transplanting on beds is best since normal herbicides used for transplanted rice can be used to control weeds. As dry seeded herbicides become available and weeds can be managed, dry seeded rice on beds will become more attractive. The use of beds also provides a way for improving fertilizer use efficiency. This is achieved by placing a band of fertilizer in the bed at planting or topdress. Using slow release formulations and experimenting with urea super granules can make further improvements. Both can be applied in the bed at the time of planting with the seed cum fertilizer drill.

Effect of conservation agriculture technologies on land, water and energy productivity

The comparative performance of zero till drill (ZT), strip till drill (ST), bed planter (BP), rotary till drill (RT) and conventional drill (CS) for rice and wheat sowing based on long term experiments at this directorate are presented in this section. The conservation agriculture technologies of zero, strip and rotary till drilling, and bed planting of rice and wheat saved 64 to 85 % resources (time, labour, cost, fuel and energy). The bed planting also saved 39 and 34 % irrigation water in rice and wheat, respectively. These technologies provided higher rice and wheat yields (2 to 8 %), B: C ratio (9 to 27 %) and energy efficiency (21 to 32 %) compared to conventional sowing. The continuous use of these technologies has also improved soil health by increasing the soil organic carbon and mean weight diameter of the soil aggregates. Also, around 70 kg ha^{-1} $year^{-1}$ CO_2 emissions to the environment could be reduced by the use of zero till drilling compared to conventional sowing which is vital to our environmental sustainability.

Farmers are adopting these technologies quickly. The adoption could be even faster if it were possible to have sufficient machinery available from small-scale manufacturers. Farmer feedback on water savings with these technologies essentially says that they save water. For zero-tillage, farmers report about 25-30% savings. This comes in several ways. First, zero tillage is possible just after rice harvest and any residual moisture is available for wheat germination. In many instances where wheat planting is delayed after rice harvest farmers have to pre-irrigate their fields before planting. Zero-till saves this irrigation. Savings in water also comes from the fact that an untilled soil has less infiltration than a tilled soil and so water flows faster over the field. That means farmers can apply irrigation much faster. Because zero tillage takes immediate advantage of residual moisture from the previous rice crop, as well as cutting down on subsequent irrigation, water use is reduced by about 10 cm-hectares, or approximately 1 million liters per hectare. One additional benefit is less water logging and yellowing of the wheat plants after the first irrigation that is a common occurrence on normal ploughed land. In zero till, less water is applied in the first irrigation and this yellowing is not seen.

Environmental impact of conservation agriculture technologies

The widespread adoption of one or several reduced tillage methods would also bring significant environmental benefits. Use of Zero Till Drill on one hectare of land would save 60-70 liters of diesel and approximately 1 million liters ha^{-1} of irrigation water. Using a conversion factor of 2.6 kilograms of carbon dioxide per liter of diesel burned, this represents about a quarter ton less emissions per hectare of carbon dioxide, the principal contributor to global warming. These benefits increase dramatically if extended across even a portion of the rice-wheat region's 13.5 million hectares. Adoption of zero-till on, say, 5 million hectares would represent a savings of 5 billion cubic meters of water each year. In addition, annual diesel fuel savings would come to 0.5 billion liters (including the pumping) - equivalent to a reduction of nearly 1.3 million tons in CO_2 emissions each year.

This lecture note has described various conservation agriculture technologies available for use in the Rice-Wheat system. The benefits include improved production at lower cost, improving the efficiency of natural resources, benefits to the environment, and improved livelihoods of farmers. The success of these technologies is dependent on rapid adoption by farmers. Conservation agriculture technologies are a key to ensuring sustainable food production in the next decade. Overcoming mindsets that hold traditional beliefs about excessive tillage and providing the enabling factors that allow exposure of the technology to all those involved in agriculture will be key factors for future success. This technology revolution is seen as one way to sustainably increase food production to meet future demands while conserving natural resources and improving farmer livelihoods.

With the rapid expansion of wheat zero tillage in the Indo-Gangetic plains, there has been, within that region a surge of interest in resource conservation technologies. In the 2004-2005 wheat season, zero tillage is estimated to have been used on nearly 2 million ha of sown area (RWC, 2005). And wheat zero tillage is seen by many as merely the first step in a broad movement towards the development and adoption of an ever richer collection of resource conserving, conservation agriculture technologies.

The assessment on the scientific, technical and institutional issues associated with rice-wheat cropping system is urgently needed. For the past 40 years, the growth in the productivity of rice-wheat cropping system was the result of technological innovations in the form of green revolution. With the result, supply exceeded demand and real prices of food such as cereals went down. However, the yield growth rate of many crops especially cereals have started declining. Reasons for declining in the productivity growth rate are multiple (Duxbury *et al.*, 2000; Ladha *et al.*, 2000; Timsina and Connors, 2000). Sustainability and profitability of rice-wheat cropping system in Indian agriculture is the lifeline and future of Indian economy with more than 60% people living in rural areas. The challenges are enormous ranging from conservation of natural resources to investment in new technologies based on biotechnology.

Factors for generating new technologies in rice-wheat cropping system

Following factors need to be taken into account for generating new technologies in rice-wheat cropping system:

1. Ideally, farmers should cut spending rather than investing more on inputs like fertilizers or pesticides.
2. We must look for extra revenues to plug the gap in net profits. The best way to increase revenue is increase yields without increase in input use.
3. It remains true that yield is primarily a time phenomenon i.e. it is a function of time for which the crop remains in the field. That means longer the crop remains in the field for its growth and development, higher the grain yield. Most parts of Indo-Gangetic Plains (IGP) have the big difference in the base line when crop is sown. For instance, in the eastern sector of IGP, sowings are delayed beyond December. This is where maximum gains in productivity will come especially from North-East plane zones where sowing of wheat is delayed beyond December.
4. Many farmers especially small farmers spend their income as soon as they receive it, in effect, they have no seed money to invest in the next season. Whatever little money they save, farmers channel most part of saving into tillage operations. Technologies like zero tillage can help scrapping such spending and provide room for diverting the spending towards input that improve yield.
5. Per capita arable land area is getting smaller. The arable land scarcity index as measured by per capita land area was 0.36 in 1960 and 0.14 in 1990 which is expected to be 0.12 in 2025 (Kaosa-Ard and Rerkasem, 2000). Similarly, the irrigated land area per capita is also declining which is 0.06 ha per capita.
6. Increase in the soil organic matter by retaining more residues on the soil is important. Promoting existing biological cycle and soil biological activity, maintaining environmental resources and using them more carefully and efficiently and reusing residues as much as possible can help sustaining the rice-wheat cropping system. Thus, minimizing only pollution both on-site and off-site is an important feature of reducing soil degradation.
7. Rice-wheat cropping system requires enormous expenditure of energy to frequently till the land and to pump the groundwater for irrigation. To sustain the energy based activities, diesel consumption will increase

in future decades. Saving of diesel consuming operations can help sustaining the import of oil for other purposes.

8. The water resources are under great stress. The real water saving will come by obtaining more crop production from same amount of water. Bed planting and laser leveling can help to reduce this stress.

9. Puddling of alkali soils further degrades the soil structure, and can facilitate formation of subsurface plough pan further restricting the percolation of water through soil profile. Reduced infiltration slows down the process of reclamation, therefore, puddling should be avoided (Gupta and Zia, 2003). Work conduced by Kumar *et al.* (2005), Malik *et al.* (2004 and 2005) and Reddy *et al.* (2005) has shown that puddling can be avoided in transplanted rice.

10. The existing practices like straw burning lead to pollution, which is spread of, from smoke of burnt straw. Farmers do not bear the cost of such pollution, which is publicly unacceptable. Zero tillage can effectively serve as an opportunity to evolve residue management technologies because management of surface residue is easier than incorporation.

Retention and management of adequate amount of crop residues (at least 30%) under conservation agriculture is the key to realize long-term benefits and also to reverse the process of soil degradation. In a soil that is not tilled for many years, the crop residues remain on the soil surface and produce a layer of mulch. Retention of crop residues improves organic carbon content, water stable aggregates, bulk density, and hydraulic conductivity and reduces runoff. But most of the farmers in Haryana and Punjab burn the crop residues to get their fields well cleaned before sowing. Therefore, to replace residue burning, and to realize benefits of residue cover under conservation agriculture, its efficient management through machinery modification is the need of time.

Water scarcity

The global water scarcity analysis has shown that up to two-third of world population will be affected by water scarcity over the next several decades. More important, wherever in the world water is scarcest, this is mostly in developing countries, irrigation for agriculture gobbles up at least 75% and sometime as much as 90% of the available water (The Economist, 17 July, 2003). The agricultural community sees continued growth of irrigation as an imperative to achieve the goals adopted to reduce hunger and poverty. International Water Management Institute, Colombo, Sri Lanka estimated that 29% more irrigated land will be required by the year 2025, but productivity gains and more efficient water use might decrease this diversion to 17% (Rijsberman, 2004). Irrigation development has impaired the ability of many eco-systems to provide valuable goods and services and therefore, more attention should be given on sustaining the existing sources of irrigation rather than alternate sources. Alcamo *et al.* (2000) projected an 8% increase in the amount of water that should be diverted to irrigation if more sustainable means of production are adopted. The difference between 17% increase and 8% decrease is on the order of 625 km^3 of water, which is close to 800 km^3 of water that is presently used globally for urban and industrial use. Therefore, there should be more emphasis on water conservation and improved efficiency of use and reallocation of water from one use to another, presumably shifting to a higher value use. Gleick (2003) calls for a soft path for water with a focus on overall productivity of water rather than seeking new supplies. That would mean a paradigm shift from supply management to demand management in the form of integrated water resources management. The most tangible proposals that have come out of this direction are: (a) to involve users more in the management of water, often through the establishment of forms of water user associations; (b) to price water and/or make it a tradable commodity; and (c) establish river basin authorities that integrate the usually fragmented government responsibilities for water into a single authority responsible for a hydrographically defined area, river basin.

The number of tube wells has grown exponentially in the North-west India. Pump irrigation now dominates gravity irrigation in many countries. In the field, the upper limit of water productivity of well managed, disease free water limited cereal crops is 20 kg ha^{-1} mm^{-1} (grain yield per ha water used). If the productivity is less than this, it is likely that major stress other than water stress such as weeds, diseases, poor nutrition or poor inhospitable soil health so, greatest advantages will come from dealing with these first (Passioura, 2004).

A big reorientation of crop and water science is needed. Development of varieties which can resist moisture stress through the use of biotechnology is necessary for increasing overall water productivity. There are no immediate prospects of producing GM crops that could greatly improve water productivity. There are hundreds of patents that claim drought tolerance but it is hard to discern any of these likely to influence water productivity in the field (Passioura, 2004).

Machinery for zero-tillage and minimum tillage cultivation

Zero-till drilling, strip till drilling and rotary till drilling of wheat after harvest of rice were compared to the conventional tillage sowing as practiced by farmers. Brief specifications of the direct drilling machines are given in Table 1.

Table 1 : Specifications of direct drilling machines

Particulars	Zero-till drill	Strip till drill	Rotary till drill
Source of power	45 hp tractor	45 hp tractor	45 hp tractor
Type / No. of furrow openers	Inverted 'T' type / 09	Shoe type / 09	Shoe type / 11
Row spacing, mm	180 (adjustable)	200 (fixed)	160 (adjustable)
Working width, mm	1600	1800	1750
Drive wheel	Angle lug - front mounted	Angle lug - side mounted	Star lug - rear hinged
Weight, kg	250	280	300
Unit price, Rs	22000	60000	70000

The sowings at shallow depth (50-60 mm) under residual moisture condition (21.8-23.6%) with 100 kg ha^{-1} seed rate and fertilizer dose of N:P:K::120:50:30 kg ha^{-1} were compared to the conventional practice of 03 tillage operations by duck-foot sweep cultiva or (size = 1800 mm) and sowing by seed-cum-fertilizer drill (09 row, row spacing = 180 mm). The time required as per actual field capacity of the machines, fuel used and cost of operations are given in Table 2.

Table 2 : Minimum tillage seeding compared to conventional tillage-sowing of wheat

Particulars	Zero tillage seeding	Strip tillage seeding	Rotary tillage seeding	Conventional tillage (03 passes) - sowing
Time, h ha^{-1}	3.23 (70.2)	4.17 (61.5)	3.45 (68.1)	10.82
Fuel used, l ha^{-1}	11.3 (67.3)	17.5 (49.4)	13.8 (60.1)	34.6
Operational energy, MJ ha^{-1}	648.9 (67.2)	1001.7 (49.3)	783.6 (60.3)	1976.1
Cost of operation, Rs ha^{-1}	639.5 (66.4)	979.9 (48.5)	807.3 (57.5)	1903.0

( ) values show percent savings over conventional practice

The results showed that zero tillage drilling was time, energy and cost-effective to the extent of 70.2, 67.2 and 66.4 percent respectively over the conventional practice. The rotary tillage seeding combined with full width shallow tillage in single pass operation was 60.3 percent energy efficient and 57.5 percent cost effective compared to the conventional tillage seeding. The strip tillage seeding was advantageous over conventional tillage seeding but for the intermittent strip tillage the operational energy and cost requirements were higher compared to the rotary tillage and no tillage seeding.

Cultural practices specific to these conservation systems were developed in terms of frequency of irrigation and fertilizer applications. First irrigation of 40-50 mm was critical for all the direct seeding systems for initial establishment especially in no tillage seeding. Performance of direct drilled wheat (Table 3) showed that in direct drilling systems although the grain yields were at par, the benefit: cost ratio were higher by 15.2 - 23.4 percent with savings in operational energy of 8.4 - 14.7 percent compared to the conventional practice.

Table 3 : Production economics and operational energy of direct drilled wheat after harvest of rice

Particulars	Zero till drilled	Strip till drilled	Rotary till drilled	Conventionally sown
Grain yield, t ha^{-1}	4.84	4.62	4.78	4.60
Cost of production, Rs ha^{-1}	8635	9114	9315	10710
Benefit: cost ration	3.64	3.29	3.34	2.79
Operational energy, MJ ha^{-1}	8114	8712	8444	9516

Making the Shift to Resource Conserving Technologies

A major bottleneck in large scale adoption of RCTs is due to mind set of the stake holders and the age-old practice of excessive tillage for establishment of rice and wheat. The shift to resource conserving technology will require a reorientation and retraining of farmers, development workers, scientists, policy makers, educators and other interested stakeholders and overcoming the 'Not Invented Here' (NIH) syndrome on the part of technocrats. The technologies are more management sensitive; they work when done properly. Farmers will need training in proper use and calibration of machinery, good after sales service and available spare parts, trained mechanics and more. Resource conserving curriculum will need to be introduced into places of learning so that extension workers, scientists and farmers can be taught the benefits and needs of this technology. Public awareness of the benefits of RCT at the farm, village, country and global level is needed. New innovative ways to upscale the technology and make it available to farmers are needed that rely on more participatory approaches of all stakeholders.

Policies needed

To enhance efficient use of inputs (agro-chemicals, water, fossil energy) and the development and use of resource conserving technologies certain policies concerning pricing, incentives, research, agricultural education, funding *etc.* have to be made. Efficient use of water will not occur if farmers are given it free. Subsidies have to be more production oriented and linked to improving the WUE. Laser assisted precision land leveling is one simple option to save on water and improve crop productivity. Farmers and service providers wish only duty of exemptions, no subsidy on equipment. For the low quality irrigation systems of South Asia, the strategy of irrigation scheduling to save on water has not worked well and need reconsideration. The same applies to pricing of fertilizers and other inputs. The subsidy instrument in fertilizer sector could be easily used to correct imbalanced fertilizer N:P:K use ratio; switch over from prilled urea to urea super granule (USG) and for developing machines that help deep placement of these costly inputs. Deep placement of urea (with or without neem formulations/ slow release materials) in rice, reduce ammonia volatilization and leaching losses of nitrogen and saves nearly 2 million tones of urea. There is a need for the new implements to experiment with the resource conserving technologies; more funds are needed for refinement and development of farm equipments to promote precision agriculture. Farmers should save sufficient funds in the first year to pay for the cost of these new drills and other equipments. A better policy would make credit more easily available although repayment schedules must be met. Subsidies on equipments are seen as a step forward in deteriorating the quality of farm implements manufactured in small-scale sectors, where quality control is difficult, and quality standards are generally missing. This whole issue of policy is complex, since there is a need to balance the needed encouragement of farmers to produce more food at lower prices without unduly degrading the environment and the resource base while still providing cheap food for the urban and rural poor.

Future research needs

For enhancing the pace of adoption of resource conserving technologies in farmer participatory mode, we need a paradigm shift in approach; the way research is planned and conducted in the country. India has a large acreage of calcareous and acidic soils. In last half a century we had not been able to extend to soils the benefits of liming for enhancing crop productivity. Similarly, the acreage of alkali soils still in need of amendments is quite large. Also, there is a very large deviation from the ideal N:P:K fertilizer use ratio indicating a very imbalanced use of fertilizers. For maintenance of high productivity in intensively cultivated systems such as R-W systems, it is essential that the nutrients mined/extracted by crops are replenished in a balanced manner. The average NPK removal by rice based cropping systems range from 554 to 814 kg $ha^{-1}yr^{-1}$. The annual deficit of NPK (removal over and above additions) is about 4-5 million tones in rice and wheat. Net depletion of nutrients brings to fore the important issue that is central to the sustainability of R-W system vis-à-vis soil resources base. Unless extracted nutrients are replenished through chemical fertilizers, releases from soil minerals and mineralization of organics etc., crop yields may decline in due course. A greater proportion of the N not used by crops is lost to air or leached to the ground water during monsoon season, out of the root zone and has little residual value unless incorporated into microbial biomass. Calcareous and acidic soils have an insatiable demand for P due to its fixation in to these soils through two different mechanisms of fixation. Inabilities of the farmers to apply soil amendments (liming materials) to acidic soils low in bases significantly reduce the ability of HYV to produce more and realize their potential productivity. We need to have a fresh look at our strategies of meeting this very large requirement of liming materials. Indigenous farmer practices in the Almorah hills show us the way. They apply pine needles to overcome P deficiency in slightly acidic soils (pH 5.5- 6.5). The role of organics, rich in silica (rice straws) affect the point of zero charge in a manner that reduces fixation and increases the availability of P to crop plants. The strategy of manipulating electrochemical properties of the soil colloids through of silica rich plant residues to enhance availability of P in acidic soils has not been tested so far and deserve immediate attention to tackle problem of non-availability of liming material and P shortages. More of such strategies in combination with RCTs have the potential of changing the face of Indian agriculture. Proper targeting of technologies in eastern IGP can reduce "rice fallows", increase crop intensification/ diversification and can nearly double the current levels of RW system productivity.

References

Alcamo, J., Henrichs, T. and Rosch, T. (2000). World water in 2025: Global Modelling and Scenario Analysis, In : *World Water Scenario Analyses*, (Rijsberman, ed.), World Water Council, Marseille.

Cassman, K.G., Dobermann, A., Walters, D. and Yang, H. (2003). Meeting cereal demand while protecting natural resources and improving environmental quality. *Annual Review Resources and Environment*, www.irri.org/publications/irrn/pdfs/ vol28no2/IRRNmini3.pdf.

Duxbury, J.M., Abrol, I.P., Gupta, R. and Bronson, K. (2000). Analysis of soil fertility experiments with rice-what rotations in south Asia. RWC Paper Series 5. Rice-Wheat Consortium for the Indo-Gangetic Plains and CIMMYT, New Delhi, India.

FAO, (1998). *FAO Yearbook: Production* Vol. 52.

Gleick, P. (2003). 'Soft Path' solution to 21st century water needs. *Science* **320**: 1524-1528.

Gupta, R.K. and Zia, M.S. (2003). Reclamation and management of alkali soils. In : *Addressing Resource Conservation Issues in Rice-Wheat Systems of South Asia: A Resource Book*. Rice-Wheat Consortium for the Indo-Gangetic Plains. International Maize and Wheat Improvement Center, New Delhi, India. 30 p.

Kaosa-Ard, M.S. and Rerkasem, B. (2000). The growth and sustainability of agriculture in Asia. Asian Development Bank, Manila, Philippines.

Kumar, V., Yadav, A. and Malik, R.K. (2005). Effect of planting methods and herbicides in transplanted rice. Project Workshop proceedings on "Accelerating the Adoption of Resource Conservation Technologies in Rice-Wheat Systems of the Indo-Gangetic Plains" held on June 1-2, 2005 at Hisar, haryana, India, pp. 122-126.

Ladha, J.K., Fischer, K.S., Hossain, M., Hobbs, P.R. and Hardy, B. (2000). Progress towards improving the productivity and sustainability of rice-wheat system: a contribution by the consortium members. IRRI Discussion Paper No. 40.

Lipton, M. (2004). Crop science, poverty and the family farm in a globalizing world. 4th International Crop science Congress held in Brisbane, Australia from September 26 to October 1. 49 p.

Malik, R.K., Yadav, A., Gill, G.S., Sardana, P., Gupta, R.K. and Pigging, C. (2004). Evolution and acceleration of no-till farming in rice-wheat system of the Indo-Gangetic Plains. 4th International Crop science Congress held in Brisbane, Australia from September 26 to October 1. 73 p.

Mailk, R.K., Yadav, A. and Singh, S. (2005). Resource Conservation Technologies in Rice-wheat cropping system of the Indo-Gangetic Plains. *In:* Conservation Agriculture: Status and Prospects (eds. Abrol, I.P., Gupta, R.K. and Malik, R.K.), Center for Advancement of Sustainable Agriculture, NASC Complex, New Delhi.

McCalla, A.F. (1998). Agriculture and food needs to 2025. *In:* International Agricultural Development, 3rd edition. Pp. 39-54 (eds. C. K. Eicher and J. M. Staatz). The John Hopkins University Press, Baltimore, Md, USA. 615 p.

Passioura, J. (2004). Increasing crop productivity when water is scarce – breeding to field management. Proc. 4th International Crop Science Congress held in Brisbane, Australia from September 26 to October 1. 54 p.

Pingali, P.L. (ed.) (1999). CIMMYT 1998-99. World Wheat Facts and Trends. Global wheat research in a changing world: Challenges and achievements, Mexico, DF, CIMMYT.

Reddy, C.V., Malik, R.K. and Yadav, A. (2005). Evaluation of double zero tillage in rice-wheat cropping system. Project Workshop proceedings on *"Accelerating the Adoption of Resource Conservation Technologies in Rice-Wheat Systems of the Indo-Gangetic Plains"* held on June 1-2, 2005 at Hisar, haryana, India, pp. 122-126.

Rice Wheat Consortium for the Indo-Gangetic Plains. (2005). "Research Highlights 2004". Presented at the 11th meeting of the Regional Steering Committee, Spectra Convention Center, Dhaka, Bangladesh. February 6-8, 2005.

Rijberman, F.R. (2004). Water scarcity: Fact or Fiction? *Proc. 4th International Crop Science Congress* held in Brisbane, Australia from September 26 to October 1. 54 p.

Timsina, J. and Connors, D.J. (2000). The productivity and sustainability of rice-wheat cropping systems: Issues and challenges. *Field Crops Research,* **69** : 93-132.

❑❑❑

System Based Integrated Nutrient Management, 2012
© B. Gangwar & V.K. Singh (eds.), pp. 275-285
New India Publishing Agency, New Delhi (India)
e-mail : info@nipabooks.com; website : www.nipabooks.com

CHAPTER **18**

Microbial Diversity – Key to Sustainable Agriculture and Soil Health

S.S. PAL

Biodiversity is the variety of life in different microbes, animals, plants and their genes present in the ecosystem. Soil provides a vital habitat and environment primarily for microflora such as bactria, actinomycetes, fungi as well as microfauna *viz.* protozoa and nematodes, mesofauna such as microarthopods, enchytraeids, macrofauna such as earthworms, beetles, termites and millipedes (Gaur 2006a).

Biodiversity and soil are interlinked which interact closely with wider biosphere. Conversely, biological activity is of prime importance in influencing the physical and chemical conditions of soil (Bardgett 2005). The level of soil biological activity is, therefore, affected by soil type and also depends on the use of management practices which contribute to crop productivity.

Soil biodiversity and soil processes

A majority of soil organisms participate in various processes which are essential for the biosphere. The main ecological function is to decompose organic matter largely derived from the plant and animals residues which consist of celluloses, hemicelluloses and lignin, and to supply plant nutrients to support growth. These organisms participate in various nutrient cycles and processes such as nitrogen mineralization, nitrification, nitrogen fixation, transformations of phosphorus, potassium, sulphur, calcium, magnesium, iron and micronutrients (Pankhurst *et al* 1997). They participate in bioremediation of harmful compounds such as heavy metals and pesticide residues (Gaur 2006 b). They not only maintain physical and chemical conditions of soil, but also regulate emissions / assimilations of green house gases and suppress soil borne diseases and pests (Table-1).

Table 1 : Soil organisms and their functions.

Groups	Numbers	Functions
Non-Mycorrhizal fungi	18-35000	Elemental immobilization; elemental mineralization; mutualistic and com-mensal associations; resource for arthropods, protozoan; soil aggregation; decomposers of agrochemicals and xenobiotics
Mycorrhizal	10,000	Mycorrhizal plants given competitive advantage by the following mechanism: mediations of transport of essential elements. and water from soil to plant roots; mediation of plant to plant movement of essential elements and carbohydrates: sequestration of essential elements present in forms not avail- able to plants: regulation of water and ion movements through plants: regulations of photosynthesis rate of plants: regulation of below ground C allocation: decreased seedlings mortality: protection from root diseases and root herbivores
Protozoa	1900	Genesis of root mycosphere for bacteria; high quality resource for mesofaunal and microfaunal grazers; grazers of bacteria and fungi; "enhance mifrobial growth; enhance C and N availability to, higher trophic levels; key components of microbial-loop systems; Prey for nemotodes and mesofauna ; host of bacterial pathogens; parasites of higher-level or- granisms.
Nematodes	5000	Grazers of bacteria and fungi; enhance C and N availability to higher trophic levels; disperse bacteria and fungi; root herbivores / plant parasites; para- sites/ predators of microfauna, mesofauna and insects; prey for meso-and macrofauna.
Mites	30,000	Grazers of bacteria and fungi; consumption of plant litter and animal car- casses ; predators on nematodes and insects; root herbivores; dispersal of helminth parasites; host for protozoan parasites and parasitoids of insects and other arthropods; prey for macrofauna; microecosystem engineers
Insects general		Grazing of rhizosphere microorganisms; dispersal of microorganisms; preda- tors of other soil organisms; decomposers of plant and animal matter.

Insects-root herbivores	40000	Modification of plant performance below ground by root herbivory (modi- ncation of plant performance above ground by root herbivory and modifi- cation of herbivore populations above ground through changes in plant physiology resulting from herbivory below ground).
Insects-collembola	6500	Grazing of microflora and microfauna especially in rizhorsphere; consump- tion of plant litter and animal carcasses; micropredartors of nematodes and rotifers; dispersal of microorganisms; dispersal of helminths and cestode parasites; host 9f parasites prey for macrofauna; microecosystem engineers
Insects-ants	8800	Bioturbators; enhancement of microbial growth; keystone species for inquilinous fauna and plants associated anthills
Insects-termites	2000	Bioturbators; enhancement of miocrobial growth, keystone species for inquilinous microorganism; fauna and plants associated mounds.
Enchytraeids	>600	Fragmentation of plant litter; enhancement of microbial growth; bioturbators; dispersal of micro-organisms
Earthworms	3627	Bioturbators; enhancement of microbial growth; dispersal of microorganism and algae; host of protozoan and other parasites

The elements of sustainable agriculture include preservation of biodiversity and protection of soil health, simultaneously ensuring soil conservation, maintenance of water quality, avoiding deforestation and maintaining healthy soil biota interactions (Fig. 1).

Nature of soil biodiversity

Soil is a natural habitat for diverse groups of macro and microorganisms related to plant and .animal kingdom. It refers to richness of life as manifested by morphological, chemical and biochemical characteristics, and their interdependence in soil as a habitat. They vary in, shape, size and morphology from microscopic to macroscopic belonging to flora and fauna. These are closely interacting groups of flora and fauna having varying energy needs, are classified as heterotrophic, chemo heterotrophic and phototrophic (Gaur 2006 b). The soil biodiversity is more extensive than any other environment when living forms are considered. Many of these are beneficial in improving soil fertility but some pathogenic organisms damage crops by causing plant diseases.

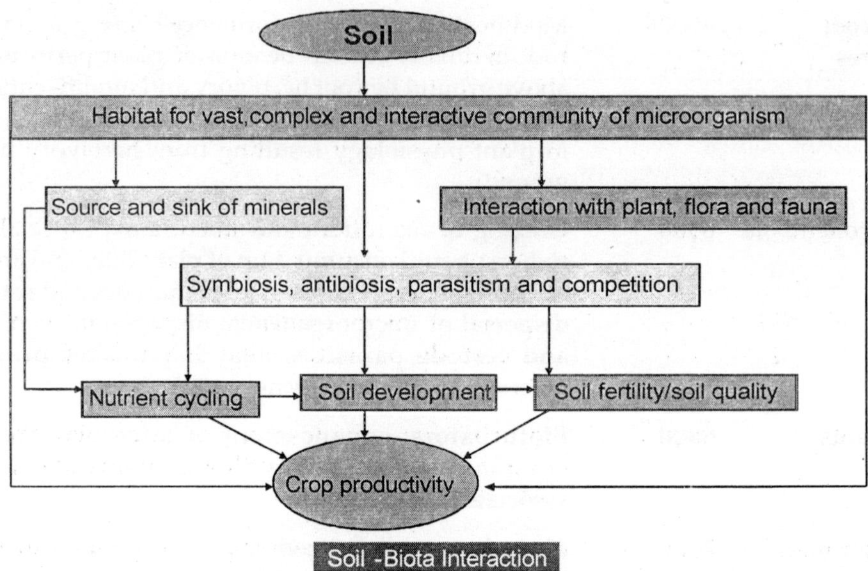

Fig. 1 : Soil Biota Interaction

Interrelated elements of biodiversity

Biodiversity can be represented by three interrelated elements based on taxonomic, genetic and functional aspects. Available knowledge on species richness of soil organisms is based on very limited inventory efforts. It has been estimated that only about 5 percent of all micro organisms present on earth, have been named and classified. A large number of unknown species are supposed to reside in the soil (Hawksworth 1991, Foissner 1994, Torsvick *et. al* 1990).

Soil algae and protozoa, like other higher plants and animals, can be identified by their morphology, while bacteria, and fungi, however, require more biochemical and genetic analysis to enable their identification. Molecular methods such as DNA techniques need to be applied to bacteria and fungi where morphological taxonomy offers limited scope. The properties of phospholipids fatty acids (PLFA), that constitute component of cell membranes, differ among different groups of microorganisms, thus showing microbial diversity (Tunlide and White 1992).

Soil health assessment and monitoring

Soil health has been defined as the continued capacity of soil to function as a vital living system, within the ecosystem and land use boundaries to sustain biological productivity, promote quality of air and water environment and sustain plant, animal and human health(Doran and Safely 1997. Soil health assessment requires not only indicators of biodiversity *i.e.* key species and functional groups, different ecosystem parameters such as water storage/ movement, structural stability, soil fertility, organic matter status and specific functions such as carbon, nitrogen, phosphorous cycling, microbial biomass activity but also establishment of target values and thresholds. Microbial diversity measurement as bioindicators have been recommended in soil health study programmes (Turco *et al* 1994). They can be grouped according to different soil health parameters such as carbon and nitrogen cycling, biomass microbial activity and bioavailability.

Besides taxonomic classification studies, molecular and biochemical techniques for estimation of abundance and number of each species must be applied (Table-2). A medium to high diversity is generally supposed to be good indicators of soil health. The genetic diversity is an indicator of genetic resources which can be measured by use of several molecular methods.

The microbial functional diversity in soil can also be measured by carbon utilization and extracellular enzyme pattern or diversity of nucleic acid within the cell (Table-3). Indicators of functional diversity are indicators of microbial activity, thus integrating diversity and function. The measurement of CO_2 production and / or O_2 consumption can also be measured as an indicator for quantification of heterotrophic microbial activity in soil (Alef 1995). Soil biomass includes bacterial, fungal and protozoan biomass. Measurement of biotic biomass is fundamental for different soil processes to occur and quantification of microbial biomass (Table-4) is generally recommended for evaluation of soil health (Visser *et al.* 1992 and Jat *et al.* 2002).

Table 2 : Soil microbial count as influenced by seed inoculation of PAS-2 (an acid tolerant strain of Phosphate Solubilizing Bacteria (PSB).

Treatment	MPN of PSB	Total bacteria	Total fungi	Total actinomycetes
Crops				
Finger millet	4.5	29.0	22.0	7.0
Amaranath	6.0	41.0	20.5	9.5
Buckwheat	6.5	39.5	21.0	6.5
Freanchbean	9.0	49.5	27.0	13.0
Maize	5.6	36.0	16.0	7.0
LSD(P=0.05)				
Sampling				
Rhizosphere	11.0	51.5	31.5	16.0
Nonrhizosphere	7.5	32.0	17.5	11.0
LSD (P=0.05)				

(PSB $X10^4 g^{-1}$soil, fungi $X10^2 g^{-1}$, soil and actinomycetes $X10g^{-1}$ soil)
Source: Pal S.S.2004

Table 3 : Phosphate solubilizing capacity and dehydrogenase activity of soil as influenced by seed inoculation of PAS 2 (an acid tolerant strain of PSB)

Treatment	Phosphate solubilizing capacity (mg P solubilized per mg $Ca_3(PO_4)_2$ per g dry soil	Dehydogenase activity (µmol formazone formed per g dry soil)
Crops		
Finger millet	0.39	79
Amaranath	0.53	88.5
Buckwheat	0.50	88
Freanchbean	0.58	105.0
Maize	0.47	86.5
LSD(P=0.05)	0.02	7.6
Sampling		
Rhizosphere	0.58	118.5
Non rhizosphere	0.40	90
LSD(P=0.05)	0.05	5.8

Source: Pal, S. S. 2000

Table 4 : Cropping system affects soil microbial biomass carbon (After 03 crop cycles) in an *Ustochrept*

Cropping system	MBC (µg g^{-1} soil)
Rice-wheat	128
Rice-potato-sunflower	165
Rice-berseem	160
Rice-potato-wheat	130
Rice-wheat-greengram	144
Sugarcane-ratoon-wheat	145
Sorghum(F)-toria-wheat	145
Sorghum(F)-wheat	170
Pigeaonpea-wheat	200
Maize-wheat	174
Initial	153

Source: Jat *et al* (2004)

Soil biodiversity and ecosystem functioning

Soil micro-organisms are clearly vital to soil functionality. Measures of the activity, biomass and diversity of soil micro-organisms, including the presence and health of root dwelling symbionts, are considered as reliable indicators of soil quality. It has been suggested that those soils harboring a greater diversity of microorganisms are more likely to be resilient to stresses. Loss of microbial diversity makes biological systems less able to adapt to environmental stresses. Various theories have been put forward to explain biodiversity-ecosystem function relationships. Based on his earlier work (Pathak and Rao, 1998), Rao hypothesized for soils, that the degree to which a particular ecological hypothesis would apply will depend on the degree of soil degradation and proposed a 'Stress Dependent Biodiversity-Functionality Relationship' (SDBFN). The main tenets of the hypothesis were recently outlined (Rao 2007b).

Measurement of soil biodiversity and soil health indicators

Soil biological quality is measured using several parameters but Rao and Manna (2005) considered it expensive and needless to devote so much time and effort for measuring dozens of soil quality indices. They considered that 4 parameters *viz* microbial biomass carbon, active (particulate) organic matter, soil respiration and N mineralization are sufficient to give a reliable picture of soil biological quality. This could be complimented with soil pH, organic matter and water holding capacity to give a complete picture of soil quality.

It is important to realize however that the concept of soil health includes the ecological attributes of the soil, which have implications beyond its quality or capacity to produce a particular crop. These attributes are chiefly those associated with the soil biota: its diversity, food web structure, activity and the range of functions it performs. Soil biodiversity per se may not be a soil property that is critical for the production of a given crop, but it is a property that may be vital for the continued capacity of the 'soil to produce that crop (including pastures and trees).

Microbial indicators would be found among the copiotrophic bacterial populations. Several species of bacteria and actinomycetes can cause plant diseases, but most damage is caused by fungi, which account for most soil-borne crop diseases, such as wilts, root rot, blights, *etc.* General suppressiveness of soils is the inhibition of pathogens as a result of a high total microbial biomass combined with a very intense com-petition for carbon and/or nutrients.

It should be cautioned however that studies on soil health must go beyond the current emphasis on only microbial activity indicators and also include soil–biota. Soil microfauna (protozoa and nematodes) may reduce or increase microbial numbers and speed the turnover of microbial biomass and thus enhances nutrient availability. Mesofauna (mites, collembola) increase substrate surface by fragmentation, soil macrofauna (ants, termites, centipedes, millipedes, earthworms) community and redistribute organic residues in soil profile which increase the surface area and substantially modify soil structure through formation of macropores and aggregates through burrowing activities and faecal pellets. Soil microbiota is sensitive to the intensification of land-use, type of crop plants, tillage and application of crop residues, fertilizers and pesticides. It has been estimated that under favourable conditions, one tenth of the organic matter in a soil is made up of soil animals. Thus, a layer of 10 cm of a hectare of soil with 1% organic matter contains roughly 1500 kg of soil fauna. Although soil fauna are present in many highly productive soils, it is difficult to make broad generalizations about earthworms or soil fauna, in general, as indicators of high quality soils. Enumeration and identification of at least the burrowing soil fauna, and measurements of faecal deposits have been recommended to be a part of minimum data set for assessing soil quality.

There are now well-documented cases to show that conversion of natural vegetation to other land- uses, including agriculture, results in change in the diversity of the soil community, e.g., microorganisms and invertebrate animals both above and below ground -lowering the biological capacity of the ecosystem for self-regulation and hence leading to further need for substitution of biological functions with agrochemical and petro-energy inputs. Furthermore, decline in soil biodiversity is expected to affect soil turnover, decrease natural soil aggregation, increase crusting, reduce infiltration rates, and thus exacerbate soil erosion. The sustainability of these systems thus comes to depend on external and market-related factors rather than internal biological resources. The biodiversity of soil under natural vegetation could therefore be taken as a baseline for monitoring changes under varying land-use. However the detection of critical thresholds for functional change is still elusive.

Managing soil biodiversity

Soil animal and microbial diversity is part of the biological resources of agro-ecosystems, and strongly impact soil processes and thereby productivity, and must be factored in the management decisions of agricultural land use practices. Microorganisms improve plant health and play a significant role in helping plants to overcome biotic and abiotic stresses. Current researches indicate that high-input agriculture, particularly tilled agro-ecosystems with narrow crop rotation/short fallow management, leads to decrease in species richness and dominance of some species. In contrast, management characterized by rotations, no-tillage, organic amendments leads to an increase in species richness and overall density. It is through the better understanding of these communities that we gain a greater understanding of soil health, and how best to intervene in agricultural systems to make them more productive. Some of the potential management approaches that were discussed above include a wide range of 'Soil biotechnologies'- direct management by inoculation with beneficial biota and indirect management through appropriate design and management of cropping system, organic input, other soil amendments and soil tillage.

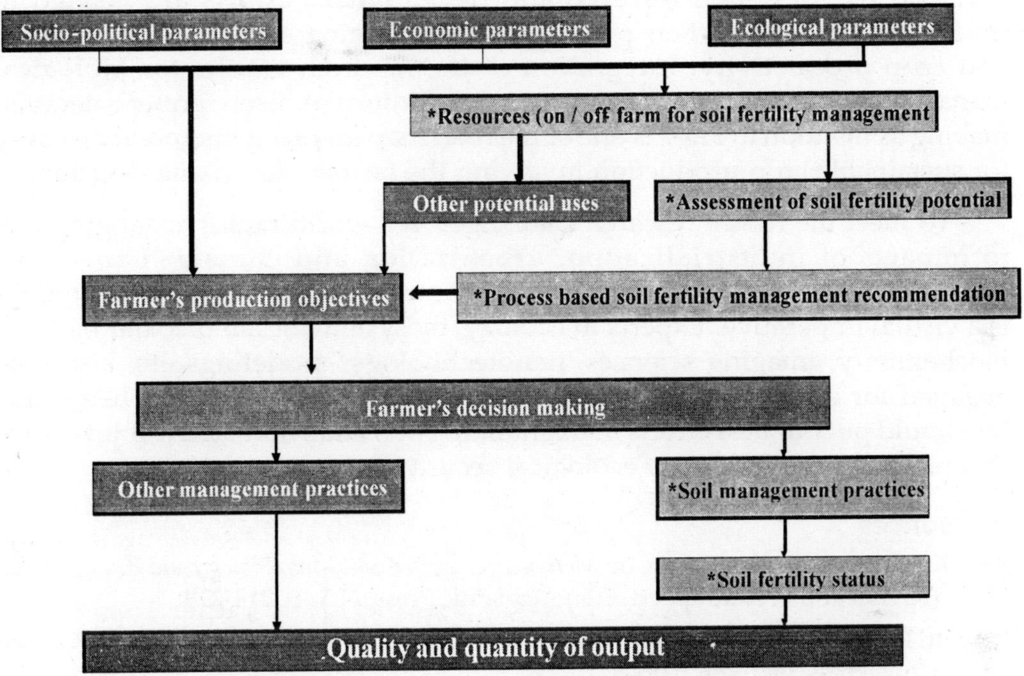

Fig. 2 : Integration of the research results for biological management of soil fertility with system level information on farmer's decision making

Conclusion

Soil is a part of terrestrial environment and supports all forms of life. It is a habitat for flora and fauna of different shapes and sizes. It is well established that soil organisms play an important role in various soil processes, maintenance of soil fertility and sustainable agriculture.

Soil health is a result of continued conservation and degradation processes, and represents the continued capacity of soil to function as a vital living system. It is concluded that the soil microorganisms appear to be excellent bio-indicators of soil health, because they quickly respond to changes in the soil ecosystem. In some instances, changes in microbial population can precede detectable changes in soil physical and chemical properties. Since soil organisms are involved in many soil processes, and these may provide an integrated measure of soil health which cannot be obtained with physical and chemical analysis alone. However, there are many gaps in scientific knowledge related to extent and quantum of soil biodiversity' and their functions. The deficiency in knowledge related to soil biodiversity may be due to limitations of available techniques to isolate and culture them as well as limited inventory efforts made so far. Systemic research and development efforts are needed to investigate and study their potential in maintaining sustainable soil health and crop productivity. Integration of the research results for biological management of soil fertility with system level information on farmer's decision making as mention in Fig-2 is one of the road map for using microbial diversity for sustainable crop production involving the farmers for taking decision.

To meet the future research challenges in a world rapidly changing due to impacts of industrialization, urbanization and climate change, our approaches need re-fashioning. Greater reintegration of the basic disciplines is a virtual imperative. Experts in ecology, biosystematics, molecular biology, biochemistry, imaging sciences, nanotechnology, modelings etc. are to be engaged for ensuring efficient land care, clean environment and better life; we should put our best efforts in maintaining this vital, non-renewable reource in a pristine state for future ecological security.

References

Alef, K. (1995). Soils respiration. In: *Methods in Applied Soil Microbiology and Biochemistry.* (K. Alef and P. Nannipeeri, Eds) Academic Press, N.Y. p. 214-218.

Bargett, R.D. (2005). The biology of soil, community and ecosystem approach. Oxford Univ. Press, Inc. New York.

Brossard, L. *et al.* (1997). Biodiversity and ecosystems functioning. Soil. *Ambio* **26**, 563-570.

Doran, J.W. and Safely S.M. (1997). Defining and assessing soil quality and sustainable productivity. In *Biological Indicators of Soil Health* (C.E. Pankhurst, B.M. Doube and V. V .S.R. Gupta, Eds). CAB International, p. 1-28.

Gaur, A.C. (2006a). *Handbook of Organic Farming and Biofertilizers.* Ambika Book Agency, Jaipur p. 291-298.

Gaur, A.C. (2006b). *Biofertilizers in Sustainable Agriculture.* Indian Council of Agricultural Research, New Delhi, p. 57-65.

Hawksworth, D.L. (1991). The fungal dimension of biodiversity magnitude, significance and conservation. *Mycology Research,* **95** : 641-655.

M.L Jat, S.S. Pal, Leevin Shukla, G.M.S. Mathur and Mohan Singh (2004). Rice Residues management using cellulolytic fungi and its effect on crop yield and soil health on rice-wheat cropping system. *Indian Journal of Agricultural Science,* **74** (3): 117-120.

Pankhurst, C.E., Doube, B.M. and Gupta, V.V.S.R. (1997). Biological indicators of soil health: Synthesis. In *Biological Indicators of Soil Health* (C.E. Pankhurst, B. M. Doube and V. V. S.R. Gupta, Eds). CAB International. p. 419-435.

Torsvick, V. Goksoyr, J. and Daae, F.L. (1990). High diversity in DNA of soil bacteria. *Applied Environmental Microbiology,* **56** : 782-787.

Tunlid, A. and White, D.C. (1992). Biochemical analysis of biomass, community structure, nutritional status and metaboliactivity of microbial communities in soil. In : *Soil Biochemistry* (G. Stotzsky and J.M. Bollag, Eds). Vol 7 .Marcel Dekker, N.Y. p. 229-262.

Turco, R.F., Kennedy, A.C. and Jawson, M.D. (1994). Mi crobial indicators of soil quality. In *Defining Soil Quality for a Sustainable Environment* (Doran *et al.,* Eds) *Soil Science Society of America,* Inc., Madison, p. 73-96.

Visser, S. and Parkinson, D. (1992). Soil biological criteria as indicators of soil quality: Soil microganisms. *American Journal of Alternative Agriculture,* **7** : 33-37.

Pathak, H and Rao, D.L.N. (1998). Carbon and nitrogen mineralization from added organic matter in saline and alkali soils. *Soil Biology and Biochemistry,* **30** : 695-702.

Manna, M.C., Swarup, A., Wanjari, R.H., Ravankar, H.N., Mishra, B., Saha, M.N., Singh, Y.V., Sahi, K.K. and Sarap, P.A. (2005). Long term effect of fertilizer and manure application on soil organic carbon storage, soil quality and yield sustainabilityu under sub-humid and semi-arid tropical India. *Field Crops Research,* **93** : 264-280.

Rao D.L.N. (2006). Maintaining the Soil Ecosystems of the Future. In *The Future of Soil Science* Ed. A. E. Hartemink, International Union of Soil Sciences, Wageningen, pp 116-118.

Rao, D.L.N. (2007). Microbial diversity, soil health and sustainability. *Journal of the Indian Society of soil science,* **55** : 392-403.

S.S. Pal (2000). Management of soil microbial population and crop yield. *Journal of the Indian Society of Soil Science,* **48** (1): 184-188.

S.S. Pal and M.L Jat (2004). Enhanced decomposition of rice straw and its effects on soil health and crop productivity in rice-wheat cropping system on *Ustochrept* of Indo-Gangetic Plain. *Journal of the Indian Society of Soil Science,* **74** (8): 409-414.

□□□

System Based Integrated Nutrient Management, 2012
© *B. Gangwar & V.K. Singh (eds.), pp. 287-294*
New India Publishing Agency, New Delhi (India)
e-mail : info@nipabooks.com; website : www.nipabooks.com

CHAPTER **19**

System Based Nutrient Budgeting

N.C. UPADHAYAY AND M.A. KHAN

A nutrient budget is a valuable indicator of the long term sustainability of a cropping system. It indicates where fertilizer application is inadequate and leading to decline in the soil nutrient status. Conversely, it can indicate excessive inputs which result in a nutrient surplus and greater potential for losses to the environment and waterways. (*Combs et al., 1996*) Thus nutrient budget is the comparison between all the sources of nutrients available to the producer and the requirement of nutrients to meet the crop and soil needs. The sources can either be from on the farm, such as livestock manure or credits from legumes, or from off the farm, such as purchased fertilizer or irrigation water. The requirement is the amount of nutrients needed by the crop to obtain the expected yields.

Most values for nutrients available from different sources (credits) and crop nutrient requirements are calculated from estimates taken from many years of long historical averages and field research. There is no "real time" method of calculating exactly what neither the crop's nutrient requirement nor the nutrients available are at any one time. More closely, both the nutrient requirements and availability are based on past performance for the climatic and soil condition. These values are given with some surety that the crop grown will be supplied with adequate nutrient during the growing season and the crops will not be limited in its growth. All incidental environment losses, such as runoff and leaching, have been accounted for. Climatic conditions, particularly temperature and soil moisture, greatly influence both the crop performance and the soil's capacity to provide nutrients to the plant. During any growing season, there can be many changes in the climate conditions that will affect both the crop growth and soil delivery of nutrients to the crop.

Although a nutrient budget is not an exact formula for supplying nutrients, it is one method for organizing the nutrient needs of the crop with the nutrients available on the farm. Nutrient budgets can easily determine if there is a gross imbalance between the nutrients that are available vs. the amount required. Nutrient budgets are one of the best methods to see the over all supply of crop nutrients available, compared to the estimated crop needs as given by historic records and field research. Continued use of soil testing, plant and water analyses and yield monitoring are essential to maintain good nutrient balance with desired results.

Nutrient budgeting is based on a soil test analysis and crop nutrient recommendations. As its basis, the nutrient requirement of the crop has been determined from historical field research for that soil and climate. The nutrient credits for nutrients supplied are taken from analysis of soil, water, plants and organic material that provide nutrients to the crop. Some of these values keep modifying by research data, to reflect the estimated supplying power of these individual sources.

Farm specific nutrient management planning tools have been developed to evaluate on-farm nutrient flow. Typically, these tools employ a system of record keeping to generate nutrient budgets, which account for nutrient movement within and among farm enterprises. Whole farm nutrient budgeting may operate from a field-by-field (*Bullington and Combs, 1995*) or farm-gate (*Kelling et al.*, 1991) level. It simulates the interconnections between crop yield, manure/fertilizer, cropping patterns and existing soil test levels. Resulting information is used to streamline nutrient flow and minimize excessive application to fields (*Bacon et al., 1990; Lanyon and Beegle, 1989*). There are a few basic elements that guide the producer in making decisions on the location, rate, timing source, form and method of nutrient application. These elements are necessary in order to be fully aware of the management obligations that are required to be undertaken to successfully manage nutrients and protect the natural resources. These basic elements are as following:

1. Site aerial photographs or maps, including a soil map

These maps are generally part of the over – all conservation plans. However, additional site information may be given for the location of fields, where nutrients will be applied and proximity to sensitive resource areas for nutrient application. Soil interpretations for nutrient application should be given.

2. Soil, plant, water and organic sample analysis results

Since nutrient management is based on crop needs and sources of nutrients, an analysis of these factors is essential to know the supplying power of the nutrients and the crop response. These are basic factors to determine the nutrient budget.

3. Current or planned plant production sequence or crop rotation

Nutrient application is based on crop requirements. The sequence of crops will determine nutrient needs as well as nutrient carried over from one crop to another.

4. Realistic yield goals

The expected crop yield is the basis for determining the nutrient requirement for that particular yield level. Generally, the higher the yield expectation, the higher the nutrient requirement to reach that yield. There are a number of methods available to calculate expected yield goals.

5. Quantification of all important nutrient sources

This could include, but not limited to, commercial fertilizer, animal manure and other organic by products, irrigation water, atmospheric deposition, and legume credits. The estimates used to determine the amount of nutrient supplied is based on the soil, plant, water and organic manure analysis.

6. A nutrient budget for the complete plant production system

A nutrient budget determines the amount of nutrients available from all the sources and compares this to the amount of nutrients required to meet the realistic yield goal. When yield requirements of nutrients exceed the available sources, then additional nutrient must be brought in, to satisfy the crop's requirement. On other hand, if nutrient supply exceeds crop needs, management measures must be taken to ensure that the excess nutrients are either reduced as inputs or that their application will not cause detrimental effects to the plants, soil or surrounding environment.

7. Recommended rates, timing and methods of application

These specifications are for individual fields or for groups of fields depending on the soil and rop rotation. The specification for rates is based on the nutrient requirement of the crop (usually taken from soil test recommendations). Timing is determined by crop growth stage, nutrient needs and by the climatic conditions that can affect the transformation and transport of nutrients. How the nutrient is applied is based on the form and consistency of the nutrient, soil conditions and potential for movement and loss to the environment.

8. Location and nutrient management restrictions within or near sensitive areas or resources

If present, sensitive resource areas should be delineated on the maps. Any restrictions of nutrient application should also be mentioned. This may include set backs required for application of animal manure, reduced application rates or timing limitations based on soil conditions, or areas with special resource concerns. One must remain aware of these areas and modify management accordingly.

9. Operation and maintenance of the nutrient management plan

A number of items are required to be assessed on a routine basis. These include calibration of application equipment, maintaining a safe working environment, review and update of plan elements, periodic soil, water, plant, and organic waste analysis, and monitoring of the resources. This element keeps the nutrient management component plan up to date.

This list of above nine elements may not be exclusive, but is the basic guidance to be put in the nutrient plant component.

Essentials of ideal nutrient budget guide

1. Planned crop or crop rotation

List the crop that will receive nutrient application. In the case of rotation, list the crops in sequence. Nutrients budgets can be calculated for a single crop or over the entire crop rotation.

2. Yield expectation

Describe the expected crop yield based on realistic soil, climate and management parameters. Yield expectation can be determined from producer or county yield records, soil productivity tables or local research.

3. Soil test analysis results

The laboratory analysis will be recorded from the information sent by the soil test laboratory with the soil test results. Generally soil tests laboratories record the lab analysis for nutrients in parts per million (ppm) or in kilogram per hectare (kg/ha.). It may be important to know the soil extraction method used by the laboratory for determination of the element.

4. Nutrient recommendations based on soil analysis and expected yields

In general, results from similar soils and climate conditions are ideal to determine additional nutrient requirements at specific soil test levels and for desired yields needed for the specific crop. After consulting the laboratory analysis and factoring in the desired yield expectation, recommendations are given for the nutrient additions. Climatic and environmental loses, such as runoff, leaching and denitrification are considered in the recommendation.

5. Nitrogen credits

Nitrogen is a very mobile nutrient and occurs in the soil and plant in many forms. It can be stored in soil's organic matter and releases as the organic matter decomposes. Nitrogen is fixed by legume plants and brought into the soil. The amount of nitrogen added by legumes varies with plant species and growing conditions. All the nitrogen applied to the crop isnot available. A percentage of last year's manure application and even smaller amount of previous applications will become plant available during this crop season. One should refer to local mineralization rates to determine the residual release of nitrogen.

Phosphorus and potassium are considered 100 % plant available the year of application, therefore no residual amounts are calculated. An analysis is necessary because the quantity of nutrients in manure varies greatly from farm to farm, depending on the diet of the animals and the amount of bedding and liquid added to the manure. The result of the analysis will give an indication of the amount of nutrients present in manure. One should obtain a sample, after complete agitation, every time the storage is emptied. The sample should comprise at least 20 sub samples. When sending a sample to the lab, fill a plastic jar half full, place in a sealed plastic bag and store in a cool place. Analysis of the manure sample should include total nitrogen, ammonium nitrogen, phosphorus, potassium and dry matter. After several analyses, a trend becomes evident and one would only need to test the manure when making major changes in livestock feeding or bedding methods.

Irrigation water, especially from shallow aquifers, contains some nitrogen in the form of nitrate nitrogen. This nitrogen is available for crop use. To calculate the amount of nitrogen applied with irrigation water, determine the concentration of nitrate nitrogen in the water (in ppm or mg/l). The application amount will equal the nitrate nitrogen concentration (in ppm) multiplied by the volume (in acre-inches). Other nitrogen credits come from atmospheric deposition from dust and ammonia in rain water. This value is recorded by a number of weather stations. Atmospheric deposition may range from a few kilograms per hectare per year to over 30 kilograms. Any other material brought on to the site, like mulch or compost can be determined by estimating the mass weight and percent concentration of nitrogen in the material.

6. Sources of nutrients available to the field

The producer has the capability to bring various sources of nutrients on to the field to supply the requirements of the crop. The nutrient budget is designed to allocate the sources of nutrients available and adjust the amounts based on the calculations to match the crop's needs. For this

Manures and biosolids can be produced either on the farm or transported to the farm with the expressed purpose of utilizing the nutrients. Manure application rates should be based on crop nutrient requirements over a broader number of fields.

Nitrogen credits are summed and carried to the calculation.

If starter fertilizers are required, such as in cases of cool, wet soils or reduced tillage systems, the amount of starter nutrients should be entered.

7. Show nutrient balance

The required amount of crop nutrient, either determined from the recommendations or from the crop removal, will be subtracted from the total nutrients available to the field. A deficiency of nutrients in the balance (shown as a positive number) will mean that additional nutrients need to be applied to the field to meet the crop requirement. This can be done with additions of higher rates of fertilizer or animal manure. There is no opportunity to increase the manure residual mineralization rate or amount of atmospheric deposition and only a slight increase with additional irrigation water is feasible. Fertilizer is considered the easiest because the exact nutrient ratio can be derived by using any of a number of fertilizer blends.

When the balance number becomes negative, there are more of one or more nutrients available in the field than required by the crop. These excess nutrients can become an environmental liability when subject to run off and leaching. It is necessary to adjust the field inputs, most likely the manure additions, to balance with the crop requirements.

Keeping all the above factors into consideration, the desired information for each crop of the cropping system has been shown in Table-1

Table 1 : Indices for nutrient budgeting of each crop of cropping system, based on soil test analysis and crop nutrient recommendation.

A. Planned Crop or Crop Rotation:				
B. Yield Expectation (goal) :				
C. Soil Test Analysis Result	N =	P =	K =	Secondary & micronutrients
D. Nutrients recommendation based on soil analysis	N =	P_2O_5 =	K_2O =	Ca = Mg = S = Zn = Cu = Fe = Mn = B = Mo =
(E.) Expected yield (qtl ha^{-1}).				
F. Nitrogen Credits:				
F 1. Legume credits from previous crop			Kg ha^{-1}	
F 2. Residual from previous manure applications			Kg ha^{-1}	
F 3. Irrigation water nitrate nitrogen			Kg ha^{-1}	
F 4. Other (e.g., mulch)			Kg ha^{-1}	
F 5. Total N Credits			Kg ha^{-1}	
G.Sources of Nutrients available to the field:	N	P_2O_5	K_2O	Secondary & micronutrients
G 1. Manure Applied				
G 2. Nitrogen Credits (E 5.)				
G 3. Starter Fertilizer				
G 4. Others				
G 5. Total Nutrient Sources				
H. Show Nutrient Balance:	N	P_2O_5	K_2O	Ca Mg S Zn Cu Fe Mn B Mo
H 1. Nutrient Recommended (D.)				
H 2. Total Nutrient Sources (F5.)				
H 3. Nutrient Balance (H1. - H2.)				

If H3 is a positive number, this amount of additional nutrient is required and supply through fertilizer or other sources of nutrient.

If H3 is a negative number, this amount of nutrient is in excess and reallocates the source of nutrient available.

Future thrust

A nutrient budget is effectively a measurement of all nutrient inputs into a farming system (e.g. through fertilizer, rain, supplements, irrigation, biological nitrogen fixation) versus what nutrient output are occurring (e.g. through production, transfer, supplements exported, gaseous losses, leaching). By helping to improve the assessment of nutrient requirements and indicating improved nutrient management strategies, farms are more profitable and have less environmental impact. In future these budgets need to be computed using the nutrient management module of the Crop Rotation Options Programme (CROP), which is a spreadsheet - based planning.

References

Bacon, S.C., Lanyon, L.E. and. Schlauder Jr. R.M. (1990). Plant nutrient flow in the managed pathways of an intensive dairy farm.*Agron. J.*, **82** : 755-761.

Bullington, S.W. and Combs, S.M. (1995). Wisconsin interactive soils program for economic recommendations (WISPer Ver. 2). In : *Proceedings of the 87th annual meeting, ASA, CSSA, SSSA, St. Louis, MO.* October 30-31.

Combs S.M., Bullington, S.W., and Herring H. (1996). Twenty years of Wisconsin soil testing, 1974-1994. *New Horizons in Soil Science,* No. 9-96.

Kelling, K.A., Schulte, E.E., Bundy, L.G., Combs, S.M., and Peters J.B. (1991). Soil test recommendations for field, vegetable and fruit crops.*Univ. Wis. Cop. Extn. Ser. Bull.* No. A 2809.

Lanyon, L.E., and Beegle, D.B. (1989). The role of on-farm nutrient balance assessments in an integrated approach to nutrient management. *J. Soil Water Cons.*, **44**: 164-168.

□□□

System Based Integrated Nutrient Management, 2012
© B. Gangwar & V.K. Singh (eds.), pp. 295-299
New India Publishing Agency, New Delhi (India)
e-mail : info@nipabooks.com; website : www.nipabooks.com

CHAPTER **20**

Soil Carbon Management for Better Soil Health

K.P. TRIPATHI

Organic C in agricultural soils contributes positively to soil fertility, soil tilth, crop production, and overall soil sustainability. Changes in agricultural management can potentially increase the accumulation rate of soil organic C (SOC), thereby sequestering CO_2 from the atmosphere. Optimizing agricultural management for accumulation of SOC can result in the sequestration of atmospheric CO_2, thereby partially mitigating the current increase in atmospheric Carbon dioxide. In addition to the environmental benefits of soil C sequestration, Carbon dioxide consideration has also been given to the implementation of a C credit trading system which may provide economic incentives for C sequestration initiatives.

Changes in agricultural practices for the purpose of increasing SOC must either increase organic matter inputs to the soil, decrease decomposition of soil organic matter (SOM) and oxidation of SOC, or a combination thereof. These practices include, but are not limited to, reducing tillage intensity, decreasing or ceasing the fallow period, using a winter cover crop, changing from monoculture to rotation cropping, or altering soil inputs to increase primary production (e.g., fertilizers, pesticides, and irrigation). Implementing practices that sequester C can reverse the loss of SOC that may have occurred under intensive cultivation thereby increasing SOC to a new equilibrium

Average global C sequestration rates, when changing land use from agriculture to forest or grassland, were estimated to be 33.8 or 33.2 g C m^{-2} yr^{-1}, respectively (Post and Kwon, 2000). Silver *et al.* (2000) estimated that reforestation of abandoned tropical agricultural land and pasture sequesters C in the soil at a rate of 130 g C m^{-2} yr^{-1} for the first 20 yr, and then at an average rate of 41 g C m^{-2} yr^{-1} for the following 80 yr.

Loss of SOC can also be reversed by using less intensive cultivation practices or by changing from monoculture to rotation cropping. In an analysis of 17 experiments (n = 38), Kern and Johnson (1993) concluded that a change from CT (Conservation tillage) to NT (No. tillage) sequesters the greatest amount of C in the top 8 cm of soil, a lesser amount in the 8- to 15-cm depth, and no significant amount below 15 cm. They also concluded that, unlike NT, no significant change in SOC was realized in response to reduced tillage (RT). Kern and Johnson (1993) assumed the duration of C sequestration to be between 10 and 20 years. Paustian *et al.* (1997) compared 39 paired tillage experiments, ranging in duration from 5 to 20 years, and estimated that NT resulted in an average soil C increase of 285 g m^{-2} with respect to CT. Using an average experiment duration of 13 years implies an approximate C sequestration rate of 22 g m^{-2} yr^{-1}.

The Intergovernmental Panel on Climate Change (IPCC) has developed guidelines for accounting of greenhouse gases, including C sinks in forest and agricultural ecosystems (Houghton et al., 1997). The IPCC suggests using a multiplication factor of 1.1 for a change from CT to NT (Houghton et al., 1997; Land-use change & forestry section), essentially corresponding to a 10% increase in SOC. Moving from CT to RT, factors of 1.05 and 1.0 are recommended for agricultural lands in temperate and tropical climate regimes, respectively. The IPCC suggests these factors be applied to a depth of 30 cm and over a period of 20 years. Additional factors are provided for residue management, soil inputs (e.g., mulching and manure), and fallow frequency.

Carbon sequestration in Indian scenario

Soil degradation is a widespread problem in India. Soil degradation through the different land degradative processes is presented in Table 1. The data indicate that total potential of carbon sequestration with the restoration of degraded lands in the country is about 10-14 Tg yr^{-1}. The largest potential lies with the restoration of soil fertility, followed by desertification control and checking the soil erosion.

Managing crop residues are one of the major tools to synthesize soil organic carbon in the agricultural system in the country. The total crop residue produced by the principal crops in the country is about 440 Tg of which cereals share is about 90%. If properly decomposed in the soil, it could contribute significantly in the soil organic carbon build-up.

Table 1 : Potential of total soil carbon sequestration through restoration of degraded soils in India

Degradation Process	Area (Mha)	SOC sequestration rate (kg ha^{-1} yr^{-1})	Total SOC sequestration (Tg yr^{-1}) potential
Water erosion	32.8	80-120	2.6-3.9
Wind erosion	10.8	40-60	0.4-0.7
Soil fertility decline	29.4	120-150	3.5-4.4
Water logging	3.1	40-60	0.1-0.2
Salinisation	4.1	120-150	0.5-0.6
Desertification control	68.1	40-60	2.7-4.1
Total			9.8-13.9

Soil carbon sequestration and food security in India

Compilation of data from different sources indicate that increase in SOC pool by 1 Mg hectare^{-1} can result in increase in yield of grain crops by 30-50 kg ha^{-1} of wheat, 100-300 kg ha^{-1} of maize, 30-50 kg ha^{-1} of rice and similar to the other crops (Table 2, Lal 2005). If such trend of increase in yield takes place in different agroclimatic conditions of the country, it would be optimistic presumptions. Realization of such potential is highly crop, soil and management specific.

Table 2 : Relation between soil carbon sequestration and yield of some crops in India

Crop	Area (Mha)	Current yield (kg ha^{-1} yr^{-1})	Projected increase in the yield(kg ha^{-1} yr^{-1} /Mg of SOC)	Total increase in production (Mt yr^{-1})
Wheat	27.3	2640	30-50	0.8-1.4
Rice	42.5	2927	30-50	1.3-2.1
Maize	14.0	670	100-300	1.4-4.2
Millet	9.4	850	30-50	0.3-0.5
Sorghum	9.2	700	100-140	0.9-1.3
Chick pea	6.5	890	20-30	0.1-0.2
Soybean	7.6	920	20-30	0.2-0.3
Rapeseed	6.8	1000	200-300	6.9-12.51

Estimation of change in the annual rate of SOC sequestration

$$\Delta SOC_R\,\%yr^{-1} = \left|\left[(NT_2 - CT_2) - (NT_1 - CT_1)\right] / (NT_1 - CT_1) / (t_2 - t_1)\right| \times 100$$

where $NT_{1\text{ and }2}$ and $CT_{1\text{ and }2}$ is SOC under NT and CT during the first and second years in which SOC was measured, respectively; SOC_R is the estimated annual rate of soil C sequestration; and t_1 and t_2 are the number of years following initiation of the experiment in which SOC was measured.

Soil carbon economy: A timely opportunity for the country

The economic value of soil carbon needs to be assessed with consideration for both onsite and offsite effects. Procedures are needed for a defensible soil carbon accounting system, and policies need to be established that provide incentives for net soil carbon sequestration at national scale. Such policies can provide financial incentives for the restoration of degraded and impoverished soils, while achieving reductions in the rate of buildup of atmospheric CO_2 levels. Enhancing the SOC pool through global and national policy incentives may be especially beneficial to improving agronomic productivity, enhancing food security and reducing poverty in countries with degraded soils, thus providing splendid benefits to humankind. The dynamics of carbon sequestration processes must be evaluated in the context of local soil and crop attributes including biogeochemical cycles and soil spatial variability. Accounting for variability of soil and plant processes is needed at a range of scales from fields to watersheds and from regional to continental scales. Spatial and temporal variability of soil C is known to be high in many areas. Any soil C commodity price structure will need to be conservatively discounted for soil C variability. Incentives and plans for adoption of sound technical practices are needed for farmers, ranchers, foresters and other land managers. Implementation of such plans will require better understanding of rural society and the market incentives needed for adopting changes in farming, grazing, and forestry practices that benefit all sectors of society. Advancement in knowledge through targeted research will permit refined quantitative assessment of the total carbon sequestration potential.

References

Houghton, J.T., L.G. Meira Filho, B. Lim, K. Tréanton, I. Mamaty, Y. Bonduki, D.J. Griggs, and B.A. Callander (ed.). (1997). IPCC Guidelines for National Greenhouse Gas Inventories. Volumes 1–3. Hadley Centre Meteorological Office, United Kingdom.

Kern, J.S. and M.G. Johnson (1993). Conservation tillage impacts on national soil and atmospheric carbon levels. *Soil Science Society of America Journal*, **57** :200–210.

Lal, R. (2005). Carbon sequestration and climate change with specific reference to India. *Indian Society of Soil Science*, ICSWEQ-Proceedings, pp 295-302.

Paustian, K., O. Andrén, H.H. Janzen, R. Lal, P. Smith, G. Tian, H. Tiessen, M. Van Noordwijk, and P.L. Woomer, (1997). Agricultural soils as a sink to mitigate CO_2 emissions. *Soil Use Manage*, **13** :230–244.

Post, W.M. and K.C. Kwon (2000). Soil carbon sequestration and land-use change: Processes and potential. *Global Change Biological*, **6** : 317–327.

Silver, W.L., R. Osterlag, and A.E. Lugo (2000). The potential for carbon sequestration through reforestation of abandoned tropical agricultural and pasture lands. *Restoratian Ecology*, **8** : 394–407.

□□□

System Based Integrated Nutrient Management, 2012
© B. Gangwar & V.K. Singh (eds.), pp. 301-310
New India Publishing Agency, New Delhi (India)
e-mail : info@nipabooks.com; website : www.nipabooks.com

CHAPTER **21**

Balance Crop Nutrition and Plant Disease Management

CHANDRA BHANU

Balanced crop nutrition is a most realized critical factor in allowing crops to give their full yield potential. The application of macro and micronutrients through manures and fertilizers to achieve this balance is an integral practice in modern crop production system. Macro and microelements have long been recognized as being associated with changes in the level of many diseases and yield and /or quality of crops. Three major elements N, P, and K are the most studied nutrients which chiefly affect relative resistance of crop plants to various diseases. Others macro/micro/beneficial nutrients including Ca, Mg, S, Mn, Fe, Zn, Mo, Si, Bo and Cu have shown varying degrees of disease suppression/induction on many crop species under various environmental conditions. Crop nutrition influence plant health through: i. affecting the growth, reproduction, virulence, and survival of pathogens; ii. resistance and tolerance, predisposition and escape in the host crop; and iii. root exudates and microbial interactions facilitating biological control. Balanced crop nutrition can protect crops from pathogens by: i. avoiding plant stress, which may allow crops to better withstand the attack of pathogens and ii. manipulating nutrients to the advantage of crops and disadvantage of the pathogen. The modern intensive agriculture which mainly relies on chemical inputs (fertilizers and pesticides) and fertilizer responsive HYVs (high yielding varieties) has led the self sufficiency in food production in India and several other countries of the world. But, the indiscriminate and imbalance use of chemical fertilizers along with exhaustive cropping initiated several second generation problems including widespread micronutrient deficiencies and increased attack of several insect-pests and pathogenic diseases. The excessive use of nitrogen and no or reduced application of phosphorus, potash and other micronutrients is creating conducive conditions for several crop pathogens and their insect vectors. This situation frequently results in serious crop losses during favourable environmental conditions for disease development. The specific role of macro and micronutrients in maintaining crop health, effect of their dose and form of application on suppression / induction of various diseases and interaction with other soil/ environmental factor(s) will be discussed in following sections.

General scenario

The application of manures and fertilizers is a universal practice in modern crop production. The advent of readily available chemical fertilizers for agriculture has brought about the demise of many diseases through disease escape, improved plant resistance, altered pathogencity and/ or microbial interactions influencing these. The physiological roles associated with the elements may be more or less understood but the research is lacking on direct linkage of biological and physiological factors such as enzymes, cell wall and membrane permeability, competing ions, amino acids, phenols, alkaloids, phytoalexins, phytoanticipins etc. with disease. The complex balance of interdependent functions of the host plant are changed and influenced by chemical elements which in turn, influence the reaction to a crop disease. However, at the same time, there is a simple and direct relationship in physiological plant pathology.

Nutrition influence plant health through: i. the growth, virulence, reproduction/ replication, and survival of pathogens; ii. predisposition, escape, tolerance, and resistance of plants; and iii. root exudates and microbial interactions important in biological control. The soil environment and severity of diseases in plants are changed by macro and microelement additions. Nitrogen stimulates the soil microflora so that competition for nutrients and space decreases the relative number(s) of propagules of the pathogen(s). Antagonists may be stimulated to produce more antibiotics. Calcium and nitrate nitrogen raise the soil pH, thereby decrease the virulence of *Fusarium oxysporum* and reduce the level of wilts caused by them whereas, NH_4-N decreases the soil and rhizosphere pH and restrict the growth of take all pathogen, *Gaeumannomyces graminis*. Nitrogen, P, K and Ca reduce damping-off. Excess N (imbalance) can cause thin cell walls, allowing fungi to penetrate more rapidly while balanced nutrition can reduce vigour and reduce damping-off disease. The level of *Fusarium* increases while *Verticillium* wilt decreases when the soil pH (or Ca) is low. Phosphorus and K can promote root growth and reduce seedling diseases, root rots etc. The effects of various nutrients in changing the level of plant diseases have been reviewed by Engelhard (1989).

Responses to a particular nutrient (N, P, K, Mn, Ca, S etc.) may be different when going from deficiency to sufficiency and then from sufficiency to excess. Metabolic systems may respond differently depending on the form of a nutrient (NH_4-N vs. NO_3-N) or availability (e.g. Mn or S, oxidized vs. reduced). Nutrient availability may vary depending on environmental conditions, the previous crop, microbial activity in the rhizosphere or ratio with other elements (N, K, Mn, S, Fe, Zn, Cu). Time of application (stage of crop growth) may also influence the disease response to a nutrient.

Present status of research

All the essential mineral elements are reported to influence disease incidence or severity. The effect of mineral nutrients on disease has been determined by: a. observing the effect of mineral amendment (fertilization), b. comparing mineral concentration in resistant and susceptible cultivars or tissues, c. correlating conditions influencing mineral availability with disease incidence or severity or d. a combination of all three. A particular element may reduce some pathogens but increase others, and have an opposite effect with modification of the environment. An example is the effect of specific forms of N on diseases of potato. Inhibiting nitrification of NH_4-N reduces *Verticillium* wilt but may increase *Rhizoctonia* canker. Other cultural modifications may be utilized to reduce *Rhizoctonia* canker so that the NH_4 form of N could be used to suppress *Verticillium* wilt in the production system.

The kind of crop host response (susceptibility, resistance and reduced receptivity) to parasitic attack is correlated with different forms of host nutrition in following manner:

i. In general, crop susceptibility to pathogens is most evident in case of potash deficiency
ii. Resistance is most evident during nitrogen deficiency and
iii. Reduced receptivity in case of phosphorus deficiency
iv. When one nutrient (e.g. nitrogen) is given in excess there is also a relative deficiency of rest other main nutrients (i.e. phosphorus and potash)

It is reported in several research that, nitrogen and potash determine the reaction of host crop to pathogens more than the phosphorus.

A. Effect of nitrogen

Since the effect of nitrogen on plants is more pronounced, it has been studied in great detail. Fertilization with nitrogen, especially at enhanced doses, causes new succulent vegetative growth of the plant and delays the maturity. Those pathogens which attack such plant organs are, therefore, favoured by high nitrogen content. It is known that excess of nitrogen predisposes the host to rusts and powdery mildew of wheat and blast of rice and many more diseases. When the nitrogen is deficient, the plant is weak, its development is incomplete and its maturity is hastened. Pathogens by slow growth of the host are thus favoured by low nitrogen.

Studies on the effects of nitrate (NO_3-N) and ammonium (NH_4-N) forms of nitrogen have revealed that pathogen-crop interactions depends frequently on the from rather than the amount of N available. These interactions are, however, complex in nature.

Mechanisms

Suppression and stimulation of parasitic diseases by specific forms of nitrogen has been attributed to preference of N- forms by plants, altered host resistance, modification of plant constituents and exudates, shift in soil and rhizosphere pH, effect on pathogens and modified microbial equilibrium. In most altered soils, NO_3-N predominates and plants adapted to such soils grow well with NO_3-N as the sole source of nitrogen. Crops that utilize NH_4-N grow well following fumigation while, those requiring NO_3-N may be adversely affected. Evidence that, nitrogen affected resistance of wheat to pathogen causing take all disease was demonstrated. Increased root growth permitting wheat to escape take all has also been reported. NH_4-N reduce bacterial canker of *Prunus* by hastening periderm formation. NH_4-N has been found to increase permeability and exudation of broad bean leaf surface resulting in higher levels of sugars and amino acids. *Botrytis cinerea* infected such leaves more frequently than supplied with NO_3-N. Ammonium forms of nitrogen fertilizers generally lower the pH while, nitrate fertilizers raise it. Nitrate nitrogen suppresses damping- off of sugar beet (caused by *Pythium ultimum*) but ammonium fails to do so. This is due to change in rhizosphere pH towards alkalinity by NO_3-N and beets being alkali tolerant plant develop more vigour under these conditions, thereby developing resistance quickly through tissue maturity. The effects of dose and form of nitrogen fertilization on increase or decrease of various plant diseases are given in Table 1-3.

Some nitrogen compounds are directly toxic to pathogens. Ammonia is known to be toxic to *Sclerotium rolfsii* and *Gaeumannomyces graminis* and some plant parasitic nematodes. Calcium cyanamide is toxic to *S. rolfsii, Fusarium oxysporum* and *Pythium* spp. Many nematodes are suppressed by urea due to direct toxicity.

Many a times, nutrient application may modify microbial equilibrium in soil and results in the suppression of many diseases.

Table 1 : Crop diseases decreased by nitrogen fertilization

Name of the disease	Causal organism
A. Fungal diseases	
1. Bunt of wheat	*Tilletia* spp.
2. Club root of cabbage	*Plasmodiophora brassicae*
3. Cotton root rot	*Phymatotriehum omnivorum*
4. Pea root rot	*Aphanomyces euteiches*
5. Wheat root rot	*Rhizoctonia solani*
6. Stem canker of soybean	*R. solani*
7. Sclerotium rot of sugarbeet	*Sclerotium rolfsii*
8. Take all of wheat	*Gaeumannomyces graminis*
9. Cotton wilt	*Verticillium albo-artum*
B. Bacterial diseases	
1. Bacterial wilt of cucumber	*Erwinia tracheiphila*
2. Bacterial spot of peach	*Xanthomonas arboricola* pv. *pruni*
3. Bacterial wilt of tomato and tobacco	*Ralstonia solanacearum*
4. Brown rot and wilt of potato	*R. solanacearum*
5. Bacterial wilt of eggplant	*R. solanacearum*
6. Wild fire of tobacco	*Pseudomonas syringae* pv. *tabaci*
C. Viral diseases	
1.Bean mosaic	*Tobacco mosaic virus*
2. Celery ring spot	*Celery ring spot virus*

Table 2 : Crop diseases increased by nitrogen fertilization

Name of the disease	Causal organism
A. Fungal Diseases	
1. Paddy blast	*Magnaporthe grisea*
2. Rusts of several crops	*Puccinia* spp.
3. Powdery mildews of cereals	*Blumeria graminis*
4. Downey mildew of cabbage	*Peronospora parasitica*
5. Wilts of tomato, melons and cabbage	*Fusarium oxysporum*

B. Bacterial diseases	
1. Citrus canker	*Xanthomonas axonopodis* pv. *citri*
2. Fire blight of apple	*Erwinia amylovora*
3. Stewart's wilt of maize	*Erwinia stewartii*
4. Stalk rot of maize	*Erwinia chrysanthemi* pv. *zeae*
5. Black leg, wilt and soft rot of potato	*E. carotovora* subsp. *carotovora* and *E. carotovora* subsp. *atroseptica*
6. Bacterial leaf blight of rice	*Xanthomonas oryzae* pv. *oryzae*
7. Bacterial leaf streak of rice	*X. oryzae* pv. *oryzicola*
C. Viral diseases	
1. Common bean mosaic	*Common bean mosaic virus*
2. Lettuce mosaic	*Lettuce mosaic virus*
3. Tobacco mosaic	*Tobacco mosaic virus*
4. Wheat mosaic	*Barley stripe mosaic virus*

Table 3 : Effect of form of nitrogen on plant diseases

Crop disease	Causal organism	Effect of NO_3-N	Effect of NH_4-N
A. Fungal and Oomycetes diseases			
1. Paddy blast	*Magnaporthe grisea*	D	I
2 Black scurf of potato	*Rhizoctonia solani*	D	I
3. Brown spot of rice	*Drechslera oryzae*	I	D
4. Charcoal rot of many crops	*Macrophomina phaseolina*	D	I
5. Eye spot of wheat	*Pseudocercosporella herpotrichoides*	D	I
6. Northern leaf blight of maize	*Drechslera turcica*	D	I
7. Root rot of pea	*Aphanomyces euteiches*	D	I
8. Root rot of pea	*Pythium* spp.	I	D
9. Root rot of tobacco	*Thielaviopsis basicola*	I	D
10. Cotton root rot	*Phymatotrichum omnivorum*	I	D
11. Stem rust of wheat	*Puccinia graminis tritici*	I	D
12. Stripe rust of wheat	*P. striiformis*	I	D
13. Stalk rot of maize	*Fusarium* sp.	D	I
14. Take all of wheat	*Gaeumannomyces graminis*	I	D
15. Wilt of bean	*Fusarium oxysporum* f.sp. *phaseoli*	D	I
16. Wilt of potato and tomato	*Verticillium albo-artum*	I	D

B. Bacterial diseases			
1. Angular leaf spot of cotton	*Xanthomonas axonopodis* pv. *malvacearum*	D	
2. Peach canker	*X. campestris* pv. *pruni*	D	
3. Canker of tomoto	*Clavibacter michiganense* sub sp. *michiganense*	I	
4. Crown gall of several plants	*Agrobacterium tumefaciens*	I	
5. Potato ring rot	*Clavibacter michiganense* sub sp. *sepedonicum*	I	
6. Stewart's wilt of maize	*Erwinia stewartii*	I	
7. Potato scab	*Streptomyces scabies*	I	D
C. Nematode diseases			
1. Lima bean root knot	*Meloidogyne incognita*	I	D
2. Soybean cyst	*Heterodera glycinea*	I	D
3. Tobacco cyst	*H. tabacum*		D
D. Viral/viroid diseases			
1. Coconut cadang-cadang	*Viroid*		D
2. *Potato virus X*			D
3. Tobacco mosaic	*Tobacco mosaic virus*		I

D = Decrease, I = Increase

B. Effect of phosphorus

Effect of phosphorus on plant diseases appear to be connected with one of the two situations:

1. Either the crop grows on soil markedly deficient in P, and correction of the imbalance contributes to its health or
2. Maturity of crop is somewhat advanced by liberal supply of P, and this helps in escape of biotrophs (pathogens which survives only on living host tissues) preferring younger tissues

The effects of phosphorus fertilization on increase or decrease of various plant diseases is given in Table 4.

C. Effect of Potassium

Usually high application of potassium has been reported to reduce the incidence of several crop diseases. The mechanism of disease reduction by potash may be:

i) Direct - which include the reduction of pathogen penetration, multiplication, survival, aggressiveness, and rate of establishment in the host, and

ii) Indirect – which include promotion of wound healing, increase in resistance to frost injury and delay in maturity in some crops.

The effects of potash fertilization on increase or decrease of various plant diseases is given in Table 5.

Table 4 : Effect of Phosphorus fertilizers on crop diseases

Name of the disease and causal organism	Effect (increase or decrease)
A. Bacterial diseases	
1. Bacterial wilt of cotton (*Pseudomonas cryophylli*)	I
2. Fire blight of apple (*Erwinia amylovora*)	I
3. Blight of lima bean (*Pseudomonas syringae*)	D
B. Fungal diseases	
1. Bunt of wheat (*Tilletia* spp.)	D
2. Club root of cabbage (*Plasmodiophora brassicae*)	I
3. Downey mildew of grape (*Plasmopara viticola*)	D
4. Late blight of potato (*Phytophthora infestans*)	D/I
5. Wilt of tomato (*F. oxysporum* f. sp. *lycopersici*)	D/I
C. Nematode diseases	
1. Root knot of pea (*Meloidogyne incognita*)	I
D. Viral diseases	
1. Bean mosaic (*tobacco mosaic virus*)	D
2. Spinach virus (Cucumber virus-1)	I
3. Tobacco/tomato mosaic (*tobacco mosaic virus*)	I

D. Effect of Calcium

The principal role of calcium in host-pathogen relationship is the formation of calcium pectate in the cell wall thus making them resistant to degradation to facultative parasites like *R. solani*, *S. rolfsii*, *Botrytis cinerea* and *Erwinia amylovora*. Besides this, calcium also provides strength to bio-membranes and slows down the efflux of nutrients. The vascular wilts caused by *F. oxysporum* are also reduced by calcium especially in conjunction with NO_3-N. However, calcium favours black shank of tobacco (*Phytophthora tnicotianae*). It also makes potato tubers prone to the attack of common scab (*Streptomyces scabies*).

Table 5 : Effect of Potash fertilizers on crop diseases

Name of the disease and causal organism	Effect (increase or decrease)
A. Bacterial diseases	
1. Bacterial blight of rice (*X. oryzae* pv. *oryzae*)	D
2. Fire blight of pear (*Erwinia amylovora*)	D
3. Blight of lima bean (*Pseudomonas syringae*)	D
4.Angular leaf spot of cotton (*X. axonopodis* pv. *malvacearum*)	D
5. Soft rot of cabbage (*Erwinia. carotovora*)	D
B. Fungal diseases	
1. Ascochyta blight of chickpea (*Ascochyta rabiei*)	D
2. Paddy blast (*M. grisea*)	D
3. Downey mildew of grape (*Plasmopara viticola*)	D
4. Late blight of potato (*Phytophthora infestans*)	D
5. Early blight of potato (*Alternaria solani*)	D
6. Wilt of tomato (*F. oxysporum* f. sp. *lycopersici*)	D
7. Wilt of cotton (*F. oxysporum* f. sp. *vasinfectum*)	D
8. Grey mold of grape (*Botrytis cineria*)	D
9. Powdery mildew of cereals (*Blumeria graminis*)	D
10. Rust of cereals (*Puccinia* spp.)	D
11. Sheath blight of rice (*R. solani*)	D
12. Wilt of melons (*F. oxysporum* f. sp. *melonis*)	D
C. Nematode diseases	
1. Root knot of many crops (*Meloidogyne* spp.)	D
2. White tip of rice (*Aphelenchoides besseyi*)	D
D. Viral diseases	
1. Bean mosaic (*tobacco mosaic virus*)	D
2. Spinach virus (Cucumber virus-1)	I
3. Tomato mosaic (*tobacco mosaic virus*)	I
4. Tobacco mosaic (*tobacco mosaic virus*)	D

E. The role of silicon

The Graminaceous plants in general and wetland rice in particular are typical silicon accumulator. An increase in the silicon supply/content to rice plant causes corresponding decline in susceptibility to fungal diseases such as paddy blast, brown spot and sheath blight. Silicon has also been found to reduce the powdery mildew of cucurbits. It has also been shown to protect plants from some sucking insects like aphids.

The mechanisms of plant resistance due to silicon application include:

i. Formation of a physical barrier in epidermal cells against penetration hyphae and insect mouth parts.

ii. Preferential accumulation and fixation of silicon around the infection peg of fungal pathogens and

iii. Manipulation of the synthesis of phenolic compounds.

Other micronutrients i.e. Bo, Mn., copper play important roles in the synthesis of lignin and phenolic compounds, thus provides resistance to hosts against many pathogens and insects.

Organic manures/amendments and management of plant disease

Organic manures and amendments on one side provide all essential nutrients for improved plant health and on other side promotes the growth of several natural enemies of plant pathogens/nematodes and insects and in general, manipulate the soil biological environment in favour of crops and whole agro-ecosystem, thus promotes the natural/biological control of pests and diseases.

Future thrusts/concerns

Disease management through nutrient manipulation and site specific integrated plant nutrient supply system in coordination with other components of integrated disease management will be the future concern. The increased nutrient availability through amendments, or by changing the physical, and/or biological environment of soil, increased nutrients uptake by improved crop cultivars, the induction of systemic resistance by nutrient solubilizing rhizobacteria and the effect of changing climate on nutrient availability and plant diseases will be thrust area on the matter.

References

Engelhard, A.W. (1989). Soilborne Plant Pathogens: Management of Diseases with Macro- and Microelements. American Phytopathological Society, Scientific Publishers, Jodhpur, India.

Lu, Z.X., Heong, K.L., Yu, X.P. and Hu, Cui (2005). Effects of nitrogen on the tolerance of brown planthopper, *Nilaparvata lugens*, to adverse environmental factors. *Insect Science*, **12** (2): 121–128.

Marschner, H. (1995). Mineral nutrition of higher plants. Sec. Ed., Academic Press, London.

Rechcigl, N.A. and Rechcigl, J.E. (1997). Environmentally safe approach to crop disease control. CRC Press, Lewis Publishers, New York.

□□□

System Based Integrated Nutrient Management, 2012
© B. Gangwar & V.K. Singh (eds.), pp. 311-326
New India Publishing Agency, New Delhi (India)
e-mail : info@nipabooks.com; website : www.nipabooks.com

CHAPTER **22**

System Based Estimation of Cost of Production and Profit Maximization with Special Reference to Nutrients

S.P. SINGH

Cost is the value of the factors of production used in producing and distributing goods and services. The cost of a factor unit equals the maximum amount which the factor could earn in alternative employment. Broadly speaking, cost is the measure of what has to be given up in order to achieve something. Production costs play an important role in the decision of the farmers. Cost of production here means the exogenesis incurred per unit of output. In general, at given level of prices, a farmer can increase his farm income in two ways i.e. (i) by increasing production or (ii) by reducing the cost of production. The cost of production involves both cash cost and non cash cost items. Cash costs are incurred when resources are purchased and used immediately in the production process such as casual labour, fertilizer fuel etc. And non cash cost like depreciation on farm building and equipment and payment made to farmer himself or family labour, management and owned capital. Before estimating the cost of cultivation/production, we should know the concepts/terms are used in estimating the cost. Concept means idea underlying or general notion.

Cost concepts and items of cost

The cost of production of a crop is considered at three different levels viz., cost A, cost B and cost C. The concept of nine costs such as is followed by the Directorate of Economics and Statistics, Ministry of Agricultures Government of india in their cost studies are given below.

- Cost A_1 = All actual expenses in cash incurred in production by the owner.
- Cost A_2 = Cost A_1 + rent paid for leased in land.
- Cost A_2 + FL= Cost A_2 + imputed value of family labour.
- Cost B_1 = Cost A_1 + interest on value of owned capital assets (excluding land).
- Cost B_2 = Cost B_1 + rental value of owned land (net of land revenue) and rent paid for leased in land.
- Cost C_1 = Cost B_1 + imputed value of family labour.
- Cost C_2 = Cost B_2 + imputed value of family labour.
- Cost C_2* = Cost C_2 estimated by taking in to account statutory minimum or actual wage whichever is higher.
- Cost C_3 = Cost C_2* + 10% of cost C_2* on account of managerial function performed by the farmers.

The inputs items included under each category of cost are given below.

1) Cost A_1: Actual paid out costs for owner cultivator, inclusive of both cash and kind expenditure, which include following cost items:
   1) Hired human labour a) Male b) Female
   2) Total bullock labour a) owned b) hired
   3) Seeds
   4) Manures
   5) Fertilizers
   6) Insecticides and pesticides
   7) Irrigation charges
   8) Land revenue, cesses and other taxes
   9) Depreciation on capital assets
   10) Transport and marketing
   11) Interest on working capital
2) Cost A_2: For the tenant cultivator the rent paid by him to the land lord is another item of actual cost. Cost on account of all the above cost items exclusive of land revenue and other taxes, but inclusive of this rent is

referred to as cost A_2. So cost A_1 = cost on account of all the above items of cost except land revenue and other taxes + rent paid by tenant.

3) Cost A_2 + FL : It is defined as the sum of cost A_2 plus the imputed value of the holding's own labour/ imputed value of family human labour.

4) Cost B_1: It is the hypothetical interest that the owned capital invested in farm business would have earned if invested alternatively, is also considered as cost. Interest on owned capital represents imputed costs which are added to Cost A_1 to give cost B_1.

5) Cost B_2: If the amount invested in purchase of land would have been put in some other long term enterprise or in a bank, it would have yielded some returns or interest. But due to the investment of the amount in purchase of land, the farmer has to part with the returns or interest that he would have otherwise gained. And as such, this loss is considered as cost. It is called rental value of land which are added to Cost B_1 to give Cost B_2.

6) Cost C_1: It is the total cost of production, which includes all cost items excluding rental value of owned land (net of land revenue) and rent paid for leased in land., actual as well as imputed. The value of holding own labour is to be imputed and added to cost B_1 to work out cost C_1.

7) Cost C_2: It is the total cost of production, which includes all cost items, actual as well as imputed. The value of holding own labour is to be imputed and added to cost B_1 to work out cost C_1.

8) Cost C_{2*} : It is the total cost of production, which includes all cost items, actual as well as imputed. The value of holding own labour is to be imputed by taking statutory minimum or actual wage whichever is higher.

Variable costs, fixed costs and the total costs

The individual cost items included in the total cost can also be grouped into variable and fixed costs.

Variable costs + fixed costs = Total cost.

1) Variable costs

Variable costs are also called operational costs or direct costs or prime costs. The costs associated with the variable inputs are called variable costs. These costs very with the level of production. Higher the production more will be the variable costs. Lower the production less will be the variable costs. The farmers have to control over only the variable costs. He can decrease or increase or total avoid the variable costs. The following are the variable costs:

1. Family human labour
2. Hired human labour
3. Owned bullock labour
4. Hired bullock labour
5. Seeds
6. Manures and fertilizer
7. Insecticides and pesticides
8. Irrigation charges
9. Transport and marketing
10. Interest on working capital

2) Fixed costs

These are also called overhead costs or indirect costs. Fixed costs are those which do not change in magnitude the amount of output of the production process changes. Fixed costs are incurred even when the production is not undertaken. These costs, when once incurred, do not vary with the changes in the output and hence have no bearing upon decision to increase or decrease production. The following are the fixed costs

1. Land improvement
2. Maintenance of farm buildings, implements, machinery wells, livestock etc.
3. Depreciation on capital assets
4. Interest on fixed capital
5. Land taxes
6. Rent or rental value of owned land.

Cash and non cash costs

In common usage, the term cost of production is used in the sense of money costs or expenses of production. In economics however, cost means total efforts, sacrifices or exertion involved in the production of a commodity. So the cost of reduction involves both the cash cost items and non cash costs items.

Cash costs

Cash costs or out of pocket costs are incurred when the resources are purchased and used immediately in the production process.

Non cash costs

Non-cash costs or imputed costs are those for which no direct cost expenses are made and for which if the money costs are to be derived, some values must be imputed.

Cost of cultivation and cost of production

The terms cost of cultivation and cost of production are used as synonyms for the purposes of cost study. However, a distinction can be made between the two, restricting the content of the cost of cultivation to include factor costs up to the stage of gathering the harvest and that cost of production to include factor costs up to the stage of market the produce.

Per unit cost of production

Cost of production is to be worked out as cost per unit of area and production i.e. per hectare and per quintal/ tones.

a) Per hectare cost of production

$$= \frac{\text{Total cost - value of by produce}}{\text{Area under the crop in ha.}}$$

b) Per quintal ton cost of production

$$= \frac{\text{Total cost- value of by produce}}{\text{Quantity of main produce in quintals/ tones.}}$$

Evaluation of farm assets, estimation and allocation of different costs.

1) Human labour : *they are of the following types.*

i) Family human labour

a) permanent labour

b) family labour

c) exchange labour

d) gift labour

ii) Hired human labour

a) Permanent labour

The total annual wages earned by a permanent labour include:

1. Cash payments
2. Kind payments evaluated at harvest price
3. Purchase prices of pre-requisites like clothes, blankets, shoes etc. and
4. Rent prevalent in the village for the same type of accommodation as is provided to 'Saldar'.

His per day wage rate can be calculated by dividing the total annual wages earned by him by his total work days on the farm.

Allocation : Cost on account of a permanent labour for a particular crop can be worked out by multiplying wage rate by total labour days he worked on that crop.

b) Family labour

Cost on account of family labour should be calculated on the basis of maintenance cost of family. But if the data are not available, the wage rates of family labour may be calculated as follows.

i) *Men*: The average wage rate per day earned by permanent labour is to be applied. In case where no such permanent labour is employed, the cost is to be calculated on the basis of operation wise wage rates for hired male labour.

ii) *Women*: At the wage rates paid to hired female labours

   a) *Exchange labour*: At the rate of family labour

   b) *Gift labour*: At the rate of family labour.

   c) *Hired human labour*: At actual wages paid.

2) Bullock labour

a) Owned bullock labour

In case of draught animals, the total useful service life is treated as of 13 years. The value of one year old calf appreciates at the rate of 1: 3: 5 upto the 3rd year, it remains constant for the 4th and 5th year and thereafter depreciates at the rate of 12% every year. In case of purchased animals, the purchase cost and remaining years of useful service are to be considered to calculate depreciation.

This depreciation on the value of drought animals and the maintenance cost constitute the total cost of owned bullock labour. The maintenance cost includes cost of feeds and concentrates, both purchased (at actual cost) and farm production (at harvest price), miscellaneous expenses e. g. veterinary charges, showing, ropes and neck-straps etc. and the wages for care of the animals.

The value of manure proucdced and wages earned on other frarms by the draught animals are to be deducted from this total cost and the per day owned bullock labour cost is worked out as follows:

Per day owned bullock labour cost =

Total cost of bullock labour- Value of manure produced + Wages earned during the year

Total number of owned bullock labour days

Allocation : The total cost is allocated to different crops on the basis of bullock labour required for each crop. In case the data on maintenance cost are not available, the hire charges may be considered.

b) Hired bullock labour

At actual hired rate. While considering the hiring rate of bullocks, the charges of human labour are to be excluded as they are already included under human labour.

3. Seeds

a) *Purchased* : At Actual cost

b) *Farm- produced* : At sowing prices
   This also includes cost of fungicides used for seed-treatment.

4) Manures

a) *F.Y. M. and Compost:*
   i) *Purchased*: At actual cost including transport charges.
   ii) *Farm produced*: At the rate prevailing in the village.

b) *Green manuring*: Total cost on all operations and inputs up to the stage of burying it into the soil, is to be considered and allocated on the crops benefited, on area basis.

5. Fertilizers

At actual cost including transport charges.

6. Insecticides and Pesticides

Actual cost incurred and hiring charges of appliances, if paid.

7. Irrigation charges

a) Well irrigation

It includes the depreciation on the original costs of sinking the well and installing oil engine or electric motor and pump-sets plus cost on account of minor repairs/ replacements, fuel and lubricants or electricity charges. Rate of depreciation:

I) Pacca construction substantial 2.5%

II) Kaccha construction (less substantial 5.0%

III) Non-substantial 10 to 20%

b) Canal irrigation

Actual charges paid to the irrigation department are to be considered. Human and bullock labour used for irrigation operation is not to be included under this item, irrespective whether crop is irrigated by well or canal.

Allocation : Total irrigation cost to be apportioned on the irrigated crops only, on hectare-irrigation basis.

Land revenue, cesses and other taxes

a) *In case of owner cultivator*: The actual amount paid as land taxes is to be accounted for.

b) *In case of tenant cultivator*: Net annual rent paid to the landlord in cash or in kind.

Allocation : This cost is allocated to different crops on the hectare-months basis.

Under this cost item, depreciation on the present values of residential building (only 1/3rd value is to be considered), farm house, cattle shed, engine shed, grain store, fencing, tools and implements and machinery, hiring charges and cost on repairs are to be taken into consideration. Depreciation rates, in general, are as follows:

Depreciation

a) Farm buildings

I) Substantial 2.5%

II) Less substantial 5.0%

III) Temporary 10 to 20%

b) Tools like sickles, baskets cropes etc.100% costing less than Rs. ten

c) Implements

i) Wooden implements 20%

ii) Iron implements 10%

iii) Bullock cart 15%

Machinery

i) Fodder cutter 15%

ii) Cane crusher 10%

iii) Oil engine 10%

iv) Tractor 7%

Allocation : The total cost of depreciation, hiring and repairing charges so calculated is to be apportioned on different crops grown during the year, on hectare months (area*length of season) basis. When a particular implement is hired for a particular crops its charges are to be allocated to that/those crops only. Depreciation on wells and drought animals is not to be considered as it is already accounted for under owned bullock labour and irrigation costs.

10) Transportation and marketing

This cost includes:

a) Expenses on packaging

b) Transportation charges

c) Marketing charges

11) Interest on working capital

The total of all the paid out costs such as costs on hired human labour, bullock labour , seeds, etc. the interest on this amount @ prevailing bank rate for the half period of crop life is to be considered.

12) Interest on fixed capital

Interest on the present value of fixed capital assets other than land has to be charged @ 10 percent /annum.

Allocation : This total amount of interest on fixed capital is to be allocated to the different crops on hectare month basis.

13) Rental Value of land

It is valued at prevailing market rate of the area.

System based cost of cultivation and net return

Table 1 gives the per hectare cropping system wise and size group wise cost and returns structure of the Rice-wheat and sugarcane–ratoon cropping systems. On an average, the per hectare cost of cultivations was Rs. 48280/- for rice-wheat and Rs. 37042/- for sugarcane cultivation (2004-05). Size group wise comparisons with respect to cost per hectare of these crops showed that the cost per hectare was relatively higher for marginal farmers in rice-wheat and large farmers in sugarcane crop. However, cost of production/quintal was lower for large farmers in rice-wheat and medium farmers in sugarcane production. The net return at cost- c was highest for marginal farmers for rice-wheat and for medium farmers in case of sugarcane-ratoon cropping system. Size group wise comparison of cost and return per hectare for individual cropping system revealed that although there was no uniform trend, with respect to size of farm but there was a tendency for these costs to decrease with increase of size of farm except sugarcane - sugarcane cropping system in case of small and large farms. Output input ratio indicated that there is no considerable difference in profitability (at cost c).

The negligible higher output input ratio for large farmers was due to lower amount of input use in rice-wheat cropping system. Similar observations were recorded for medium farmers of sugarcane-ratoon cropping systems. The return per rupee invested were Rs. 1.31 in case of rice-wheat and Rs. 1.38 for sugarcane-sugarcane cropping system. The estimate of returns of the farmer's family labour (imputed value) ranged from Rs. 1711.5 in sugarcane (Medium farmers) to 3074.4 in rice-wheat (Marginal farmers).

Profit maximization : Production function approach

In order to measure the contribution of specific factor in combination with other factors which are responsible for the change in the level of output, multiple regression analysis was used in the form of Cobb Douglas production function. The function was used in this study since the elasticity coefficients are free from the unit of measurement and factor ratio. This function is preferred because of the computational ease and theoretical fitness to agriculture data. Before fitting the function, the zero order correlation matrix were estimated to test multicolinearity. The algebraic form of the Cobb-Douglas production function used in the study was as under.

$$Y_i = b_o\, x_1^{b1}\, x_2 b^2{}_4{}^{b4}\, x_5^{b5}\, x_6^{b6}\, x_7^{b7}\, x_8^{b8}\, u_i$$

The variables considered for the production function analysis are:

Y = Annual output of cropping system in quintals.

X_1= Area in ha.

X_2= No. of human labour employed (mandays)

X_3= Tractor use in hours

X_4= No. of irrigation

X_5= Seed in Kg for rice wheat and in quintals for sugarcane -sugarcane

X_6= FYM in quintals

X_7= Fertilizer in kg (n+p+k and zink sulphate)

X_8= Insecticides/pesticides in rs.

BI = 1,2,3,........8 are parameters, b_0 = Intercept and ui = error term

The function was estimated using the ordinary least square (OLS) method. The step down method was used for final retention of variables.

Optimum level of resource use

The profit maximization criteria enable us to work out the optimum level of input by solving the first order marginal condition and the formula worked out is given below

$$MVP_i = \frac{bi \ \bar{Y} \ P(Y)}{\bar{X}_i}$$

Optimum value = $MVP_i = P(X_i)$

$$X_i = \frac{[b_i(\bar{Y})] \ P(Y)}{[P(X_i)]}$$

$$t = \frac{MVPi - price(i)}{SE \ of \ MVPi}$$

$\bar{X}$ = Bar indicates inputs at geometric mean level

$\bar{Y}$ = Estimated output when all factors are fixed at their geometric mean

bi = regression coefficient of x_i

MVP = Marginal value product of X_i

In this study we measure both dependent variables in the function in physical unit. One advantage of this measurement is that resource productivities worked out from the function will be free from the impact of fluctuation in prices. Yield of rice (main produce and by produce) and wheat straw were converted in to wheat equivalent yield in quintals. Accordingly Sugarcane top was also converted into sugarcane main produce in quintals.

Table 1 : Cost of Cultivation, output/return from Rice-wheat and Sugarcane –sugarcane cropping System in mid western plain of U.P. (2004 - 05) (ha^{-1})

Particulars	Rice - Wheat					Sugarcane - ratoon				
	Marginal	Small	Medium	Large	All	Marginal	Small	Medium	Large	All
I. Costs (Rs)										
(i) Cost A	30744.0	28718.9	27615.4	25293.3	28092.9	17240.0	17353.0	17115.0	17388.4	17313.4
(ii) Cost B	47676.4	46240.2	45172.6	42792.9	45470.5	35240.0	35300.0	35089.9	35380.6	35310.1
(iii) Cost C	50750.8	49112.1	47934.2	45322.2	48279.8	36964.0	37035.3	36801.4	37119.4	37041.5
II. Output/Return										
(i) Main produce (qtls)	76.9	76.6	72.7	73.2	74.9	469.8	458.5	460.9	450.3	455.8
(ii) Main produce income (Rs.)	57013.8	55389.2	52683.5	53019.2	54526.4	48802.4	47056.4	48901.9	47082.3	47715.0
(iii) By produce Yield in (qtls)	76.9	70.7	70.1	73.2	72.7	91.2	88.0	91.4	88.0	89.2
(iv) By produce income (Rs.)	9288.7	8787.1	8823.2	7120.6	8504.9	3648.8	3518.2	3656.2	3520.2	3567.5
(v) Gross Return (Rs.)	66302.5	64176.3	61506.8	60139.8	63031.4	52451.2	50574.6	52558.1	50602.5	51282.5
(vi) Gross Yield in qtls	91.5	88.2	84.8	82.7	86.8	490.2	472.7	491.2	472.9	479.3
(vii) Net Return at cost A (Rs.)	35558.5	35457.4	33891.4	34846.5	34938.5	35211.2	33221.7	35443.1	33214.1	33969.1
(viii) Net Return at cost B (Rs.)	18626.1	17936.1	16334.2	17346.9	17560.8	17211.2	15274.6	17468.2	15221.9	15972.3
(ix) Net Return at cost C (Rs.)	15551.7	15064.2	13572.6	14817.6	14751.5	15487.2	13539.3	15756.7	13483.1	14241.0
III. Per qtl cost of prod at B (Rs.)	521.1	524.5	532.9	517.6	524.0	71.9	74.7	71.4	74.8	73.7
IV. Per qtl cost of prod at C (Rs.)	554.7	557.1	565.5	548.2	556.4	75.4	78.4	74.9	78.5	77.3
V. Output - Input ratio	1.31	1.31	1.28	1.33	1.31	1.42	1.37	1.43	1.36	1.38
VI. Intensity of investment (Cost B-A)	16932.4	17521.4	17557.2	17499.6	17377.7	18000.0	17947.1	17975.0	17992.2	17996.8
VII. Imputed value Of family labour C-B	3074.4	2871.9	2761.6	2529.3	2809.3	1724.0	1735.3	1711.5	1738.8	1731.4

Source : S.P. Singh *et al.*

Resource productivities

It is seen from the table 2 that the adjusted coefficient of multiple determination of productions functions in all the cases were high and thus the independent variables included in the production function explained about 94-99 percent variations in the yield of all size categories across cropping system. The observed ratio indicated that all the production function were highly significant. From the analysis it can be inferred that the important input which made a positive and significant contribution to the crop yield were land for marginal farmer of rice-wheat cropping system and human labour (man days) for medium and over all level in sugarcane-ratoon cropping system. Further the influence was highly significant for all categories of rice-wheat system except marginal farmers, tractor (hrs) for medium farmers. Irrigations have positive and significant influence on small farms of sugarcane - sugarcane and large farms of rice-wheat system. Quantity of seed has positive and significant influence for all categories of farmers in both the systems. FYM applied had positive and significant influence on small and over all farmers of rice-wheat cropping system. Fertilizer application was having positive and significant influence for all farms except medium and large farms of sugarcane-ratoon cropping system and large farms of rice-wheat cropping system.

The non significant input variables in some cases may be the reflection of lack of timely availability of irrigations water, timely non availability of inputs and lack of managerial skill, soil fertility etc. The return to scale was also compared and the same are presented in Table 2. They generally indicated constant returns to scale.

Productivity and resource use efficiency

Under perfect competition a firm will be maximizing its profit by equating marginal value product (MVP) of each factor with its cost. The MVP of various inputs namely human labour, seed manures and fertilizers were worked out at their geometric mean level in order to compare their productivity. Since there is open competition, nobody will accept less than marginal productivity. For the estimations of MVP the annual average wholesale market prices were used. The price paid for a unit of input was taken as the factor cost. These costs facilitate the shifting of resources from the less remunerative uses to the more profitable ones. A dynamic farming cannot function without this shifting.

The geometric mean level of input with respect to all farms, the estimated MVP of different inputs in the functions and the factor prices, their ratio and value are given in table 3. In general the MVP of variable inputs ranged from Rs. 12.56 for fertilizer to Rs. 272.86 for seed in case of sugarcane-sugarcane and Rs. 7.03 for fertilizer to Rs. 209.03 for manures in case rice-wheat cropping system.

Table 2 : Regression co-efficient, standard error and co-efficient of multiple determination (R^2adj) for production function of Rice-wheat and Sugarcane –sugarcane cropping System in western plain zone of U.P.

Type of farm	Sample size	Inter-cept	Area in ha X1	Labor M. days X2	Tractor Hrs X3	No. of Irrigation X4	Seed in Kg X5	FYM Qtls x6	Ferti lizer Kg X7	Insec-ticides Rs. x8	Adj R^2
				Sugarcane - Ratoon cropping System							
Marginal	12	1.9411***	NS	0.1565	0.1948	NS	0.3746***	NS	0.2932**	NS	0.994
		-0.3348		-0.119	-0.0908		-0.0996		-0.0966		
Small	19	2.0115***	NS	NS	NS	0.1089**	.7372***	NS	0.1563**	0.1183	0.998
		-0.343				-0.0343	-0.616		-0.569	-0.0075	
medium	21	1.6487**	NS	0.4812**	0.0822	NS	0.3092**	NS	0.145	NS	0.993
		-0.4298		-0.1251	-0.0646		-0.1751		-0.1357		
Large	18	1.3265**	NS	NS	NS	NS	0.3409*	NS	0.258	0.0071	0.995
		-0.4142		-0.1181			-0.1659		-0.1431	-0.0088	
Overall	70	1.5982***	NS	0.2000**		NS	0.3007**	NS	0..1400**	0.0083	0.997
		-0.1636		-0.0517			-0.0789		-0.0556	-0.0052	
				Rice - Wheat cropping System							
Marginal	22	0.5486	0.1572*	NS	NS	0.1361	0.1940***	NS	0.5388***	NS	0.965
		-0.3525	-0.058			-0.1221	-0.0455		-0.0666		
Small	27	1.3548**	NS	0.6298***	NS	NS	0.2498**	0.0216*	0.5320**	NS	0.945
		-0.4252		-0.152			-0.0783	-0.0104	-0.1449		
medium	28	0.268	NS	0.3549***	0.2100*	NS	0.2880**	NS	01048*	0.0137*	0.974
		-0.352		-0.0924	-0.0902		-0.0805		-0.0382	-0.0052	
Large	25	1.929***	NS	0.4260**	NS	0.4842**	0.3851***	NS	0.1806	NS	0.968
		-0.4552		-0.1492		-0.1381	-0.0798		-0.0971		
Overall	102	1.0477***	NS	0.5905***	0.1402	NS	0.1201**	0.0138*	0.1390**	NS	0.973
		-0.2151		-0.085	-0.0888		-0.0534	-0.0058	-0.0449		

Source : S.P. Singh et al

Figures in parenthesis are standard error of respective variable ***significant at 0.01 percent level, ** significant at 1 percent level, significant at5 percent level and NS Non significant

Table 3 : Regression co-efficient, Standard error, Marginal value product (MVP) to factor cost ratio Optimum value of Resources used for different cropping systems at over all level in western plain zone of U.P.

Resources	Regression coefficient	Standard error	G. Mean	Price per unit	MVP	Ratio of MVP	t value	Present Level	Optimum level	Input Requirement ha^{-1}
Sugarcane – ratoon cropping system (0.81)										
Land in ha										
Human labour (Man days)	0.2000**	0.0517	117.39	80	50.66	0.63	-42.843	117.39	74.34	92.17
Seed in Qtls	0.3007**	0.0789	32.77	140	272.86	1.95	23.526	32.77	63.87	79.18
Fertilizer Kg	0.1400**	0.0556	331.50	7.75	12.56	1.62	17.401	331.50	537.11	665.90
Rice - Wheat cropping System (2.11)										
Land in ha										
Human labour (Man days)	0.5905**	0.085	238.91	80	150.01	1.88	75.543	238.90	447.94	211.91
Seed in Kgs	0.1201**	0.0534	164.28	26	44.38	1.71	25.356	164.28	280.41	132.66
Manures in Qtils	0.0138*	0.0058	3.91	30	209.03	6.97	343.203	3.91	27.27	12.90
Fertilizer in Kgs	0.1390**	0.0449	1200.78	7.75	7.03	0.91	-0.071	1200.78	1088.50	514.95

Source: S.P. Singh et al,***significant at 0.01 percent level, ** significant at 1 percent level and significant at5 percent level.
Figures in parenthesis is the average area per household under the system

The ratio of MVP of human labour and the labour wage was less than one for sugarcane-sugarcane and more than 1 for rice-wheat cropping system indicating sufficient scope for profit maximization by increasing the use of this resource at over all level in Rice -wheat system. As regards to seeds, the ratio of MVP to its price was more than 1 in sugarcane and rice-wheat cropping system indicating a good scope to increase the use of this input. In case of manures of rice-wheat cropping system the ratio MVP to its price is more than 6 indicating very high scope for using this input for profit maximization. The MVP of fertilizer use and its prices ratio was more than 1 for sugarcane -ratoon system indicating very good scope for increasing the use of NPK in sugarcane production for profit maximization, but the ratio was less than one in case of rice-wheat system indicating scope to reduce the use of this input in rice-wheat cropping system.

Optimum level of input

The estimated optimum levels and their present value in respect of human labour seed, manures and fertilizer for both the systems are presented in table 3. The optimums levels in case of other resources have not taken as their regression coefficient at over all level were not significant. It is evident from the table that the present level of human labour was much lower in case of rice-wheat cropping system than the optimum level. But the human labour use in sugarcane was higher to optimum level. In case of seed its use is significantly lower than the optimum level. Manures and fertilizer were also used below optimum level.

Reference

S. P. Singh, B. Gangwar and B. M. Garg (2007). Input use efficiency - A comparative study of rice-wheat and sugarcane-ratoon cropping system in mid western plain of U.P. *Agricultural. Situations in India* **LXIV** (8): 377-383.

□□□

System Based Integrated Nutrient Management, 2012
© B. Gangwar & V.K. Singh (eds.), pp. 327-338
New India Publishing Agency, New Delhi (India)
e-mail : info@nipabooks.com; website : www.nipabooks.com

CHAPTER **23**

Integrated Nutrient Management in Seed Spice Based Cropping System

O.P. AISHWATH

Introduction

Seed spices include group of annuals whose dried fruit or seeds are used as spices. These are characterised by pungency, strong odour, sweet or bitter taste. Coriander, cumin, fennel and fenugreek occupy largest area among the seed spices. However, ajowan, dil, celery, anise, nigella and caraway having secondary place as far as area and production is concerned. Major area under seed spices is in Rajasthan and Gujarat, and other seed spices producing states are Chattisgarh, Madhya Pradesh, Andhra Pradesh, Tamil Nadu and Uttar Pradesh. The seed spices account for about 36% and 17% of the total area and production of spices in country. The export of seed spices from India during 2008-09 was 4.71 thousand tonnes worth Rs. 54914.8 million rupees. The per cent export growth rate during 1995-96, 2000-01, 2005-06, 2007-08 and 2008-09 was 12.0, 11.0, 8.0, 36.0 and 19.0% in terms of value and 13.0, 3.0, 8, 13.0 and 6.0 in respect of quantity, respectively.

Integrated plant nutrient management includes use of organic manure/ compost, bio-fertilizer, chemical fertilizer, green manuring, residue management, legume based cropping system, use of nutrient-responsive varieties, proper method and time of organic manure and fertilizer application, soil and water management to minimize the nutrient losses occurring through volatilization, denitrification, runoff and leaching. Application of plant nutrients in proper balance form is also a part of Integrated Plant Nutrient Management (IPNS) system. Supply of nutrients to seed spices in appropriate quantities and at the correct times is essential for economically and environmentally sustainable agriculture. Soil organic matter, crop residues and manures play a vital role in the supply of macro and micronutrients and the transformation between the various organic and inorganic forms often control availability, both for plant uptake and loss to the environment (Aishwath and Vashishtha, 2008; Lal, *et al.*, 2009).

Distribution, area production and productivity trend in seed spices

The seed spices are grown in different parts of the world covering mainly Mediterranean region, South Europe and Asia. The seed spices producing countries other than India are Morocco, Russia, Bulgaria for Coriander; Turkey, Iran, Egypt for Cumin; Egypt, China, Romania, Russia for Fennel; Morocco, Bulgaria for Fenugreek; Iran, Egypt for Ajowain ; Germany, Hungary for Dill; Southern France, China for Celery; Bulgaria Cyprus, Germany, Russia for Aniseed; Pakistan, Sri Lanka, Egypt for Nigella. Many of the countries in South America, Europe, South Africa and Asian continent are also likely to enter in the production of seed spices. Almost all of the seed spices crops are cultivated in India and has got the privilege to be called as the largest seed spices producing country in the world. The prominent states where seed spices produced are Rajasthan and Gujarat while other states where commonly grown are Haryana, Panjab, M.P., Bihar, U.P., West Bengal, Orissa, Tamil Nadu and Karnataka. Almost all the states in India grow one or more seed spices. But major growing belt spread from arid to semi-arid regions covering large areas in Rajasthan and Gujarat. Last five years data for area, production and productivity of coriander, cumin, fenugreek and ajwain did not indicate any definite trend from 2002-03 to 2006-2007. Moreover, it seems in increasing trends with inconsistent rise and fall in area, production and productivity. However, area production and productivity of fennel increased with increase in time. In the recent past (2008-09), there is a quantum jump in area, production and productivity of coriander, cumin, fennel and fenugreek. However, the productivity of fenugreek was lower in 2008-09 than 2007-08 (Aishwath *et al.*, 2010).

Major prevailed cropping system for seed spices

Research on integrated nutrient management on seed spices in cropping system is very merger. However, farmers used to take these crops in various crpping system and crop rotations. Besides that, slow growth rate of seed spices, successfully has been intercropped with chilli, potato, sugarcane and pigeon pea (cv. Sharad), pea for pods, radish, palak and methi for leaves. Pareek and Gupta (1984) carried out several trials on crop rotation of anise, coriander and methi with medicinal plants and they reported that senna-basil-coriander based crop rotation was more profitable (Table 1). However, some major prevailing cropping systems are as follows.

Punjab conditions

- Maize-potato-coriander,
- Summer moong-coriander
- Rice-coriander.
- Senna-anise
- Senna- coriander
- Senna-fenugreek

Rajasthan, Gujarat and other conditions

- Green gram/black gram-coriander
- Cluster bean/cowpea- coriander
- Cluster bean-coriander-summer maize
- Maize/pearl millet-fenugreek
- Sesame-fenugreek
- Sesame-fenugreek-summer maize.

Table 1 : Performance of anise, coriander and menthi insenna and sacred basil based cropping system.

Crop-sequence	Yield of crops (q ha^{-1})		Cost of cultivation (Rs ha^{-1})		Grooss income (Rs ha^{-1})		Net profit (Rs ha^{-1})		
	Kharif	Rabi	Kharif	Rabi	Kharif	Rabi	Kharif	Rabi	Total
Senna-basil-anise*	69.25	5.80	2500	2000	6925	4640	4425	2625	7050
Senna-basil-methi	72.60	10.75	2500	2500	7260	4030	4760	1530	6290
Senna-basil-Coriander	72.70	12.25	2500	2500	7270	6125	4770	3625	8395

(Adopted from Pareek and Gupta, 1984)

Components of nutrient management

Aim of integrated nutrient management to develop and promote options for efficient nutrient supply to the seed spices for their production maximization with minimal environmental impact. Integrated nutrient management includes- chemical fertilizers, composts, FYM, vermicompost, green manuring, crop rotations, biofertilizers, crop residues, intercropping, carpet, wool and farm waste and other industries by-products etc as its components.

Response of seed spices for various integrated inputs

a) Coriander (Coriandrum sativum)

Coriander is a tropical crop and generally taken as a *Rabi* crop. A dry and cold weather favors high seed production. Irrigated coriander can be cultivated in almost all types of soils by giving sufficient organic manure. Unirrigated crop survives only on heavier types of soils having good water retention capacity. The effects of N fertilizer (0, 30, 60 or 90 kg ha^{-1}) and biofertilizer (*Azotobacter*, *Azospirillum* and *Azotobacter* + *Azospirillum*) on the yield and quality of *Coriandrum sativum* (cv. RCr-435), was positive up to 60 kg N ha^{-1} and biofertilizers (Kumar, *et al.*, 2002). Inoculation with *A. chroococcum* or *A. brasilense*, combined with *G. mossese* (VAM), gave significant increase in vegetative growth, total carbohydrates, photosynthetic characteristics, essential oils, seed parameters, N, P and K contents and linalool (Abou and Gomaa, 2002). Manure *et al.* (2000) reported that yield attributes, se :d, essential oil yield and oil content of coriander (*Coriandrium sativum* L.) enhanced by the graded levels of nitrogen, sulphur and zinc nutrition in red sandy loam soils. Sharma *et al.* (1996) recommended application of 10-20 tonnes ha^{-1} of well decomposed FYM or compost at the time of field preparation. In addition to this, 20 kg N, 30 kg P_2O_5 and 20kg K_2O ha^{-1} has been recommended for un-irrigated coriander in the form of fertilizers at the time of sowing. Choudhary and Jat (2004) reported higher yield and net return in coriander at 100% inorganic N + *Azospirillum* + 5 t FYM (Fig.1). Whereas, Prabu *et al* (2000) reported that seed yield of coriander was significantly higher with 25% recommended dose of fertilizer + FYM @ 10 t ha^{-1}+ *Azospirillum*+ VAM over other combinations of nutrient sources and individual source of nutrient at recommended dose of fertilizer. Integrated use of fertilizer improved the yield of coriander, cumin fennel and fenugreek (Sharma, 2010).

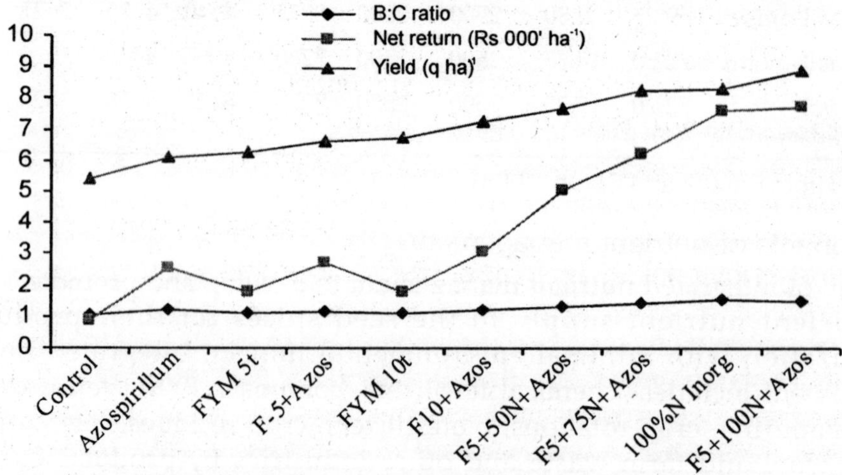

Fig. 1 : Yield net return and B:C ratio in coriander with manures and fertilizer inputs

b) Cumin (Cuminum Cyminum)

Cumin is a tropical plant can be cultivated as a *Rabi* crop in areas where atmospheric humidity is low during flowering and seed formation. Cumin thrives well in deep friable, medium and well-drained soil under mild climate. For cumin application of 15-20 tones FYM is sufficient under rain-fed condition whereas additional 30 kg N is recommended under irrigated conditions (Peter, K.V. *et al.*, 2000). Patel *et al.* (2004) reported that application of recommended dose of nitrogen through mustard-cake and fertilizer at ratio of 1:1 recorded the maximum cumin seed yield (869 kg ha^{-1}), net return (Rs. 70,765 ha^{-1}) and benefit: cost ratio (4.4:1) as give in Table 2. Cumin grown on loamy sand soils with 30 kg N+ 20 kg P_2O_5+ 30 kg K_2O ha^{-1} reduced the wilt incidence (Champawat and Pathak, 1988). Cumin crop grown in sequence with pearl millet responded significantly to the recommend dose of 30 kg N and 15 kg P_2O_5 ha^{-1} in loamy sand soils at Jodhpur (Yadav and Poonia, 1996). Application of increasing level of N up to 30 kg ha^{-1} and $ZnSO_4$ up to 20 kg ha^{-1} significantly increased the seed yield of cumin in loamy sand soils at Jobner (Meena, and Chaudhary, 1998).

Table 2 : Yield and economics of cumin with organic manures and fertilizers

Treatments	Seed yield (q ha^{-1})	Net return (Rs ha^{-1})	B:C ratio
RD-N in the form of caster-cake	6.53	49,070	3.02
RD-N in the form of FYM	6.70	50,890	3.16
RD-N in the form of caster-cake + inorganic fertilizer @ 1:1	7.23	56,585	3.60
RD as in organic fertilizer (NP)	7.59	60,383	3.89
RD-N in the form of FYM + inorganic fertilizer @ 1:1	7.69	61,245	3.19
RD-N in the form of inorganic (N)	7.46	59,372	3.90
RD-N in the form of mustard-cake	7.74	60,330	3.53
RD-N in the form of mustard-cake + inorganic fertilizer @ 1:1	8.69	70,765	4.39
CD (P = 0.05)	78.00	--	

(Adopted from Patel *et al.*, 2004)

c) Fennel (Foeniculum vulgare)

Fennel is a cold weather crop and grows well in fairly mild climate either by seeding or transplanting. Moreover, transplanting is most successful in various cropping system and relay cropping. A cold whether favours higher seed production and it thrives well in fertile and well drained loamy clay soil. In fennel, application of 5 tones ha^{-1} of well decomposed FYM or compost at the time of field preparation, 90 kg N ha^{-1} in three equal splits viz. first as basal dose with 40 kg P_2O_5 ha^{-1} at the time of sowing, second and third at 30 and 60 days after sowing with irrigation which produced maximum seed yield of 13.77 q ha^{-1} as against 9.30 q ha^{-1} with no nitrogen (Sharma *et al.*, 1996). (Chaudhary and Jat, (2004) reported that combine application of 100 % inorganic N (90 kg ha^{-1}) + *Azospiriullum* + 5 t FYM ha^{-1} gave maximum seed yield of 12.0 q ha^{-1} which remained at par with 75 and 50% N through inorganic source + *Azospirillum* + 5 t FYM ha^{-1}, *Azospirillum* + 10 t FYM and 100% inorganic N alone but proved superior on applied 5 t FYM+ *Azospirillum,* 5 and 10 t FYM alone, *Azospirillum* alone and control (Fig. 2). On loamy sand soils at Jobner, soil application of 20 kg $ZnSO_4$ ha^{-1} along with recommended NPK increased the fennel seed yield by 8.14%over control (Sharma, 1998).

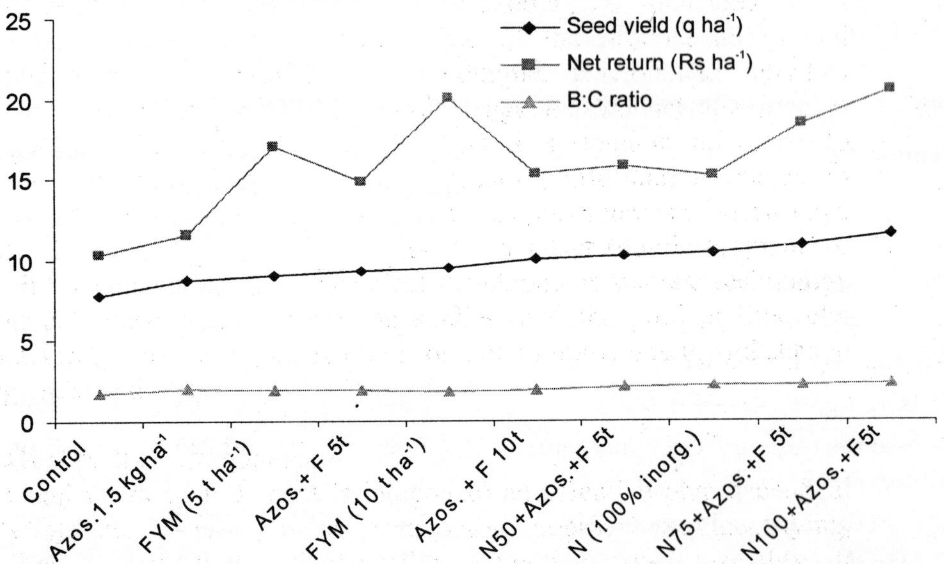

Fig. 2 : Yield, net return (in thousand) and B:C ratio in fennel with various inputs.

d) Fenugreek (Trigonella foenumgraecum)

Fenugreek is grown on well-drained loamy soils with irrigation. The crop can be grown on heavy black cotton soils as rainfed crop. It is fairly tolerant to salinity, acidity and responds favourably to fertilizer application in all types of soil. Growth and seed yield enhanced with 15 t ha^{-1} of FYM and 40 kg ha^{-1} phosphorus also proved superior for growth parameters and yield of fenugreek (Khiriya, *et al.*, 2001). Combined inoculation of Rhizobium+PSB resulted in significantly higher number of nodules plant^{-1}, plant height and dry matter accumulation over individual inoculation and control (Purbey, and Sen 2004). Highest fenugreek seed yield (12.56 q ha^{-1}), net returns (Rs 9761 ha^{-1}) and B:C ratio (1.76) recorded with 100% inorganic N alone closely followed by 100% inorganic N+ Rhizobium @1.5 kg ha^{-1}+ 5 t FYM ha^{-1} as given in Table 3, whereas the highest straw and biological yields of 33.91 and 46.43 q ha^{-1}, respectively were recorded in inorganic N (100%)+ Rhizobium @ 1.5 kg ha^{-1} 5 t FYM ha^{-1} (Jat and Chaudhary, 2004. and Chaudhary and Jat, 2004). For achieving optimum yield of fenugreek, it was recommended to apply 15 tonnes of FYM ha^{-1} and 25 kg N, 25 kg P_2O_5 and 50 kg K_2O ha^{-1} are optimum for sustainable production of fenugreek (Rethinam and Sadanandan, 1994).

Table 3 : Yield and economics of fenugreek with organic and inorganic fertilizers

Treatments	Seed yield (q ha^{-1})	Net return (Rs ha^{-1})	B:C ratio
Control	9.64	4,905	1.39
Rhizobium (@ 1.5 kg ha^{-1})	10.27	6,014	1.48
FYM (5 t ha^{-1})	10.59	5,365	1.39
Rhizobium + FYM (5 t ha^{-1})	11.01	6,096	1.44
FYM (10 t ha^{-1})	11.05	4,943	1.33
N (75%) *Rhizobium* + FYM (5 t ha^{-1})	11.76	7,146	1.51
N (50%) *Rhizobium* + FYM (5 t ha^{-1})	11.77	7,264	1.52
Rhizobium + FYM (10 t ha^{-1})	11.78	6,232	1.29
N (100%) *Rhizobium* + FYM (5 t ha^{-1})	12.52	8,414	1.60
N (100% inorganic)	12.56	9,761	1.76
CD (P = 0.05)	0.94	--	--

Source: Choudhary and Jat, 2004

e) Ajwain (Trachyspermum ammi)

Ajwain is considered as Rabi crop in most of the growing areas. However, Ajwain can be grown successfully in Kharif season also. Some of the pockets in Rajasthan ajwain cultivation is going on as a *Kharif* crop. Besides the light soils of Rajasthan, it gives better yield in heavy textured soils of Andhra Pradesh. Ajwain responds to both N and P and the maximum seed yield of 15.48 g and 11.11 g plant^{-1} was registered with 50 kg phosphorus and 100 kg nitrogen ha^{-1}, respectively (Krishnamoorthy and Madalageri, 2002). Whereas in Egypt calcium superphosphate was applied to the soil at 100 kg feddan^{-1}, 50 kg feddan^{-1} + 500 litres of a 2% foliar spray or as 1000 litres of 2% foliar spray to ajowan (*Carum copticum*) and compard with control plots. The fertilizer was applied on 16 March (soil) and/or 1 April (foliage), neither fruit yield nor essential oil content differed significantly between these treatments (Haggag and Hilal, 1976).

f) Celery (Apium graveolens L.)

Celery is a temperate crop and it cultivation could also be done in semiarid areas in *Rabi* season. Its potential yield can be taken in northern plains and hill regions. Celery responds to N and highest seed yields were obtained with 80 kg N ha^{-1} and a 45 cm row spacing (Mital, *et al.*, 1975). In a two years trial, N was applied at 160 kg ha^{-1} and FYM at 50 t ha^{-1} to sandy loam soil. The highest celery yields (21.2 and 18.1 q ha^{-1}) were obtained from plots receiving 50 t FYM ha^{-1} + 160 kg N ha^{-1} in the first year while at 50 t FYM ha^{-1} + 120 kg N ha^{-1} in the second year (Mahajan *et al.*, 1974).

Nutrient interaction impact on seed spices yield and quality

Individual and synergistic effects between N and P, and N and S was observed and found positive with respect to seed and essential oil yield in coriander (Sivakumaran, *et al.*, 1996). Application of graded levels of N in coriander gave higher yield and did not influenced oil content and quality (Rao, 1983). In cumin, application of 250 kg AS and 120 kg SSP were found most suitable for crop yield and quality (Chandola, *et al.*, 1970). Application of 5t FYM + 100% inorganic + *Azospirillum* and micronutrient improves growth, yield and quality in fenugreek (Jat, and Chaudhary, 2004). Interaction effect of P, S and growth regulators found positive in fenugreek (Rethinam, and Sadanandan, 1994). Use of *Azotobacter* and *Azospirillum* not only enhance the yield but also controls diseases in fennel (Jat *et al.*, 1998). Combine use of fertilizers and manures gave better performance of fennel (Pareek, 2005). Combine use of N and Mg reported higher seed and oil yield and quality of ajwain (El-Wahab and Mohamed, 2007). Beneficial effect of sulphur and micro-nutrient (Fe, Zn and Mn) observed in anise (Gangrade, *et al.*, 1989). Ammonium sulphate is better N source of N than others for caraway (Schroder, 1964). Micro and secondary nutrient are equally important in growth and yield of celery as their deficiency symptoms appears on various plant parts like: Boron- Cracked stem, Ca - Black heart and Mg-Chlorosis.

Conclusions and future research needs

Seed spices responds well with individual or combine application of nutrients as well as with manures. Under integrated nutrient management, application of 50-75% recommended dose of fertilizer and 25-50% through organic or along with bio-fertilizers gave higher yield and good quality of seed as compared to application of 100% recommended dose of inorganic fertilizers or organics. An application of bio-fertilizer (symbiotic and nonsymbiotic) in seed spices were also found beneficial. *Rhizibum* inoculants gave considerably good yield in fenugreek, whereas *Azospirillum* and/or *Azotobacter* gave in rest of the seed spices crops. In some of these crops inoculation with PSB and VAM also gave encouraging results in these crops. Micronutrients play significant role in improvement of growth, yield and quality of seed spices.

Future research needs on - exploration of area-wise assessment of nutrient deficiencies to be done and INM plan should be prepared accordingly. For sustainable crop production, long-term experiments to be conducted and modelling needs to be done based on results. Usage of bio-fertilizers in crops and cropping system for management of specific and multi-nutrient-deficiency needs to study in seed spices. Incorporation of organic inputs (manures, cakes, crop residue, stubble etc.) in arid and semi arid areas must be done for conserving moisture and nutrients.

References

Abou, A.H.E. and Gomaa, A.O. (2002). Influence of combined inoculation with diazotrophs and phosphate solubilizers on growth, yield and volatile oil content of coriander plants (*Coriandrum sativum* L.). *Bulletin of Faculty of Agriculture Cairo University*, **53** : 93-113.

Aishwath, O.P. and Vashishtha, B.B. (2008). Integrated Nutrient Management in Seed Spices. In: Proc. *National Seminar on Integrated Nutrient Management in rainfed Agro-ecosystems*, held on 3-4 March, 2008 at CRIDA, Hyderabad, Andhra Pradesh. pp 13.

Aishwath, O.P., Mehta, R.S. and Anwer, M.M. (2010). Integrated Nutrient Management in seed spice crops. *Indian Journal of Fertilisers*, **6**: 132-139.

Champawat, R.S. and Pathak, V.N. (1988). Role of nitrogen, phosphorous and potassium fertilizers and organic amendments in cumin (*Cuminum cyminum*) wilt incited by *Fusarium and oxysporum* f.sp. *cumini*. *Indian Journal of Agricultural Sciences*, **58**: 728-730.

Chandola, R.P., Mathur, S.C. and Srivastava, V.K. (1970). Cumin cultivation in Rajasthan. *Indian farming*, **20** : 13-16.

Chaudhary, G.R. (1999). Response of fenugreek for N, P and *Rhizobium* inoculation. *Indian Journal of Agronomy*, **44**: 424-426.

Chaudhary, G.R. (2004). Integrated nutrient management in fenugreek. In : *National seminar on new perspectives in commercial cultivation, processing and marketing of seed spices and medicinal plants*, Jobner, Rajasthan, March 25-26, 2004. pp 46-55.

Chaudhary, G.R. and Jat, N.L. (2004). Integrated nutrient management in fennel. In : *National seminar on new perspectives in commercial cultivation, processing and marketing of seed spices and medicinal plants*, Jobner, Rajasthan, March 25-26, 2004. pp 44-45.

Chaudhary, G.R. and Jat N.L. (2004). Integrated nutrient management in coriander. National seminar on new perspectives in commercial cultivation, processing and marketing of seed spices and medicinal plants, Jobner, Rajasthan, March 25-26, 2004. pp 56.

El-Wahab A. and Mohamed A. (2007). Effect of Nitrogen and Magnesium Fertilization on the Production of *Trachyspermum ammi* L (Ajowan) Plants under Sinai Conditions. *Journal of Applied Sciences Research*, **3**: 781-786.

Gangrade, S.K., Srivastava, R.D., Sharma, O.P., Iyer, B.G. and Trivedi, K.C. (1989). Influence of micro-nutrient on yield and quality of *Pimpinella anisum* L. *Indian Perfumer*, **30** : 142-146.

Haggag, M.Y. and Hilal, S.H. (1976). A preliminary study of the effect of phosphorus fertilizers on the yield of ajowan fruits and on their volatile oil content. *Egyptian Journal of Pharmaceutical Sciences*, **17** : 199-205.

Halesh, D.P., Gowda, M.C., Forooqi, A.A., Vasundhara, M. and Srinivasappa, K.N. (2000). Influence of nitrogen and phosphorus on growth, yield and nutrient content of fenugreek (*Trigonella foenumgraecum L.)*. Spices and aromatic plants: challenges and opportunities in the new century. Contributory papers. Centennial conference on spices and aromatic plants, Calicut, Kerala, India, 20-23 September 2000, 191-194.

Jat, M.L., Sharma, O.P., Shivran, D.R., Jat, A.S., Baldev, R. and Ram, B. (1998). Effect of phosphorus, sulphur and growth regulator on yield attributes, yield and nutrient uptake of fenugreek (*Trigonella-foenum graecum* L.). *Annals of Agricultural Research*, **19** : 89-91.

Jat, N.L. and Chaudhary, G.R. (2004). Effect of N, FYM and Biofertilizer on seed yield of fenugreek. National seminar on new perspectives in commercial cultivation, processing and marketing of seed spices and medicinal plants, Jobner, Rajasthan, March 25-26, 2004. pp 44.

Khiriya, K.D., Sheoran, R.S. and Singh, B.P. (2001). Growth analysis of fenugreek (*Trigonella foenum-graecum* L.) under various levels of farmyard manure and phosphorus. *Journal of Spices and Aromatic Crops*, **10** : 105-110.

Krishnamoorthy, V. and Madalageri, M.B. (2002). Effect of nitrogen and phosphorus nutrition on growth of ajowan genotypes (*Trachyspermum ammi*). *Journal of Medicinal and Aromatic Plant Sciences*, **24** : 45-49.

Kumar, S., Choudhary, G.R. Chaudhari, A.C. and Kumar, S. (2002). Effect of nitrogen and biofertilizers on yield and quality of coriander (*Coriendrum sativum* L.). *Annals of Agricultural Research*, **23** : 634-637.

Lal, G., Aishwath, O.P. and Meena, S.S. (2009). Integrated nutrient and water management in seed spices. In : *Research needs for seed spices: Issues and strategies.* (Eds. Anwer *et al*) Pub. Director NRCSS, Ajmer. pp 49-63.

Manure, G.R., Shivaraj, B., Farooqii, A.A. and Surendra, H.S. (2000). Yield attributes, seed, essential oil yield and oil content of coriander (*Coriandrum sativum* L.) as influenced by the graded levels of nitrogen, sulphur and zinc nutrition in red sandy loam soils. Spices and aromatic plants challenges and opportunities in the new century. Contributory papers. *Centennial conference on spices and aromatic plants*, Calicut, Kerala, India, 20-23 September 2000. pp 139-144.

Meena, N.L. and Chaudhary, G.R. (1998). Response of cumin to N and Zn fertilization. Extended summaries, First International Agronomy Congress, November 23-27, 1998. New Delhi.

Mital, S.P., Singh, R.P., Bhagat, N.R. and Maheshwari, M.L. (1975). Response of celery to nitrogen and row spacings. *Indian Journal of Agronomy*, **20**: 278-279.

Pareek, O.P. (2005). Organic production of seed spices in India. In : Course manual of winter school on '*Advances in Seed Spices Production*'. Pub. NRCSS, Ajmer. pp 6-11.

Pareek, S.K. and Gupta, R. (1984). Exploratory studies on yield and comparative economics of medicinal plants based cropping systems in north-western India. *Annals of Agricultural Research*, **5**: 169-177.

Patel, B.S., Amin, A.U. and Patel, K.P. (2004). Response of cumin (*Cuminum cyminum*) to integrated nutrient management. *Indian Journal of Agronomy*, **49** : 205-206.

Peter, K.V., Srinivasan, V., and Hamza, S. (2000). Nutrient management in spices. *Fertiliser News*, **45** : 13-28.

Prabu, T., Narwadkar, P.R., Sajindranath, A.K. and Rathod, N.G. (2000). Effect of integrated nutrient management on growth and yield of coriander (*Coriandrum sativum* L.). *South Indian Horticulture*, **50** : 680-684.

Purbey, S.K. and Sen, N.L. (2004). Effect of bioinoculation and growth regulators on nodulation, growth and physiological attributes of fenugreek. In: *National seminar on new perspectives in commercial cultivation*. Processing and marketing of seed spices and medicinal plants. Jobner, Rajasthan, march 25-26, 2004. pp 25.

Rao, E.V. S.P., Singh, M., Narayana, M.R. and Rao, R.S.G. (1983). Fertilizer studies in coriander (*Coriandrum sativum* L.). *Journal of Agricultural Science U.K.*, **100** : 251-252.

Rethinam, P. and Sadanandan, K. (1994). Effect of P, S and growth regulators on fenugreek. In: *Advances in Horticulture* (Eds.) K.L. Chadha and P. Rethinam, **9** : 499-510.

Schroder, H. (1964). Arznei-und Greurzpflanzen Bernburg/Saale.

Sharma, B.D. (2010). Integrated water and nutrient management in seed spices. In : Book of Lead papers '*National Consultation on Seed Spices Biodiversity and Production for Export-Perspective, Potential, Threats and their Solutions*', held on July 7th 2010 at Ajmer (Eds. G. Lal, Aishwath, O.P., Saxena, S.N., Sharma, Y.K, Mehta, R.S. and Meena, S.S.). Pub. Director, NRCSS, Ajmer, Rajasthan and President Indian Soc. Seed Spices, Ajmer, Rajasthan, India. pp 39-49.

Sharma, R. (1998). Effect of crop geometry with and without zinc on growth, yield and quality of fennel. Thesis M.Sc. (Ag.), Department of Agronomy, S.K.N. College of Agriculture, Jobner.

Sharma, R.K., Dashora, S.L., Chaudhary, G.R., Agrawal, S., Jain, M.P. and Singh, D. (1996). Status of spices research in Rajasthan. In: *Seed spices research in Rajasthan* (Eds. Pundir . P. and Dharma. A.K.). Pub. Director Research. Raj. Agri. Univ., Bikaner. pp. 9-44.

Sivakumaran, S., Gururaj, H., Basavaraju, H.K., Sridhara, S. and Hunsigi, G. (1996). Study on nutrient uptake, seed and oil yield of coriander as influenced by nitrogen, phosphorus and sulphur. *Indian Agriculturist*, **40** : 89-92.

□□□

System Based Integrated Nutrient Management, 2012
© B. Gangwar & V.K. Singh (eds.), pp. 339-352
New India Publishing Agency, New Delhi (India)
e-mail : info@nipabooks.com; website : www.nipabooks.com

CHAPTER **24**

Crop Simulation Modelling and its Applications for Integrated Nutrient Management

M. SHAMIM

Agricultural systems are complex, and understanding this complexity requires systematic research under shrinking agricultural research resources. Field experimentation can only be used to investigate a very limited number of variables under a few site-specific conditions. The crop growth simulation models simulate crop growth, development and yield taking into account the effects of weather, genetics, soil water, carbon and nitrogen, planting, irrigation and fertilizer (e.g. Nitrogen, Phosphorus) management. Of the many factors that potentially affect the growth of plants, the two that are most responsive to management are water and nitrogen. It is therefore not surprising that most crop models include routines to handle responses to water and nitrogen. Less frequently do models attempt to describe limitations due to other soil constraints (Probert and Keating, 2000). In the context of integrated nutrient management, the performance of a crop model depends mainly on its ability to adequately describe the release of nutrients from diverse inputs and their uptake by the crop.

Origins and genesis of a crop simulation model

Crop simulation model approach was started near about 50 years ago, in early 1960s when early simple crop models were developed for estimation of transpiration and photosynthesis. The first step towards crop modeling was the development of simple models to estimate light interception and photosynthesis (Loomis and williams, 1963; De wit, 1965 and Duncan *et al.*, 1967). There are many crop growth simulation models. Some are more generic in nature while others are built for specific purposes. Most of these models simulate crop growth and soil processes using daily time steps. All models are developed with some assumptions and hypotheses, and all have strengths, weaknesses and limitations for appropriate application. Well-known crop modelling groups across the world include IBSNAT/IFDC (International Benchmark Sites Network for Agro technology Transfer/International Fertilizer Development Centre) in USA, WAU/AB-DLO (Wageningen Agriculture University/Centre for Agro-biological Research) in the Netherlands, and APSRU (Agriculture Production Systems Research Unit) in Australia. The IBSNAT project was initiated in 1982, and over the past 20 years it has developed more than 15 models, including CERES (Crop Estimation through Resource and Environment Synthesis) Rice and CERES Wheat, which are available either as stand-alone models or within the DSSAT (Decision Support System for Agrotechnology Transfer) shell (Uehara and Tsuji, 1993).

All the DSSAT models are continuously being refined, calibrated, validated, and applied across the world by the scientists who developed the models and their collaborators and others. The Wageningen Group is probably the strongest modelling group in the world in terms of concentration and strength of modelers and the range of models being developed. Most of the models that are available today across the world have some sort of origin from Wageningen. The APSRU group has developed and applied APSIM (Agriculture Production Systems SIMulator) with a range of models initially targeted for rainfed cropping, but also applicable to irrigated conditions.

Calibration and evaluation of crop simulation models

Model calibration or parameterization is the adjustment of parameters so that simulated values compare well with observed values. The genetic coefficients that influence the occurrence of developmental stages in the models can be derived iteratively, by manipulating the relevant coefficients to achieve the best possible match between the simulated and observed number of days to the phenological events. Other coefficients can be derived from determinations of non-limited grain weight, number of grains per panicle (rice) and rate of grain filling (wheat). Alternatively, genetic coefficients can be determined using the GENCALC software that uses the observations of phenological events from one or several experiments from a range of environments (Hunt *et al.*, 1993).

Evaluation or Validation is the comparison of the results of model simulations with observations from crops that were not used for the calibration.

A model should be rigorously validated under widely differing environmental conditions to evaluate the performance of major processes in addition to its ability to predict yield. CERES-model has been validated and evaluated for many locations across the rice and wheat growing regions in the world. Models can be validated in terms of growth and phenophases, grain and biomass yields and yield components, crop and soil N, methane emissions and evapotranspiration etc.

CERES- models

The CERES models simulate crop growth, development and yield taking into account the effects of weather, management, genetics and soil water, carbon and nitrogen. But only the main features of official versions of CERES-models components and processes are described here.

Phenology: The phenology components of CERES-Rice and CERES-Wheat have been described by Ritchie and Ne Smith (1991), Singh (1994) and Ritchie *et al.* (1998). The models describe the progress through the crop life cycle using degree-day accumulation (heat sum). The duration of growth stages in response to temperature and photoperiod varies between species and cultivars and genetic coefficients are used as model inputs to describe these differences (Hunt and Boote, 1994; Singh *et al.*, 2002). In wheat, the duration from emergence to terminal spikelet formation is influenced by the vernalization (chilling) requirement and photoperiod. In rice, a developmental phase (juvenile stage) occurs when the crop is not sensitive to photoperiod. After this stage sensitivity to both photoperiod and temperature determine the time to panicle initiation.

Shamim and Sheikh (2009) revealed that CERES-Rice model performed well for simulation of days to anthesis. The simulated days to anthesis by the model were (97.0 ± 3.6 d) near to the mean observed days to anthesis (96.0 ± 3.1 d). Highly positive significant association was observed between simulated and observed days to anthesis (r = 0.95**, ** significant at 0.001 level). The results also revealed that days to anthesis decreased date wise irrespective of cultivars from first (July 8th) to third (August, 8th) date of transplanting, both in case of observed and simulated values. This decreasing trend was due to the prolonged period of favourable temperature and lower photoperiod in addition to higher mean vapour pressure during both crop seasons and could be explained in terms of the rate of development (inverse of duration) as related to temperature and photoperiod (Marcellos and Single, 1971 and Saini *et al.*, 1986).The mean simulated days to maturity (132:63±4.31 d) was matched with the observed mean value (131.88± 4.58 d) (Table 1).

Table 1 : Test criteria for evaluation of the model with respect to simulation of days to anthesis and days to maturity of aromatic genotypes of rice

Parameters	Days to anthesis	Days to maturity
OM	96.0	131.88
SDo	3.1	4.58
SM	97.0	132.63
SDs	3.6	4.31
r	0.95**	0.74**
Student 't'	0.00975 NS	0.01525 NS
MAE	1.33	3.75
MBE	1.08	0.75
RMSE	1.58	4.28
PE	1.65	3.25
Index of agreement (D)	0.99	0.98

where,
OM: Observed mean ; SM: Simulated mean ; SDo: Observed of standard deviation
SDs: Simulated standard deviation

The phenology component also simulates the effect of water or N deficit on the rate of life cycle progress (Singh *et al.*, 1999). These effects may vary with life cycle phase; for example, water deficit may slow the onset of reproductive growth but accelerate reproductive growth after the beginning of grain filling.

Growth: The models predict daily photosynthesis using the radiation-use efficiency approach as a function of daily irradiance for a full canopy, which is then multiplied by factors ranging from 0 to 1 for light interception, temperature, leaf N status, and water deficit. There are additional adjustments for CO_2 concentration, specific leaf weight, row spacing, and cultivar (Ritchie *et al.*, 1998). Growth of new tissues depends on daily available carbohydrate and partitioning to different tissues as a function of phenological stage which is modified by water deficit and N deficiency stress indices. Leaf area expansion depends on leaf appearance rate, photosynthesis and specific leaf area. Leaf area expansion is quite sensitive to temperature, water deficit, and N stress. During seed fill, N is mobilized from vegetative tissues to the grains.

Shamim and sheikh (2009) reported that observed as well as model simulated LAI was found maximum at the anthesis stage. Slightly low peak in simulated and observed values was noticed at beginning of grain filling stage (BGF) due to additive effect of senescence of lower leaves started and vegetative growth had ceased at that stage. The values of LAI for various genotypes of rice were highest in 22nd July transplanting and lowest in and 8th August for both observed as well as simulated by the model. The CERES-Rice model underestimated the LAI from emergence to maturity in all four genotypes and dates of transplanting except at few points (Figure 1). The highest LAI was recorded in GR-104 in all the corresponding dates of transplanting than that of others genotypes of aromatic rice. The plants with high LAI intercepted maximum solar radiation, which was vital for photosynthetic activity of the leaves. In cereals, under most circumstances, majority of the carbohydrates in the grain are derived from carbon dioxide fixation (photosynthesis) after flowering stage. This was the reason for higher grain yield in respect of D_2 transplanting treatments of all four genotypes of rice.

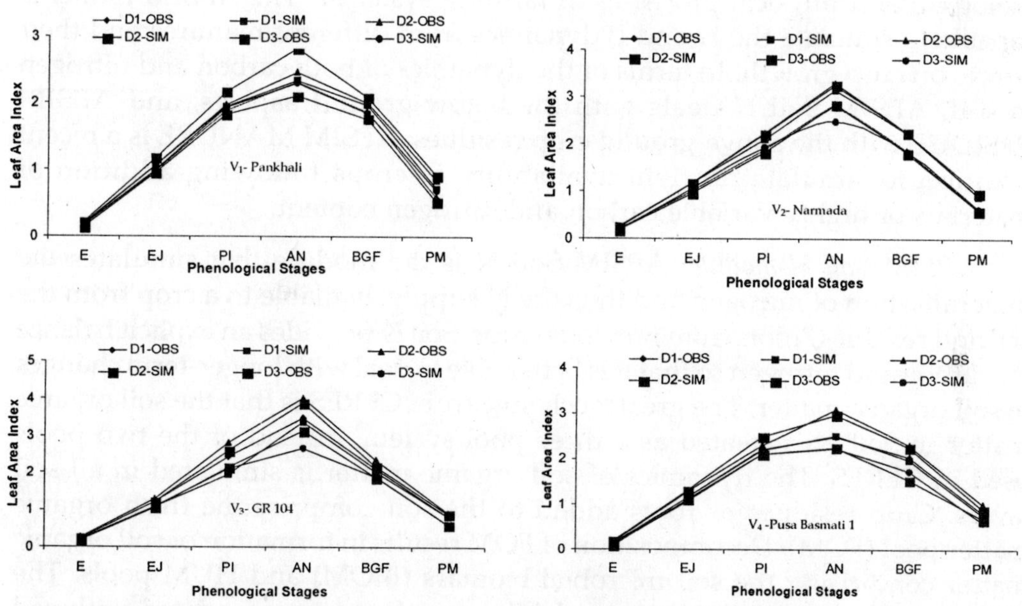

Fig. 1: Comparison of observed (OBS) and simulated (SIM) temporal leaf area index of four genotypes of aromatic rice under three dates (D_1: 8th July, D_2: 22nd July and D_3: 8th August) of transplanting at six crop phenophases (EM = Emergence, EJ = End of juvenile phase, PI = Panicle initiation, AN = Anthesis, BGF = Beginning of grain filling and PM = Physiological maturity).

Soil organic matter and N dynamics: The DSSAT model has options to simulate the soil organic matter (SOM) pools and dynamics (Godwin and Jones, 1991; Godwin and Singh, 1998) which has a variable C: N ratio and can mineralize or immobilize nutrients. The SOM decomposition rate and residue flows also vary with soil texture. The N module of CERES-Rice simulates hydrolysis of urea, nitrification, ammonia volatilization, nitrate leaching, denitrification, algal activity and floodwater pH changes, plant N uptake and partitioning under continuously flooded, intermittently flooded and non-ponded conditions (Singh, 1994). The floodwater N chemistry component of the model (Godwin *et al.*, 1990; Godwin and Singh, 1991 ; 1998) uses an hourly time step to calculate rapid N transformations and to update soil-floodwater-atmosphere equilibria.

The APSIM-model

The Agriculture Production System Simulator (APSIM) Shell includes a wheat model which was developed from a combination of the approaches used in the N-Wheat and I-WHEAT models. The APSIM modelling framework contains a set of biophysical modules that can be configured to simulate biological and physical processes in farming systems. The APSIM model is capable to simulate the N and P dynamics from different manures and their effects on crop growth. In terms of the dynamics of both carbon and nitrogen in soil, APSIM Soil N deals with the below-ground aspects, and APSIM RESIDUE with the above-ground crop residues. APSIM MANURE is a recent addition to simulate nutrient availability to crops following addition of materials of highly variable carbon and nitrogen content.

APSIM Soil N module: APSIM Soil N is the module that simulates the mineralisation of nitrogen and thus the N supply available to a crop from the soil and residues/roots from previous crops. Soil N provides an explicit balance for carbon and nitrogen so that it is better able to deal with longer-term changes in soil organic matter. The greatest change from CERES is that the soil organic matter in Soil N is treated as a three pool system, instead of the two pools used in CERES. The dynamics of soil organic matter is simulated in all soil layers. Crop residues or roots added to the soil comprise the fresh organic matter pool (FOM). Decomposition of FOM results in formation of soil organic matter comprising the soil microbial biomass (BIOM) and HUM pools. The BIOM pool is notionally the more labile organic matter associated with soil microbial biomass; while it makes up a relatively small part of the total soil organic matter, it has a higher rate of turnover than the bulk of the soil organic matter. The reasons for introduction of an additional soil organic matter pool were to better represent situations where 'soil fertility' improves following a legume ley (Dimes, 1996). A single soil organic matter pool cannot deal realistically with the changes in mineral N supply following the cumulative additions of high N material to soil organic matter (Figure 2).The release of nitrogen from the decomposing organic matter pools is determined by the mineralisation and immobilisation processes that are occurring. The carbon that is decomposed is either evolved as CO_2 or is synthesized into soil organic matter. Soil N assumes that the pathway for synthesis of stable soil organic matter is predominantly through initial formation of BIOM, though some carbon may be transferred directly to the more stable pool (HUM). The model further assumes that the soil organic matter pools (BIOM and HUM) have C:N ratios that are unchanging through time. The C:N ratio of the BIOM pool is typically set at 8, while that of the HUM pool is based on the C:N ratio of the soil, which is an input at initialisation of a simulation. The formation of BIOM and HUM thus creates an immobilisation demand that has to be met from the N released from the decomposing pools and/or by drawing on the mineral N (ammonium and nitrate) in the layer. Any release of N above the immobilisation demand during the decomposition process results in an increase in the ammonium-N. The FOM pool is assumed to comprise three sub fractions (FPOOLs), sometimes referred to as carbohydrate-,cellulose- and lignin-like, each with a different rate of decomposition. In this manner, the decomposition of added plant material under conditions of constant moisture and temperature is not a simple first-order process. The rates of decomposition of the various soil organic matter pools are dependent on the temperature and moisture content of the soil layers where decomposition is occurring. In circumstances where there is inadequate mineral N to meet an immobilisation demand, as can occur where the C:N ratio of the FOM pool is high, the decomposition process is limited by the N available to be immobilised. Other processes dealt with in Soil N are nitrification, denitrification and urea hydrolysis (following application of urea as fertiliser). Fluxes of nitrate-N between adjoining soil layers are calculated based on movement of water by the water balance module. Ammonium-N and all soil organic matter pools are assumed to be non-mobile (Delve and Probert, 2004).

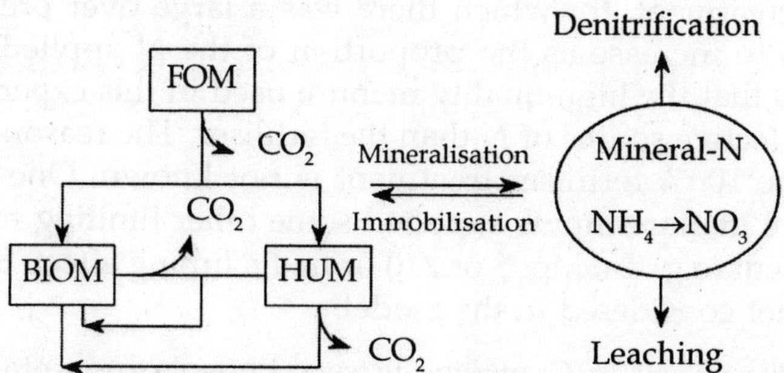

Fig. 2 : Schematic representation of the processes affecting soil organic matter andnitrogen transformations in the APSIM SoilN module. FOM = fresh organic matter, i.e. roots and incorporated crop residues; BIOM = labile soil organic matter pool; HUM = remainder of soil organic matter (Adapted from Delve and Probert, 2004).

Chivenge *et al.*, 2004 revealed that grain yield response to various combinations of manure and fertiliser applying a total 100 kg N ha^{-1} to maize at Murewa is shown in Figure 3. Combinations gave higher maize yields than sole fertiliser or sole manure. The 100% manure and 100% fertiliser treatments gave almost identical maize yields, and these were more than double the yield of the control treatment that received no inputs of N.

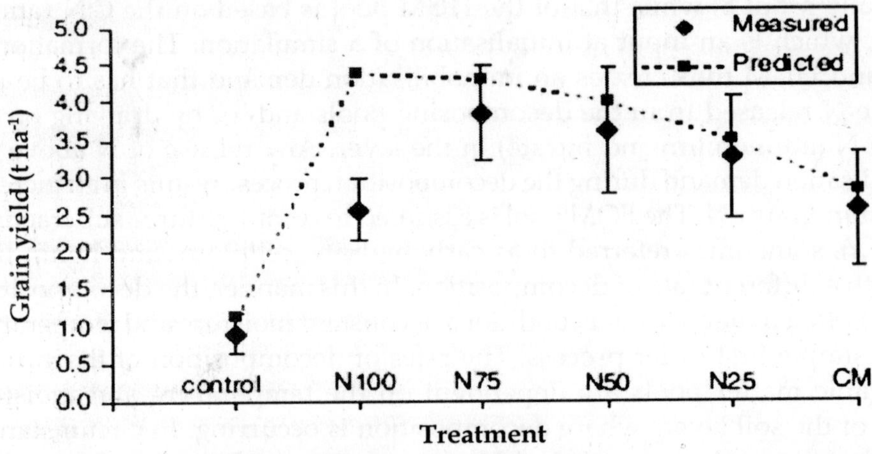

Fig. 3 : Measured and predicted effects of 100 kg Nha^{-1} applied as combinations of manure and fertiliser N on maize grain yield at Murewa in the 1997–1998 cropping season. (Control = no N inputs, N100 = 100 kg N ha^{-1} applied as fertiliser, CM = 100 kg N ha^{-1} applied as manure). Error bars denote standard deviations of measured means.

The model predictions agreed closely with the observed yields for the control and in response to the manure and fertiliser inputs, except for the 100% fertiliser treatment, for which there was a large over prediction. The trend for yields to increase as the proportion of the N applied as fertiliser increased shows that the high-quality manure used in this experiment was a relatively less effective source of N than the fertiliser. The reason for the poor prediction of the 100% fertiliser treatment is not known. One explanation could be that the manure inputs supplied some other limiting resource such as another nutrient (e.g. Ca, Mg, S or Zn) or had a liming effect. Such benefits of manure are not considered in the model.

Modelling Phosphorus in Cropping Systems: Phosphorus uptake by plants involves diffusion of phosphate to roots, and is increased by the presence of mycorrhizae. Models of diffusion to plant roots (Claassen and Barber, 1976) show that root density is a controlling factor in P uptake. But models of the diffusion process are at a greater level of detail (in both time and space) than what is found in most crop models. Crop models typically assume that water and nitrogen are homogeneous throughout each soil layer with a dimension of centimetres, in marked contrast with the diffusion models where concentration gradients exist around individual roots with dimension of fractions of a millimetre. More general system models, like EPIC (Jones *et al.*, 1984) and CENTURY (Parton *et al.* 1988), have included P routines, but the generic crop routines in these models have limited ability to address crop management issues requiring accurate simulation of crop growth in response to weather, genotype, soil and management practices. They have not been widely used to explore management strategies involving P. It has been reported that the P routines in CENTURY were not able to describe the dynamics of P in tropical soils (Gijsman *et al.*, 1996). Management of soil P (especially in high-input agricultural systems) has focused on issues like whether to apply fertiliser, at what rate, evaluating placement and residual effects, and

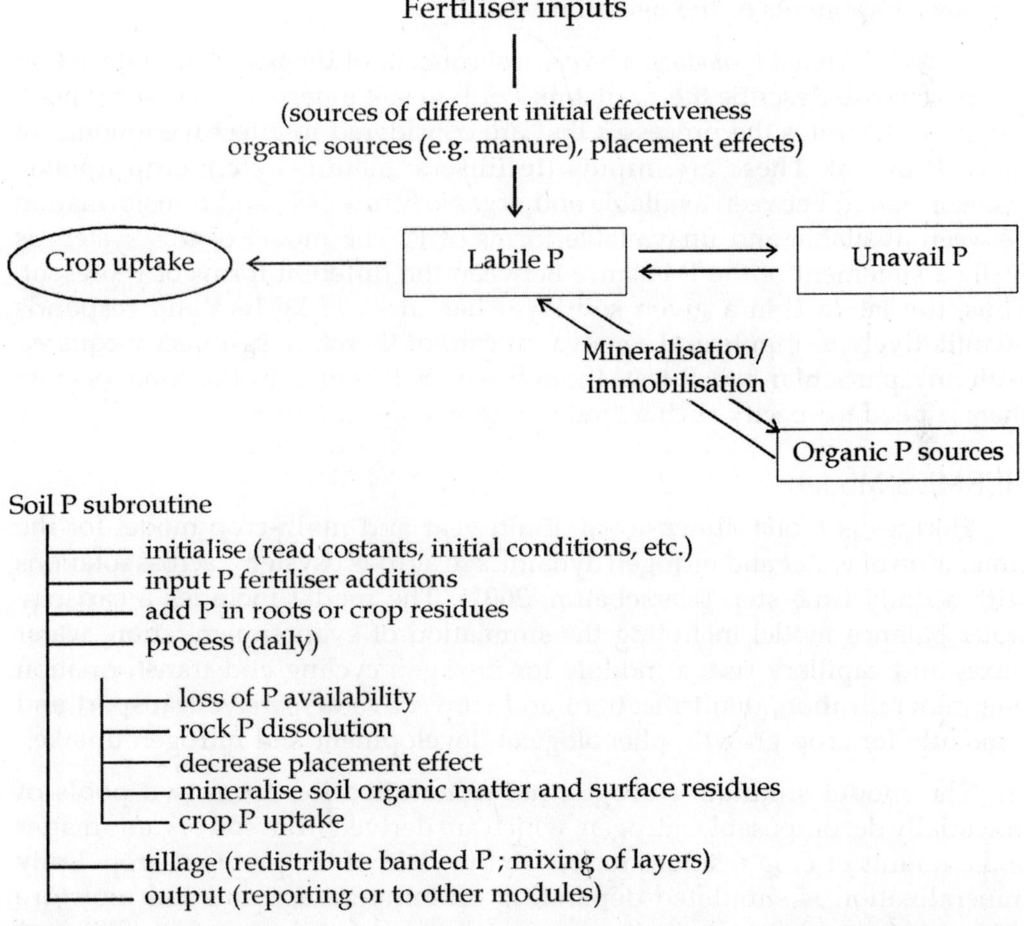

Fig. 4 : The APSIM Soil P module. The upper part of the figure shows in diagrammatic form the processes that are considered in the module. The lower part shows the simplified subroutine structure of the model where some actions are event based (e.g. initialisation, add fertiliser, tillage) whereas the 'process' activities occur on a daily time step (Adapted from Delve and Probert, 2004).

comparing relative effectiveness of water-soluble versus insoluble sources. Because P is immobile in soil (at least over the time scale of an annual crop) interactions with climate are of little importance. Unlike the management of N, there has been no need for a detailed crop model to evaluate alternative strategies for management of P. Models operating with a time-step of a growing season and an empirical relationship between yield and soil P status are adequate to gain insights into crop responsiveness to alternative fertiliser P sources and their residual effects (Probert, 1985). However, if there is a need for crop models to simulate response to manures and other organic sources in low-input systems, it is important that they respond to both N and P. Crop models tend to perform best when there is a similar degree of detail for the various components of the overall model.

The APSIM Soil P module: The central concept of the Soil P module is that it is possible to describe the availability of P in soil in terms of a labile P pool. Figure 4 illustrates the processes that are considered to affect the amount of labile P in soil. These are: inputs (fertilisers, manures etc.); crop uptake; transformation between available and organic forms of P; and transformation between available and unavailable forms of P. The model of this system is really a statement of the P balance between the different forms of P present. Thus, the labile P in a given soil layer has units of kg ha^{-1} and responds quantitatively to inputs and removal. It cannot therefore be directly equated with any particular soil P test, though we shall return to the topic of how there is need to specify such a model in terms of soil P tests.

HERMES-Model

Hermes is a one dimensional, multi year and multi-crop model for the simulation of water and nitrogen dynamics in agro-ecosystems across rotations with a daily time step (Kersebaum, 2007). The model includes a capacity water balance model including the simulation of evapotranspiration, water fluxes and capillary rise, a module for nitrogen cycling and transformation (net mineralization, denitrification) and convective-dispersive transport and a module for crop growth, phenological development and nitrogen uptake.

The model simulate nitrogen net mineralization from two pools of potentially decomposable nitrogen, which are derived from soil organic matter and amounts of crop residues related to the yield of the previous crop. Daily mineralization is simulated depending on temperature and soil moisture (Mayers *et al.,* 1982). Denitrification is modeled for the top soil (0-30 cm) using a Michalis-Menten kinetic dependent on nitrate content modified by reduction functions depending on water filled pore space, and temperature. The sub model for crop growth follows a generic approach in the style of SUCROSE model, simulating different crops by using external crop parameters file. The daily net dry matter production as a result of photosynthesis and respiration is simulated driven by global radiation and temperature. Daily dry matter is portioned between the different plant organs depending on the crop development stages, which are in turn calculated from thermal sum (°C days). Thermal sums are specifically modified for each stage by day length and vernalization, if applicable. Yield is estimated at harvest from the weight of the storage organ and nitrogen stored in crop residues is calculated from the difference between simulated crop nitrogen uptake and nitrogen exported with yield and removed by-products. Subsequently, crop residues undergo a decay process similar to soil organic matter. Crop growth is limited by water and nitrogen stress. Drought stress is indicated by the ratio between actual and potential evapotranspiration. Temporary limitation of soil air by water logging is considered through reducing transpiration and photosynthesis. Water and nitrogen uptake is calculated from crop transpiration and crop nitrogen status, depending on the simulated root distribution and water and nitrogen availability in different soil layers. An empirical exponential distribution function is used to calculate the root length density from the simulated root biomass. The concept of critical nitrogen concentration in plant as a function of crop developmental stage is applied to assess the nitrogen shortage. Climate change has an effect on nitrogen availability and the nitrogen use efficiency. Increasing temperature would enhance nitrogen mineralization if soil moisture is not limiting. On the other hand, if water is not sufficiently available, the efficiency of applied nitrogen fertilizer may decrease due to reduced diffusive transport in dry soils. The combined effect of temperature and precipitation affects the nitrogen use efficiency (Kersebaum *et al.*, 2009).

Applications of crop simulation models

Matthews and Stephens (2002) categorized the main applications or needs of models into 3 categories: as tools in research, in decision support, and in education and training. Examples of research applications included identification of desired crop genotype characteristics, investigation of management options, cropping or farming system analysis, investigations of impact of climate change on crop productivity, and prediction of greenhouse gas production. Models can be used to assist both tactical decision making (such as irrigation scheduling, and fertilizer and pest management), or in strategic decision making, such as planning for climate change or to avoid salinisation, yield forecasting and planning for national food requirements. Models can also be useful in teaching crop and soil processes and crop system behaviour in response to weather, management and site conditions.

The major applications of the models are yield forecasting, yield gap and yield trend analysis, evaluating agronomic management strategies, evaluation of cropping options in new locations, impacts of climate change on yields, prediction of greenhouse gas emissions, pest and disease management, and in farming government planning. However, the impact of the application of the models on decision making by farmers and their advisors and policy makers is generally less clear.

Usefulness of models is enhanced if they can be operationalized and harnessed directly or indirectly for farmers' benefit. In production-oriented agriculture, optimum sowing or transplanting time and suitable cultivars on local area basis are vital aspects that not only determine farm economy and profitability but also intimately associated with sustainable crop productivity.

References

Chivenge, P., Dimes, J., Nhamo, N., Nzuma, J.K. and Murwira, H.K., (2004). Evaluation of APSIM to Simulate Maize Response to Manure Inputs in Wet and Dry Regions of Zimbabwe. In : *Modelling Nutrient Management in Tropical Cropping Systems*, ACIAR Proceedings No. 114, 138p.

Claassen, N. and Barber, S.A. (1976). Simulation model for nutrient uptake from soil by growing plant root system (maize). *Agronomy Journal*, **68** : 961-964.

de Wit, C.T. (1965). Photosynthesis of leaf canopies. *Institute of Biological Chemistry Research Field Crops, Herb. Agriculture Research Report*. 663. Wageningen, The Netherlands.

Delve, R.J. and Probert, M.E. (2004). Modelling Nutrient Management in Tropical Cropping Systems, ACIAR Proceedings No. 114, 138p.

Dimes, J.P. (1996). Simulation of mineral N supply to no-till crops in the semi arid tropics. PhD. Thesis, Griffith University, Queensland.

Duncan, W.G.; Loomis, R.S.; Williams, W.A. and Hanau, R. (1967). A model for simulating photosynthesis in plant communities. *Hilgardia*, **38** : 181-205.

Gijsman, A.J., Oberson, A., Tiessen, H. and Friesen, D.K. (1996). Limited applicability of the CENTURY model to highly weathered tropical soils. *Agronomy Journal*, **88** : 894-903.

Godwin D., Singh U. (1998). Nitrogen balance and crop response to nitrogen in upland and lowland cropping systems. In: Tsuji, G.Y., Hoogenboom, G., Thornton, P.K. (Eds.), Understanding Options for Agricultural Production. Kluwer Academic Publishers, Dordrecht, The Netherlands, pp 55-78.

Godwin, D., Jones, C.A. (1991). Nitrogen dynamics in soil-crop systems. In: Hanks, R.J., Ritchie, J.T. (Eds.), Modeling Plant and Soil systems. Agronomy Monograph 31, American Society of Agronomy, Madison, Wisconsin, USA, pp. 287-321.

Godwin, D., Singh, U. (1991). Modelling nitrogen dynamics in rice cropping systems. In: Deturck P, Ponnamperuma FN (Eds.), Rice Production on Acid Soils of the Tropics'. Institute of Fundamental Studies, Kandy, Sri Lanka, pp 287-294.

Godwin, D.C., Meyer, W.S., Singh, U. (1994). Simulation of the effect of chilling injury and nitrogen supply on floret fertility and yield in rice. *Australian Journal of Experimental Agriculture*, **34** : 921-926.

Godwin, D.C., Singh, U., Buresh, R.J., De Datta, S.K. (1990). Modelling N dynamics in relation to rice growth and yield. In: Transactions of the *14th International Congress of Soil Science*, Kyoto, Japan. *International Society of Soil Science*, Japan, pp. 320-325.

Hunt, L.A., Boote, K.J. (1994). Data for model operation, calibration, and validation. In: Tsuji, G.Y., Hoogenboom, G., Thornton, P.K. (eds.), IBSNAT: A system approach to research and decision making. University of Hawaii. Honolulu, Hawaii, USA, pp. 9-40.

Hunt, L.A., Pararajasingham, S., Jones, J.W., Hoogenboom, G., Imamura, D.T. and Ogoshi, R.M. (1993). Gentale : Software to facilitate the use of crop models to analyze field experiment. *Agronomy*, **85** : 1090-94.

Jones, C.A., Cole, C.V., Sharpley, A.N. and Williams, J.R. (1984). A simplified soil and plant phosphorus model: I. Documentation. *Soil Science Society of America Journal*, **48** : 800–805.

Kersebaum, K.C., Mirschel, W., Wenkel, K.O., Manderscheid, R., Weigel, H. J. and Nendel, C. (2009). Modelling climate change impacts on crop growth and management in Germany. In: *Climate Variability, Modelling Tools and Agricultural Decision-Making*. (Edi) Angel Utset, Nova Science Publishers, pp. 183-194.

Kersebaum, K.C. (2007). Modelling nitrogen dynamics in soil-crop systems with HERMES. *Nutrient Cycling in Agroecosystems*, **77** : 39-52.

Loomis, R.S. and Williams, W.A. (1963). Maximum crop productivity: An estimate. *Crop Science*, **3**: 67-72.

Marcellos, H. and Single, W.V. (1971). Quantitative responses of wheat to photoperiod and temperature in the field. *Australian Journal of Agricultural Research*, **22** : 343-357.

Matthews, R. and Stephens, W. (Eds.) (2002). Crop-Soil Simulation Models Applications in Developing Countries. CABI Publishing, CAB International, Wallingford, UK, 277 p.

Myers, R.J.K., C.A. Campbell and K.L. Weier (1982). Quantitative relationship between net nitrogen mineralization and moisture content of soils. *Canadian Journal of Soil Science*, **62** : 111-124.

Parton, W.J., Stewart, J.W.B. and Cole, C.V. (1988). Dynamics of C, N, P and S in grassland soils: a model. Biogeochemistry, **5** : 109–131.

Probert, M.E. (1985). A conceptual model for initial and residual responses to phosphorus fertilizers. *Fertilizer Rcsearch*, **6** : 131–138.

Probert, M.E. and Keating, B.A. (2000). What soil constraints should be included in crop and forest models? *Agriculture, Ecosystems and Environment*. pp. 273-281.

Ritchie J.T. and NeSmith D.S. (1991). Temperature and crop development. In: Hanks, R.J., Ritchie, J.T. (Eds.), Modeling Plant and Soil systems. Agronomy Monograph 31, American Society of Agronomy, Madison, Wisconsin, USA, Pages 5-29.

Ritchie, J.T. (1998). Soil water balance and plant stress. In: Tsuji GY, Hoogenboom G, Thornton PK (Eds), Understanding Options for Agricultural Production. Kluwer Academic Publishers, Dordrecht, The Netherlands, pp. 41-54.

Ritchie, J.T.; Singh, U.; Godwin, D.C. and Bowen, W.T. (1998). Cereal growth, development and yield. In: Tsuji, G.Y.; Hoogenboom, G.; Thornton, P.K. (eds.). Understanding Options for Agricultural Production, Kluwer Academic Publishers, Dordrecht, The Netherlands. pp. 79-98.

Saini, A.D.; Dadwal, V.K.; Phadnawis, B.N. and Nanda, R. (1986). Thermal and photoperiodic effects on phase durations of four wheat varieties grown on different sowing dates. *Indian Journal Agricultural Science*, **56** : 646-656.

Shamim, M. and Sheikh, A.M. (2009). Simulation of environmental and varietal effects on growth and yield of aromatic rice using CERES-Rice model in middle Gujarat Agro-climatic region. Ph.D. Thesis submitted to Anand Agricultural University, Anand-388 110, Gujarat.

Singh, U. (1994). Nitrogen management strategies for lowland rice cropping systems. In: *Proceedings of the International Conference on Fertilizer Usage in the Tropics* (FERTROP). Malaysian Society of Soil Science, Kuala Lumpur, Malaysia, pp. 110-130.

Singh, U., Timsina, J., Godwin, D.C. (2002). Testing and applications of CERES-rice and CERES-wheat models to rice-wheat cropping systems. In: Humphreys, E, Timsina, J (Eds.), Modelling irrigated cropping systems with special attention to rice-wheat sequences and raised bed planting". *Proceedings of a workshop at CSIRO Land and Water*, Griffith (Australia) 25-28 February 2002, pp. 17-32.

Singh, U., Wilkens, P.W., Chude, V., Oikeh, S. (1999). Predicting the effect of nitrogen deficiency on crop growth duration and yield. In: *Proceedings of the Fourth International Conference on Precision Agriculture*. ASA-CSSA-SSSA, Madison, Wisconsin, USA, pp. 1379-1393.

Singh, U.; Wilkens, P.W.; Chude, V. and Oikeh, S. (1999). Predicting the effect of nitrogen deficiency on crop growth duration and yield. In: *Proceedings of the Fourth International Conference on Precision Agriculture*. ASA-CSSA-SSSA, Madison, Wisconsin, USA. pp. 1379-1393

Uehara, G. and Tsuji, G.Y. (1993). The IBSNAT project. In: Penning de Vries, F.W.T., Teng, P., Metsellar, K. (eds.) Systems Approaches for Agriculture Development. Kluwer, Dordrecht, pp. 505-514.

□□□

System Based Integrated Nutrient Management, 2012
© *B. Gangwar & V.K. Singh (eds.), pp. 353-371*
New India Publishing Agency, New Delhi (India)
e-mail : info@nipabooks.com; website : www.nipabooks.com

CHAPTER **25**

Soil Management Options for Sequestering Carbon

S.P. MAZUMDAR

Carbon is found in all living organisms and is the major building block for life on earth. It exists in many forms, predominantly as plant biomass, soil organic matter, and as the gas CO_2 in the atmosphere and dissolved in sea water. Carbon cycles globally among five distinct pools among which the largest oceanic pool is estimated at 38000 Pg (Pg= petagram =$1x10^{15}$g=billion tonnes) and is increasing at the rate of 2.3 Pg C yr^{-1} (Fig. 1). The geological C pool, comprising fossil fuels, is estimated at 4130 Pg, pedologic, estimated at 2500 Pg to 1m depth, atmospheric pool comprising 760 Pg of C with nearly all of it as CO_2, and the smallest among the global C pools is the biotic pool estimated at 560 Pg. The pedologic and biotic C pools together are called the terrestrial C pool estimated at approximately 2860 Pg.

Soil carbon consists of two distinct components: soil organic carbon (SOC) pool estimated at 1550 Pg and soil inorganic carbon (SIC) pool at 950 Pg (Batjes, 1996). The SOC pool includes highly active humus and relatively inert charcoal C. It comprises a mixture of: (i) plant and animal residues at various stages of decomposition; (ii) substances synthesized microbiologically and/or chemically from the breakdown products; and (iii) the bodies of live micro-organisms and small animals and their decomposing products (Schnitzer, 1991). The SIC pool includes elemental C and carbonate minerals such as calcite, dolomite and gypsum, and comprises primary and secondary

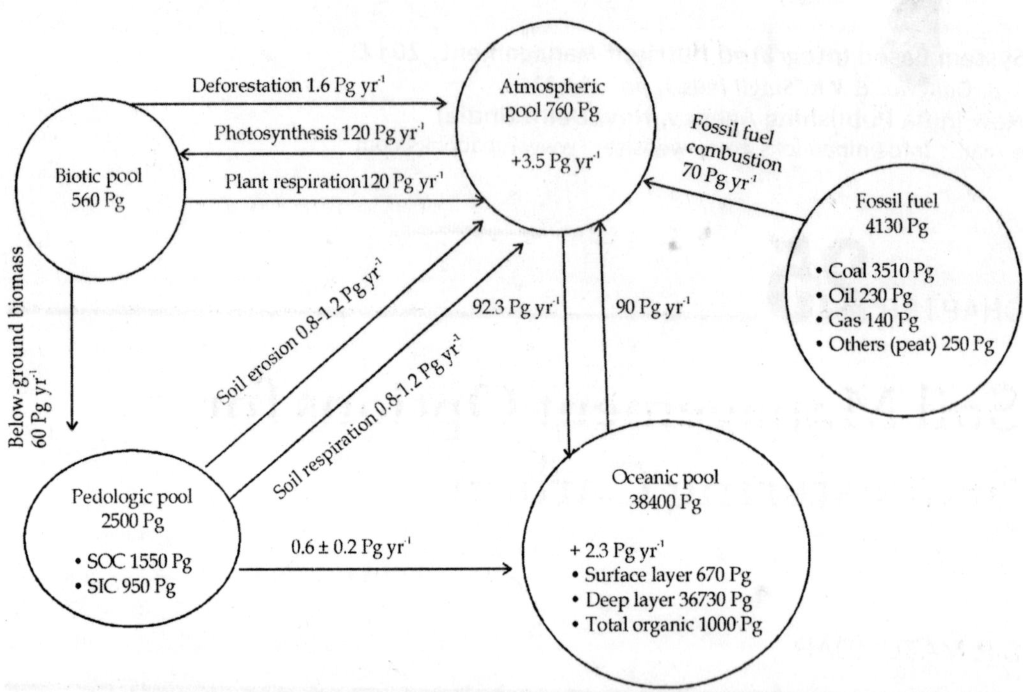

Fig. 1 : Principal global C pools and fluxes between them. The data on C pools among major reservoirs are from Batjes (1996), Fallowsi et al. (2000) and Pacala and Socolow (2004), and data on fluxes are from IPCC (2001)

carbonates. The primary carbonates are derived from the weathering of parent material. In contrast, the secondary carbonates are formed through the reaction of atmospheric CO_2 with CaC_2 and MgC_2 brought in from outside the local ecosystem (e.g. calcareous dust, irrigation water, fertilizers, and manures). The SIC is an important constituent in soils of arid and semi-arid regions.

Over the past few decades, the man-induced changes in the climate of the earth have become the focus of scientific and social attention. The most imminent of the climatic change of the earth is the increase in the atmospheric temperatures due to the increased levels of carbon dioxide (CO_2) and other greenhouse gases. Carbon di-oxide is the major greenhouse gas which contributes about 50% global warming. During the past 250 years the atmospheric CO_2 concentration has increased from a pre-Industrial value of about 275 ppm to 380 ppm in 2008 and increasing the atmospheric pool at the rate of 3.5 Pg C yr^{-1} or 0.46% yr^{-1}. Concentration of CO_2 is governed by global Carbon cycle. If present trend continues it will increase to 600ppm in 2100 and 1300 ppm in 2200. Climate change is the major global issue of 21st century and global warming can affect global hydrological cycle and alter the distribution and productivity of terrestrial biota. During the last century global surface temperature is increased by 0.6± 0.2°C at an average rate of increase of 0.17°C per decade since 1900.

Following the agreement of the Kyoto protocol in 1997 efforts are being made by various agencies to examine sinks of carbon in different areas. The key parameters in assessment of mitigation strategies are: the amount of carbon per unit of land that can be conserved or sequestered in vegetation and soil by a given management option, the time period over which this can be conserved or sequestered, the amount of available land for this purpose and the mitigation costs. These parameters vary by land use type, soil type, environmental and socio-economic factors, which in turn also vary by region.

Although agriculture is not a large emitter of greenhouse gases (GHGs), it can assist in reducing the overall carbon emissions by sequestering this in large area of land. Therefore, carbon sequestration implies transferring atmospheric carbon di-oxide into long lived pools and storing it securely, so it is not immediately reemitted. Thus, soil carbon sequestrations means increasing SOC and SIC stocks through judicious land use and recommended management practices. The SOC pool represents a dynamic equilibrium of gains and losses. Conversion from natural to managed ecosystems, extractive farming practices based on low external input and soil degrading land use, which occurred during the last 200 years has transformed soil into important sources of atmospheric CO_2, contributing about 20% to the global emission of CO_2. The prediction of global climate change has focussed attention on soil as major source and sink for atmospheric CO_2. Besides, disturbing the earth's heat budget, emission of carbon di-oxide from soil diminishes soil organic carbon pool resulted in degrading soil quality, reduces biomass productivity, and adversely impacts water quality. Conversion of natural to agricultural ecosystems causes depletion of the SOC pool by as much as 60% in soils of temperate regions and 75% or more in the cultivated soils of the tropics.

Climate change may affect global carbon cycle thereby distorting structure and functions of ecosystems. Organic matter content, which is quite low (<1.0%) in the tropical soils would become still lower and climate change may affect its quality (Lal, 2004). Soil biology and microbial populations are expected to change under changed climatic conditions (Baker, 2004). The amount of carbon stored in agricultural soils is approximately 170 Gt C (Paustian *et al.*, 2000) while vegetation contains 550 Gt C. Soils of India have lower SOC and hence there is a large sink capacity for atmospheric CO_2 sequestration. Many of the management practices that are effective in increasing SOC in agricultural soils also improve productivity and profitability, conserve the resource base and protect the environment.

Soil carbon sequestration

Carbon sequestration in Indian soils assumes a special significance towards sustainable management of resources for achieving food security through improvement in soil and environmental quality. Estimates of the total potential of C sequestration in world soils vary widely from a low of 0.4 to 0.6 Gt C yr^{-1} (Sauerbeck, 2001) to a high of 0.6 to 1.2 Gt C yr^{-1} (Lal, 2003). India is having large land area and diverse eco-regions and there is a considerable potential of soil carbon sequestration. Total SOC pool in soils of India is estimated at 21 Pg up to 30 cm depth and 63 Pg up to 150 cm depth and SIC pool estimated at 196 Pg up to 1 m depth. (Pal *et al.*, 2000) (Table1).

Table 1. Organic carbon pool in soils of India and the world (adapted from Velayuthan *et al.*, 2000; Eswaran et al., 1993)

Soil order	India		World	
	0-30 cm (Pg)	0-150 cm (Pg)	0-25 cm (Pg)	0-100 cm (Pg)
Alfisols	4.22	13.54	73	136
Andisols	-	-	38	69
Aridisols	7.67	20.30	57	110
Entisols	1.36	4.17	37	106
Histosols	-	-	26	390
Inceptisols	4.67	15.07	162	267
Mollisols	0.12	0.50	41	72
Oxisols	0.19	0.49	88	150
Spodosols	-	-	39	98
Ultisols	0.14	0.34	74	101
Vertisols	2.62	8.78	17	38
Total	20.99	63.19	652	1555

Potential of soil C sequestration in India is estimated at 7 to 10 Tg C/yr^{-1} for restoration of degraded soils and ecosystems, 5 to 7 Tg C/yr^{-1} for erosion control, 6 to 7 Tg C/yr^{-1} for adoption of recommended management practices (RMPs) on agricultural soils, and 22 to 26 Tg C/yr^{-1} for secondary carbonates (Table2). Thus, total potential of soil C sequestration in India is 39 to 49 Tg C/yr^{-1} (44 ± 5).

Table 2. Total potential of carbon sequestration in soils of India (Source: Lal, 2004)

Process	Potential (Tg Cyr^{-1})
A. Soil organic Carbon (SOC)	
· Restoration of degraded soils	7.2 - 9.8
· Agricultutal intensification	5.5 – 6.7
B. Secondary carbonates	21.8 – 25.6
C. Erosion control	4.8 – 7.2
Total	39.3 – 49.3

Several studies have been initiated in Indian National Agricultural Research System to determine the effect of various agricultural management strategies such as zero tillage, mulching, incorporation of green manure, silvi-pastoral systems and watershed management for enrichment of soil carbon. The prominent role of forestry & agroforestry systems in carbon sequestration has also increased global interest in these land use options to stabilize greenhouse gas emission. Agroforestry system especially agri-silvicultural system can be carbon sinks and temporarily store carbon. In one of the studies in Uttar Pradesh, approximately 20 million tons of carbon has been estimated to be sequestered by the farm forestry plantations. The carbon sequestration potential of agroforestry practices is however, quite variable, depending on choice of trees and associated crops, the planting density and production objective of the system.

Agricultural soils represent a potential sink for carbon although they also serve as a carbon source through soil respiration. Cultivation practices such as tillage can also enhance soil aeration, mineralization of soil organic carbon (SOC) and the flux of CO_2 from soils. Indian soils are largely carbon-depleted but can be brought back to their native carbon- carrying capacity by reforestation and adoption of recommended management practices. The current stock of organic carbon in Indian soils (24.3 Gt) can be increased to 34.9 Gt, the difference representing the potential for sequestering additional carbon in soils. The soil constitutes a significant reservoir of carbon in both mineral and organic forms. The carbon in the soil has a special feature through humus formation. The rate of SOC sequestration in agricultural and restored ecosystems depends on soil texture and structure, rainfall, temperature, farming system and soil management (West and Post, 2002 and Lal, 2004).

Soil carbon sequestration is a natural, cost-effective and environmental friendly process. Important basic strategies to enhance carbon sequestration in soil through sustainable management practices include conservation tillage, use of crop residues, manure, compost, bio-solids, mulch farming, diverse cropping, integrated nutrient management, improved grazing, forestry and growing energy crops on spare lands (Table 3). Thus, the management systems that cause minimal soil disturbance by reducing tillage intensity and frequency, addition of higher amount of biomass, conserve soil and water, improve soil biodiversity and aggregation will lead to increased SOC sequestration.

Management practices for enhancing sequestration soil carbon

No-tillage/Conservation tillage/residue management

Long term experiments provide opportunities for assessing long term changes in SOC and estimating C sequestration potential of agricultural lands (Ladha *et al.*, 2003). Although some gives net losses of carbon from soil to the

Table 3 : Technological options for soil carbon sequestration (Source Lal, 2004)

Technology	Cropping system	Region	Reference
1. Green manuring	Sugarcane	Tropical	Yadav (1995)
	Rice - wheat	North western	Aulakh *et al.* (2001)
	Rice	Tropical	Kumar *et al.* (1999)
	Rice - wheat	Northern	Joshi *et al.* (1994)
2. Mulch farming / conservation tillage	Rice - wheat	Punjab	Aulakh *et al.* (2001)
	Soybean -wheat	Central	Kundu *et al.* (2001)
	Arable land	Northern	Biswas and Narayanasamy (1998)
	Sugarcane	Tropics	Yadav and Verma (1995)
3. Afforestation / agroforestry	Silviculture	Northern	Singhal *et al.* (1975)
	Acacia nilotica	Central	Pandey *et al.* (2000)
	Agroforestry	Tropical	Chander *et al.* (1998)
4. Grazing management/ ley farming	Grassland	M.P.	Chaubey et al. (1986)
	Mixed Farming	Arid	Rao et al. (1997)
5. Integrated nutrient management / manuring	Rice - Wheat	North West	Duxbury (2001)
	Cotton	Central India	Venugopalan *et al.* (1999)
	Arable land	North east	Chakrabarti *et al.* (2000)
6. Cropping systems	Mint-Mustard	U.P.	Patra *et al.* (2000)

atmosphere, others show that by good soil management practices sequestration of carbon in soil can be increased under arable crop management and even greater carbon sequestration can be achieved by afforestation. According to Lal, (2000), soil degradation and its mismanagement in relation to SOC is the main concern to the human being. Intensive tillage, especially ploughing accentuates mineralization and CO_2 emissions by mixing crop residues into the soil, bringing it closer to microbes, increasing O_2 concentrations in the soil and disrupting aggregates and exposing physically protected organic matter to microbial and enzyme activity. Therefore, conservation tillage offers a strategy to conserve and sequester carbon in soils by reducing the intensity and frequency of ploughing and leaving crop residues on the soil surface as mulch for enhancing SOC content. This helps, improves in soil organic matter quantity and quality and provides food for micro-organisms leading to improved biological health and microbial diversity. Adoption of no-till conserves carbon and improves soil health. Doran (1980) reported that organic C concentration in the surface 15 cm of non tilled soils is greater than for tilled soil. According to Campbell *et al.* (1995) continuous wheat cultivation under no tillage condition gained about 1.5 t ha^{-1} more C than under conventional tillage condition. Organic C in the top 7.5 cm soil were 36% higher in stubble mulch than in conventional tillage, with no differences below 7.5 cm (Rasmussen and Rhode, 1988). Management practice, such as no-till coupled with crop residue management improve SOC and enhance C sequestration in soil. (Gebhart *et al.*, 1994; Havlin *et al.*, 1990; Robinson *et al.*, 1996; Paustian *et al.* 2000; West and Post 2002; Wright and Hons 2005). Pacala & Socolow (2004) estimated that conversion of plough tillage to no-till farming on 1600 M ha of cropland along with adoption of conservation-effective measures could lead to sequestration of 0.5–1 Pg C yr^{-1} by 2050. Lupwayi *et al.* (1999) observed zero tillage (ZT) systems contributed less to atmospheric CO_2 than CT and SOM accumulated more under ZT. Several authors reported that conservation tillage and leguminous based cropping sequences have the potential to enhance SOC and improve soil aggregation (Bhattacharya *et al.*, 2009). Reduced tillage increases the standing stock of macro-aggregate protected soil organic C compared with conventional tillage (CT) mostly at the soil surface (Campbell *et al.*, 1995; Franzluebbers and Arshad, 1997). Surface residue management and reduced tillage provide a favourable for soil environment and native micro flora and fauna. Among the intermediate products of decomposition, many polysaccharides including microbial gum are especially significant in the formation of soil aggregates by their binding action on soil particles. In a 23 year field study, Dolan *et al.* (2006) found that no-tillage managed soil had 30% more soil organic C (SOC) and soil organic N (SON) than the mould board plow and chisel plows soils. McCarty *et al.* (1998) showed that a soil profile typical of plow tillage management can transform to a soil profile characteristic of no tillage (NT) within a 3-year period of transition. In this time period, stratification of organic matter in the profile progressed significantly toward that which occurs after 20 years of no tillage. A similar pattern was reported by several workers who also identified increased accumulation of C and N with greater periods of no-till management.

In India under rice-wheat system until, 1998, results in northern India show that no-tillage is not a successful technology and is only an option for late sown wheat (DWR, 1996-97; 1997-1998). But the importance of no-tillage has now been realized when the increase in CO_2 content in earth's atmosphere is voiced loudly. In India the area under zero-tillage under rice-wheat system has been increased tremendously over the years. Long-term sites maintained at farmers' field in Teek, Uchani and Nangla in Haryana revealed an improvement in the NPK and organic matter status in soil with no-till (Kumar *et al.*, 2002). Similarly, long-term studies in other parts of the world indicate that organic matter, nutrient and enzymes accumulate at soil surface (Dick *et al.*, 1991). Purakayastha *et al.*, 2008 reported that greater SOC stock in 0-20 cm Paulose silt loam soil occurred in Native Prairire (63.7 Mg ha^{-1}) and CT has lowest SOC stock which is 56% less than NP (Table 4).

Table 4 : SOC, MBC stock in 0-20 cm Palouse silt loam under different soil management options (*Purakayastha et al., 2008*)

Management practices	SOC (Mg ha^{-1})	MBC (Mg ha^{-1})
NP	63.7	4.0
CRP	37.0	2.11
NT 28	52.9	2.27
BGNT24	30.2	1.46
NT4	50.0	1.85
NTR	58.4	3.36
CT	27.9	2.02

Cover crops and cropping system

The effectiveness of conservation tillage in SOC sequestration is enhanced by the use of cover crops, such as clover and small grains. Frequent use of sod type legumes and grasses in rotation with food crops is an important strategy to enhance SOC and soil quality (Entry *et al.*, 1996). Gains in SOC by growing cover crops in rotation with food crops have been reported throughout the world including Haryana, India (Chander *et al.*, 1997), southwestern Nigeria (Juo *et al.*, 1996), Syria (Jenkinson *et al.*, 1999), and Argentina (Demmi *et al.*, 1986). Research was conducted to evaluate the potential of cover crops to accumulate SOC (Amado *et al.*, 2001). Soil under maize + velvet bean (cover crop) system had 5.42 t ha^{-1} more C than the fallow/maize system in the 0 to 20 cm soil layer. Cover crops can be grazed and well maintained pastures with controlled grazing at appropriate stocking rate can enhance SOC (Grace *et al.*, 1995). Neill *et al.* (1997) observed that SOC increased in 14 of 18 pastures and the magnitude of increase in some cases was as much as 18 Mg C ha $^{-1}$ to a 30 cm depth. Strategies that increase the cropping intensity such as the use of rotations with winter cover crops to increase the amount of biomass C returned to the soil can affect the size, turnover and vertical distribution of both active and passive pools of SOC (Franzluebbers *et al.*, 1995). Sainju *et al.*, (2002) reported that concentrations of soil organic C was greater with rye, hairy vetch and crimson clover cover crops than without a cover crop.

For sustainability of intensive cropping system, it is desirable not to grow a particular crop or a group of crops on the same soil for a long period. Varying the types of crop grow can increase the level of soil organic matter. However, effectiveness of crop rotation depends on the type of crops and crop rotation times. Cropping systems provide an opportunity to produce more biomass C than in a monoculture system and to thus increase SOC sequestration. Chander *et al.,* (1997) studied soil organic matter under different crop rotations for 6 years and found that the inclusion of green manure crop of *Sesbania aculeata* in the rotation improved soil organic matter status and microbial C increased from 192 mg kg^{-1} soil in pearl millet-wheat fallow rotation to 256 mg kg^{-1} soil in pearl millet-wheat green manure rotation (Table 5). Franzluebbers *et al.,* (1995) observed that mineralizable C and microbial biomass carbon (MBC) averaged 18% greater in rotation than in monoculture, probably due to greater C input via crop roots and residues in rotation and a shorter fallow. Benefits of crop rotations in managing SOC are documented by several long-term experiments in the United States (Anderson *et al.,* 1990). Legume-based cropping systems could help to increase crop productivity and soil organic matter levels, thereby enhancing soil quality as well as having the additional benefit of sequestering atmospheric C (Gregorich *et al.,* 2001). The quantity of C below the plough layer in legume-based rotation was 40% greater than that in monoculture. The soil organic matter below the plough layer in soil under the legume-based rotation appeared to be in a more biologically resistant form (i.e., higher aromatic C content) compared with that under monoculture. Effects of rotation with legume crops had the highest total C and biomass levels reported by Granatstein *et al.,* (1987) and Lupwayi *et al.,* (1999). Similar evidence of the effects of crop rotations on soil aggregation and SOC storage comes from studies on fine textured soils producing annual row crops on cool and humid soils of USA (Wright and Hons, 2004), Eastern Canada (Angers and Carter, 1996; Whalen *et al.,* 2003) Europe (Lopez Fando *et al.,* 2007; Alvaro-Fuentes *et al.,* 2008) and soils of Indian Himalayas (Verma and Sharma, 2007; Bhattacharya *et al.,* 2009).

Table 5 : Effect of crop rotations on SOC contents in Inceptisols (*Chander et al., 1997*)

Cropping system	SOC (g kg^{-1})	MBC (mg kg^{-1})
Pearl-millet-wheat-fallow	4.83	192
Pearl-millet-fodder-cowpea-fallow	4.78	231
Pearl-millet-potato-tomato	5.04	221
Pearl-millet-potato-sunflower	4.54	184
Pearl-millet-mustard-sunflower	5.02	144
Pearl-millet-wheat-green manure	5.14	256
LSD(P=0.05)	0.09	11

In India, most of the commonly practiced cropping systems are cereal-cereal (rice-rice),cereal-cereal-cereal (rice-wheat-maize), cereal-cereal-legume (maize-wheat-green gram), finger millet-wheat-gram, or pearl millet-wheat-gram. Singh *et al.* (1996) show that over a period of five years the net change in SOC is negative under cereal-cereal sequences, whereas in other sequences having legume component the changes are positive. Mixed or intercropping systems are also advantageous in terms of OC sequestration. In a typical black soil, continuous cropping and manuring increase organic carbon content by 20-40% over a period of three years (Mathan *et al.*, 1978) or introducing summer green manure crop *dhaincha* after harvest of wheat and before planting of rice (Prasad and Goswami, 1992).

The maintenance of organic matter in rice-wheat cropping system is extremely important. In Indo-Gangetic plains rice-wheat sequence needs to be tested by introducing legume crops in the system in different ways, viz., replacing rice or wheat crop by legume i.e. rice by pigeon pea in summer and wheat by lentil in winter. The beneficial effect of growing summer mung as a catch crop in rice-wheat rotation on enhancing SOC pool and the total productivity of the crops in the system in Mollisols of Pantnagar has been reported by Ghosh and Sharma (1996).

Soil fertility and nutrient management

Integrated nutrient management is also essential for C sequestration. On a long-term basis, increased crop yield and organic matter returned to the soil with judicious fertilizer application result in higher SOC content and biological activity than under controlled conditions (absence of fertilizers). Lal *et al.*, (1998) summarized the results of a number of studies and concluded that fertility management practices can enhance the SOC content at the rate of 50-150 kg ha^{-1}yr^{-1}. A study conducted by Ryan *et al.* (1998) reported that amount of organic matter increased with increasing rates of N fertilizers (0-90 kg ha^{-1}). In a dry land annual cropping study after 11 crops Halvorson *et al.* (1999) observed that fertilizer N rates of 67 and 134 kg N ha^{-1} yr^{-1} resulted in 140 and 182 kg ha^{-1} yr^{-1} more SOC, respectively, in the 0-15 cm depth than for the no N treatment. Phosphorus fertilizer also has a beneficial impact on organic soil C (Kumar and Yadav, 2001). They observed that after 20 years of continuous rice-wheat cropping, soil organic C increased by 35, 27 and 20% in plots receiving 120-35-33, 120-35-0, and 80-35-33 kg ha^{-1} NPK, respectively, relative to the initial value of 4.5 g C kg^{-1} soil in 1977. In contrast, organic C declined in the unfertilized treatment by 62% compared with the initial level and by 49% in the treatment receiving 120 kg ha-^{1} alone. Integrated addition of inorganics and organics through farmyard manure, green manure and crop residues increased the organic carbon content significantly over control followed by recommended dose of NPK (Sharma *et al.*, 2001). According to Rasmussen and Rohde (1988) applied N increased soil C linearly on plots.

Manure ammendment is a practice that can improve the nutrient status of the soil and increase SOC levels. A long-term (16 years) field experiment was conducted in Jiangsu, China, to investigate the effects of manure application on soil microbial biomass carbon (MBC). Although the application rate of N, P and K in each sub-treatments were equal, the contents of MBC, in 0-5 and 5-10 cm depth layer followed the treatments in the order: fertilizers + pig manure > fertilizers + straw > fertilizers + green manure > chemical fertilizers > control (Xu *et al.,* 2002). Long-term fertility experiments (LTFE) play an important role in understanding the complex interaction involving plants, soils, climate and management practices and are the primary source of information to determine the effects of cropping systems, soil management, fertilizer use, and residue utilization on the quantitative and mechanistic changes soil quality as well as on SOC pools (Rasmussen *et al.,*1998).Long-term manurial experiment on vertisol of central India under soybean-wheat cropping system showed an enrichment of organic carbon ranged from 85-739 kg C ha^{-1} at 0-15 cm and 54-149 kg C ha^{-1} at 15-30 cm soil depth (Kundu *et al.,* 2001). They also reported that a minimum of 888 kg C ha^{-1} yr^{-1} is required to maintain the SOM content of the experimental soil at equilibrium level.

Rice (*Oryza sativa* L.)-wheat (*Triticum aestivum* L.) cropping is the dominant cropping sequence in the Indo-Gangetic Plains, and occupies nearly 50% of its cropped area. Use of organic amendments such as FYM, rice straw and green manure is known to improve soil productivity in rice-wheat cropping (Ghosh *et al.,* 2009) and has the capacity to add SOC and to improve soil condition (Majumder *et al.,*2008; Benbi and Senapati, 2010). Balanced application of fertilizer and manure substantially improve the SOC under different soils and cropping systems (Kukal *et al.,* 2009, Manna *et al.,* 2005, Rudrappa, 2006) (Table 6 and Table 7). Invariably green manure crops restore organic carbon status of soil and crop productivity (Swarup, 1998). In semi-arid regions of India, utilizing the wastes through compost, amended with minerals such as rock phosphates, pyrites and N application have been recognized for improving the crop yields and SOC (Hazra *et al.,* 1989; Hazra *et al.,* 1994; Manna *et al.,* 1998). Liebig *et al.* (2002) observed that high N rate treatments increased SOC sequestration rates by 1.0–1.4 Mg C ha^{-1} yr^{-1} compared with unfertilized controls. Similar observations were made by Dumanski *et al.* (1988), Gregorich *et al.* (1996), Bowman & Halvorson (1998), Studdert & Echeverria (2000), and Jacinthe *et al.* (2002). Malhi *et al.* (1997) reported that SOC sequestration depends both on the rate and the source of N application.

Water management and elimination of summer fallows

Judicious application of water particularly in arid and semi arid regions where crop growth is severely affected by water deficit and nutrient availability can increase biomass production and improve SOC. It has been estimated that conversion of dry land farming to irrigated agriculture may increase SOC content in the soil profile by 50 to 150 kg ha^{-1}. Elimination of summer fallow, which is often practised in semi arid regions, improves carbon sequestration in dry and cropping system.

Table 6 : Long term manuring and fertilization effects on TOC (0-45cm) contents in Inceptisols (*Rudrappa et al., 2006*)

Management practices	TOC (Mg ha^{-1})
50% NPK	51.5
100% NPK	54.1
150% NPK	63.5
100% NP	53.0
100% N	52.1
100% NPK + FYM	72.1
Control	48.7

Table 7 : Soil organic pool (Mg ha^{-1}) and calculated soil organic carbon sequestration rate kg ha^{-1} yr^{-1}) as affected by inorganic fertilizers and FYM in rice-wheat and maize-wheat systems (0-60 cm profile) (*Kukal et al., 2009*)

Treatments	Rice - wheat		Maize - wheat	
	SOC pool	C sequestration	SOC pool	C sequestration
Control	21.4		18.7	-
FYM	31.4	310	23.3	140
N_{120}	25.5	130	19.8	30
N_{120} P_{30}	27.4	190	20.0	40
N_{120} P_{30} K_{30}	29.6	260	20.6	70
Mean	27.1	220	20.6	70
LSD (0.05)		40	-	30

Land conversion and restoration

Out of the total degraded land in India (about 167 M ha), 90 M ha degraded through the process of water erosion. Restoration of degraded soils is a high priority for economic and environmental reasons. Problem of water erosion is more severe in hills especially in northeast hill states, Bihar and Central India. Deforestation is at a faster rate in hill regions due to practice of *jhum* cultivation (Singh 1980, 1986, 1987; Singh and Pazo, 1981). Soil erosion management by alternate land use systems viz., agro-forestry, agro-horticulture, and agro-silviculture, are important strategies for SOC restoration. In north-east hill state of India, existence of three land use systems reduces soil erosion and SOC loss considerably (Munda *et al.*, 1996).

Agro-forestry agro-horticultural systems and silviculture systems

Agroforestry systems have very high potential for carbon sequestration because of increase in organic carbon storage, both in soil and above ground biomass (IPCC, 2000). Conversion of long-term arable crop land to agro-horticulture results in a significant increase in SOC .Under a system of different intercropped fruit trees, coconut (*Cocus nufifera* L) intercropped with guava (*Psidium guavaja* L.), the SOC enhanced from 3.4 to 7.8 and 2.4 to 6.2 g kg^{-1} after 38 years and 10 years, respectively due to greater recycling of bio-litters. In a five year study the effect of *Dalbergia sisso, Pongamia spp., Leucaena Leucocephala, Acacia nilotica and Dalbergia latifolia* on SOC enhancement has been studied extensively and SOC increased from 0.44 to 0.95%. Similarly, doubling of organic carbon content has been observed in agro-horticultural and agro-forestry systems as compared to shifting cultivation (sole cropping) (Das and Itnal, 1994). The build up of SOC in an alkali soil in agro-forestry systems in six years has been in the order: Acacia based > Poplar based > Eucalyptus based > sole crop based system. It has also been reported that 6-10 fold increases in SOC status of a sodic soil when the field has been occupied by trees like *Prosopis juliflora, Acacia nilotica, Eucalyptus tereticornis, Terminalia arjuna and Albizia lebbek* for more than 20 years. Solanki *et al.* (1999) compared the enrichment of C, N, P and K under various agroecosystems and reported highest enrichment of C (377%) under silvi-pastoral system and highest increase in available N, P and K under agric-horticultural system.

Table 8 : Impact of prevalent Silviculture systems on soil carbon storage and carbon sequestration in Shivalik region of Lower Himalayas *(Bhattacharyya et al. (2009)*

Parameters	Eucalyptus hybrid	Acacia catechu	Acacia nilotica
Plantation age (years)	35	35	15
Aboveground biomass (t ha^{-1})	409.9	18.3	21.5
Soil Carbon (t ha^{-1})	35.3	50.3	28.5
Litter BC (t ha^{-1})	9.5	0.7	0.1
Root BC (t ha^{-1})	45.3	2.7	3.2
Herbaceous shrubs BC (t ha^{-1})	3.2	0.4	1.2
Total C (t ha^{-1})	503.2	72.4	54.5
C sequestration (C ha^{-1} yr^{-1})	14.4	2.1	3.6

Bhattacharyya *et al.* (2009) have shown that *Acacia* and *Eucalyptus* species based silvicultural systems prevalent in degraded *Shivalik* region had the potential to sequester carbon and carbon sequestration was found to be more under fast growing *Eucalyptus* hybrid coppice rotation (14.4 t C ha^{-1} yr^{-1}) as compared to *Acacia* species (Table 8).

Benefits of soil carbon sequestration

Benefits of Soil Carbon sequestration to soil quality are (i) improvement in soil structure, (ii) reduction in soil erosion, (iii) decrease in non-point source pollution,(iv) increase in plant-available water reserves,(v) increase in storage of plant nutrients, (vi)denaturing of pollutants, (vii) increase in soil quality,(viii) increase in agronomic productivity of food security, (ix) moderation of climate, and (x) increase in aesthetic and economic value of soil.

References

Alvaro-Fuentes, J.L., Gracia, A.R., Lo´ Pez, M.V. (2008). Tillage and cropping intensification effects on soil aggregation: temporal dynamics and controlling factors under semiarid conditions. *Geoder,* **145** : 390–396.

Amado, T.J.C., Bayer, C., Eltz, F.L.F. and Brum, A.C.R. (2001). Potential of cover crops to accumulate carbon and nitrogen in soil in direct planting and to improve environmental quality. *Revista-Brasileira-de-Ciencia-do-Solo,* **25** (1): 189-197.

Anderson, S.H., Gantzer, C.J. and Brown, J.R. (1990). Soil physical properties after 100 years of continuous cultivation. *Journal of Soil and Water Conservation,* **45** : 117-121.

Angers, D.A., Carter, M.R. (1996). Aggregation and organic matter storage in cool, humid agricultural soils. In: Carter, M.R., Stewart, B.A. (Eds.), Structure and organic matter storage in agricultural soils. *Advances in Soil Science.* Lewis/CRC Press, Boca Raton, Florida, pp. 193–211.

Baker, J.M. (2004). Yield responses of southern US rice cultivars to CO_2 and temperature. *Agriculture and Forest Meterology,* **122** : 129-137.

Batjes, N.H. (1996). Total Carbon and Nitrogen in Soils of the World. *European Journal of Soil Science,* **47** : 151–163.

Benbi, D.K. and Senapati, N. (2010). Soil aggregation and carbon and nitrogen stabilizationin relation to residue and manure application in rice–wheat systems in northwest India. *Nutrient Cycling in Agro Ecosystems,* **87** : 233–247 (DOI 10.1007/ s10705-009-9331-2).

Bhattacharyya, R., Prakash V., Kundu S., Srivastva A.K.and Gupta H.S. (2009). Soil aggregation and organic matter in a sandy clay loam soil of the Indian Himalayas under different tillage and crop regimes. *Agriculture Ecosystems and Environment* **132** : 126–134.

Bhattacharyya, P., yadav, R.P. and Agnihotri, Y. (2009). Impact of prevalent silviculture systems on soil stability and potential carbon storage in Shivalik Region of Lower Himalayas. *Journal of the Indian Society of Soil Science,* **57**: 71-75.

Bowman, R.A. and Halvorson, A.D. (1998). Soil chemical changes after nine years of differential N fertilization in no-till dry land wheat–corn–fallow rotation. *Soil Science,* **163** : 241–247. (doi:10.1097/00010694-199803000-00009).

Campbell, C.A., McConkey, B. S.G., Zenter, R.P., Dyck, F.B., Selles, F. and Curtin, D. (1995). Carbon sequestration in a Brown Chernozem as affected by tillage and rotation. *Canadian Journal of Soil Science,* **75** : 449-458.

Chander, K., Goyal, S., Mundra, M.C. and Kapoor, K.K. (1997). Organic matter, microbial biomass and enzyme activity of soils under different crop rotations in the tropics. *Biology and Fertilizer of Soils*, **24** : 306-310.

Das,S.K. and Itnal, C.J. (1994). Capability based land use systems ; role in diversifying dry land agriculture. In: soil management for sustainable agriculture in dryland areas. *Bulletin Indian Society of Soil Science*, **16** : 92-100.

Demmi, M.A., Puricelli, C.A. and Rosell, R.A. (1986). El efecto del pasto iloron en la recuperacion de los suelos. INTA-San Luis Agricultural Experimental Statistic Chemical Bulletin, **109** : 23.

Dick, W.A., McCoy, E.L.Edwards, W.M., Lal, R. (1991). Continuous application of no tillage in Ohio soils. *Agronomy Journal*, **83** : 83-65.

Doran, J.W. (1980). Soil microbial and biochemical changes associated with reduced tillage. *Soil Science Society of America Journal*, **47** : 102-107.

Dumanski, J., Desjardins, R.L., Tarnocai, C.G., Monreal,D., Gregorich, E.G., Kirkwood, V. and Campbell, C.A. (1998). Possibilities for future carbon sequestration in Canadian agriculture inrelation to land use changes. *Climate Change*, **40** : 81–103. (doi:10.1023/ A:1005390815340).

Entry, J.A., Mitchell, C.C. and Backman, C.B. (1996). Influence of management practices on soil organic matter, microbial biomass and cotton yield in Alabama's "old rotation". *Biology and Fertility of Soil*, **23** : 353-358.

Eswaran, H., Van Den Berg, E and Reich, P. (1993). *Soil Science Society of America Journal*, **57** : 192–194.

Falkowski, P. (2000). The global carbon cycle: a test of our knowledge of earth as a system. *Science*, **290** : 291–296.

Franzluebbers, A.J., Hons, F.M. and Zuberer, D.A. (1995). Soil organic carbon, microbial biomass, and mineralizable carbon and nitrogen in sorghum. *Soil Science Society of America Journal*, **59** (2): 460-466.

Franzluebbers, A.J., and Arshad M.A., (1996). Soil organic matter pools during early adoption of conservation tillage in northwestern Canada. *Soil Science Society of America Journal*, **60** : 1422–1427.

Gebhart, D.L., Johnson, H.B., Mayeux, H.S. and Polley, H.W. (1994). The CRP increases soil organic carbon. *Journal of Soil Water Conservation*, **49** : 488–492.

Ghosh, P.K. and Sharma, K.C. (1996). Direct and residual effect of green manuring on rice-wheat rotation. *Crop Research*, **211** : 133-136.

Grace, P., Ladd, J. and Bryesson, K. (1995). Management options to increase carbon storage in cultivated dryland soils. In: V.R. Squires, E.P. Glenn and A.T. Ayoub (eds.) "Combating global climate change by combating land degradation". UNEP, Nairobi, Kenya. pp. 160-173.

Granatstein, D.M., Bezdicek, D.F., Cochran, V.L., Elliott, L.F. and Hammel, J. (1987). Long-term tillage and rotation effects on soil microbial biomass, carbon and nitrogen. *Biology and Fertility of Soils*, **5** : 265-270.

Gregorich, E.G., Drury, C.F. and Baldock, J.A. (2001). Changes in soil carbon under long-term maize in monoculture and legume-based rotation. *Canadian Journal of Soil Science,* **81** (1): 21-31.

Gregorich, E.G., Ellert, B.H., Dury, C.F. and Linang, B.C. (1996). Fertilization effects on soil organic matter turnover and crop residue carbon storage. *Soil Science Society of America Journal,* **61** : 1159–1175.

Hajra, J.N., Manna, M.C. and Kole, S.C. (1989). Enrichment phosphor-compost and its response on rice-wheat system in Alluvial soil. *Journal of Indian Agricultural Science,* **67** : 540-544.

Hajra, J.N., Sinha, N.B., Manna, M.C., Islam, N. and Banerjee, N.C. (1994). Comparative performance of phosphocomposts. and single super phosphate and response of green gram (Vigna radiata L. Wilezek). *Tropical Agriculture (Trinidad).* **71**: 147-149.

Halvorson, A.D., Reule, C.A. and Follett, R.F. (1999). Nitrogen fertilization effects on soil carbon and nitrogen in a dryland cropping system. *Soil Science Society of America Journal,* **63** : 912-917.

Havlin, J.L., Kissel, D.E., Maddux, L.D., Claassen, M.M. and Long, J.H. (1990). Crop rotation and tillage effects on soil carbon and nitrogen. *Soil Science Society of America Journal,* **54** : 448-452.

IPCC, (2000). IPCC Special Report on Land Use, Land-Use Change and Forestry. A special report of the Intergovernmental Panel on Climate Change. Approved at IPCC Plenary XVI (Montreal, 1-8 May, 2000). IPCC Secretariat, c/o World Meteorological Organisation, Geneva, Switzerland. At http:/ /www.ipcc.ch*J.* **66** : 596–601.

Jacinthe, P.A., Lal, R. and Kimble, J.M. (2002). Effects of wheat residue fertilization on accumulation and biochemical attributes of organic carbon in central Ohio Luvisol. *Soil Science,* **167** : 750–758. (doi:10.1097/00010694-200211000-00005).

Jenkinson, D.S., Harris, H.C., Ryan, J., McNeill, A.M., Pibeam, C.J. and Coleman, K. (1999). Organic matter turnover in a calcareous clay soil from Syria under a two-course cereal rotation. *Soil Biology adn Biochemistry,* **31** : 643-796.

Juo, A.S.R., Franzluebbers, K., Dabiri, A. and Ikhile, B. (1996). Soil properties and crop performance on a Kaolinitic Alfisol after 15 years of fallow and continuous cultivation. *Plant and Soil,* **180** : 209-217.

Kukal,S.S., Rasool Rehana and Benbi D.K. (2009). Soil organic carbon sequestration in relation to organic and inorganic fertilizationin rice–wheat and maize–wheat systems. *Soil and Tillage Research,* **102** : 87–92.

Kumar, A. and Yadav, D.S. (2001). Long-term effects of fertilizers on the soil fertility and productivity of a rice-wheat system. *Journal of Agronomy and Crop Science,* **186** : 47-54.

Ladha, J.K., Dawe, D., Pathak, H., Padre, A.T., Yadav, R.L., Singh B., Singh Y., Singh, Y., Singh, P., Kundu, A.L., Sakal, R., Ram, N., Regmi, A.P., Gami, S.K., Bhandari, A.L., Amin, R., Yadav, C.R., Bhattarai, E.M., Das, S., Aggarwal, H.P., Gupta, R.K., and Hobbs. P.R. (2003). How extensive are yield declines in long-term rice-wheat experiments in Asia. *Field Crops Research* **81**:159-180.

IPCC, (2001). Climate change (2001). The scientific basis. Intergovernment Panel on Climate Change. Cambridge, UK: Cambridge University Press.

Lal, R. (2003). Global potential of soil carbon sequestration to mitigate the greenhouse effect. *Critical Reviewes in Plant Sciences,* **22** : 151-184.

Lal, R. (2000). Controlling greenhouse gases and feeding the globe through soil management. University distinguished lecture presented on February 17, 2000, at Wexner.

Lal, R. (2004). Soil carbon sequestration in India *Climate Change,* **65** : 277-296

Lupwayi, N.Z., Rice, W.A., Clayton, G.W. (1999). Soil microbial biomass and carbon dioxide flux under wheat as influenced by tillage and crop rotation. *Canadian Journal of Soil Science,* **79** : 273–280.

Majumder B., Mandal B., Bandyopadhyay P.K. and Chaudhury J. (2007). Soil organic carbon pools and productivityrelationships for a 34 year old rice-wheat-jute agroecosystemunder different fertilizer treatments. *Plant Soil,* **297** : 53–67.

Majumder B., Mandal B., Bandyopadhyay P.K., Gangopadhyay, A., Mani, P.K. and Kundu, A.L. (2009). Organic amendments influence Soil organic carbon pools and rice-wheat productivity. *Soil Science of America Journal,* **72** (3).

Kundu, A.L. and Mazumdar, D. (2008). Organic Amendments Infl uence Soil Organic Carbon Pools and Rice–Wheat Productivity. *Soil Science of America Journal,* **72** (3) : 775-785.

Malhi, S.S., Nyborg, M., Harpiak, J.T., Heier, K. and Flore, N.A. (1997). Increasing organic carbon and nitrogen under bromegrass with long-term N fertilization. *Nutrient Cycling in Agroecosystems,* **49** : 255-260. (doi:10.1023/ A:1009727.

Manna, M.C., Hajra, J.N. Sinha, N.B. and Ganguly, T.K. (1997). Enrichment of compost by bioinoculants and mineral amendments.*Journal of the Indian Society of Soil Science,* **45** : 831-833.

Manna, M.C. and Ganguly, T.K. (1998). Recycling of organic wastes- its potential, turn-over and maintenance in soil-a review. *Agriculture Reviews,* **19** : 86-104.

Manna, M.C., Swarup, A, Wanjaria, R.H., Ravankar H.N., Mishra, B., Sahae, M.N., Singh Y.V., Sahid D.K., Sarap, P.A. (2005). Long-term effect of fertilizer and manure application on soil. Organic carbon storage, soil quality and yield sustainability under sub humid and semi-arid tropical India. *Field Crops Research,* **93** (2-3) : 264-280.

Mathan, K.K., Sankaran K., Kanakabushan N. and Krishnamoorthy K.K., (1978). Effect of continuous rotational cropping on the organic carbon and total nitrogen content in a black soil. *Journal of the Indian Society of Soil Sciences,* **26**: 283-285.

Munda, G.C., Ghosh, P.K. and Prasad, R.N. (1996). Adopt alternative land-use system to shifting cultivation. *Indian Farming,* **40** (5): 10-14.

Neill, C., Melillo, J.M., Steudler, P.A., Cerri, C.C., De Moraes, J.F.L., Piccolo, M.C. and Brito, M. (1997). Soil C and N stocks following forest clearing for pasture in the southwestern Brazilian Amazon. *Journal of Applied Ecology,* **7**: 1216-1225.

Pacala, S. and Socolow, R. (2004). Stabilization wedges: solving the climate problem for the next 50 years with current technologies. *Science,* **305** : 968–972. (doi:10.1126/ science. 1100103).

Pal, D.K., Dasog, G.S., Vdivelu, S., Ahuja, R.L., and Bhattacharya, T. (2000). 'Secondary CalciumCarbonate in Soils of Arid and Semi-Arid Regions of India', in Lal, R., Kimble, J.M., Eswaran, H., and Stewart, B.A. (eds.), Global Climate Change and Pedogenic Carbonates, CRC/Lewis Publishers, Boca Raton, FL, pp. 149–185.

Paustian, K., Six, J., Elliott, E.T., Hunt, H.W., Rustad, L.E., Huntingdon, T.G. and Boone, R.D. (2000). Management options for reducing CO_2 emissions from agricultural soils. *Biogeochemistry,* **48** (1): 147-163.

Prasad, R. and N.N. Goswami, (1992). Soil fertility restoration and management for sustainable agriculture in South Asia. *Advances in Soil Science,* **17** : 37-77.

Purakayastha, T.J., Huggins, D.R. and Smith, J.L. (2008). Carbon sequestration under native Prairie, conservation reserve and no tillage in Paluose region. *Soil Science Society of America Journal,* **72** : 534-540.

Rasmussen, P.E. and Rohde, C.R. (1988). Long-term tillage and nitrogen fertilization effects on organic nitrogen and carbon in a semiarid soil. *Soil Science Society of America Journal,* **52** : 1114-1117.

Robinson, C.A., Cruse, R.M., Ghaffarzadeh, M. (1996). Cropping system and nitrogen effects on mollisol organic carbon. *Soil Science Society of America Journal,* **60** : 264–269.

Rudrappa, L., Purakayestha T.J., Singh D., and Bhadraray S., (2006). Long-term manuring and fertilization effects on soil organic carbon pools in a Typic Haplustept of semi-arid sub-tropical India. *Soil and Tillage Research,* **88** : 180–192.

Ryan, J., Lal, R. (ed), Kimble, J.M., Follett, R.F. (ed) and Stewart, B.A. (1998). Changes on organic carbon in long term rotation and tillage trials in Northern Syria. *Management of carbon sequestration in soil,* pp.285-286.

Sainju, U.M., Singh, B.P. and Whitehead, W.F. (2002). Long-term effects of tillage, cover crops and nitrogen fertilization on organic carbon and nitrogen concentrations in sandy loam soils in Georgia, USA. *Soil and Tillage Research,* **63** : 167-179.

Sauerbeck, D.R. (2001). CO_2 emissions and C sequestration by agriculture – perspectives and limitations. *Nutrient Cycling in Agroecosystems,* **60** : 253-266.

Schnitzer, M. (1991). Soil organic matter—the next 75 years. *Soil Science,* **151** : 41–58. (doi:10.1097/00010694-199101000- 00008).

Sharma, M.P., Bali, S.V. and Gupta, D.K. (2001). Soil fertility and productivity of rice (Oryza sativa)-wheat (Triticum aestivum) cropping system in an Inceptisol as influenced by integrated nutrient management. *Indian Journal of Agricultural Sciences,* **71** (2): 82-86.

Singh, G.B. (1980). Shifting cultivation and its control in the North Eastern Hill Region-A critical review. Proceeding of *National symposium on soil conservation and water management,* CSWCRI, Dehradun, pp. 9.

Singh, G.B. (1986). Present status of agro forestry research in India. In: Agroforestry system-A new challenge (Khosla, P.K. and Khurana, D.K. eds.), *Indian Society of Tree Scientists*, New Delhi, pp. 25-31.

Singh, G.B. (1987). Agroforestry in the Indian sub-continent: past present and future. In: Agroforestry-A decade of development (Steppler, H.A. and Nair, P.K.R. eds.), ICRAF, Kenya, pp. 117-138.

Singh, G.B. and Pazo, P.O. (1981). Agroforestry in the Eastern Himalayas. In : *Proceeding of Agro forestry seminar* held at Imphal, ICAR, New Delhi.

Singh, Y., Chaudhary D.C., Singh S.P., Bhardawaj A.K. and Singh D. (1996). Sustainability of rice (Oryza sativa) – wheat (Triticum aestivum) sequential cropping through introduction of legume crops and green manure crops in the system. *Indian Journal of Agronomy*, **41** : 510–514.

Solanki, K.R., Bisaria, A.K.and handa, A.K. (1999). In : *Sustainable rehabilitation of Degraded lands through Agroforestry*. NRC for Agroforestry, Jhansi.

Studdert, G.A. and Echeverria, H.E. (2000). Crop rotations and nitrogen fertilization to manage soil organic carbondynamics. *Soil Science Society of America Journal*, **64** : 1496–1503.

Swarup, A. (1998). Emerging soil fertility management issues for sustainable crop production in irrigated systems. In: Swarup A., Reddy, D.O. and Prasad, R.N. (eds.), Long-term soil fertility management through integrated plant nutrient supply. Indian Institute of Soil Science, Bhopal, India, pp. 54-68.

Velayutham, M., Pal, D.K. and Bhattacharya, T.K. (2000). Organic carbon stock in soils of India. In: Advances in Soil Science: Global climatic change and tropical ecosystems (Lal, R., Kimble, J.M. and Stewart, B.A. eds.)., Lewis Publishers, Boca Raton, Fl, pp. 71-95.

Verma,S. and Sharma, P.K. (2007). Effect of long-term manuring and fertilizers on carbon pools, soil structure, and sustainability under different cropping systems in wet-temperate zone of northwest Himalayas. *Biology and Fertility of Soils*, **44** : 235–240.

West, T.O. and Post, W.M. (2002). Soil organic carbon sequestration rates by tillage and crop rotation: a global data analysis. *Soil Science Society of America Journal*, 1930-1946.

Wright, A.L., Hons, F.M. (2004). Soil aggregation and carbon and nitrogen storage under soybean cropping sequences. *Soil Science Society of America Journal*, **68** : 507–513.

Wright, A.L., Hons, F.M. (2005). Soil carbon and nitrogen storage in aggregates from different tillage and crop regimes. *Soil Science Society of America Journal*, **69** : 141–147.

Xu, Y. C., Shen, Q. R. and Ran, W. (2002). Effects of zero-tillage and application of manure on soil microbial biomass C, N, and P after sixteen years of cropping. *Acta-Pedologica-Sinica*, **39** (1): 89-96.

❑❑❑